# Modern War and the Utility of Force

This book investigates the use and utility of military force in modern war.

After the Cold War, Western armed forces have increasingly been called upon to intervene in internal conflicts in the former Third World. These forces have had to carry out missions that they traditionally have not been trained and equipped for, in environments that they often have not been prepared for. A number of these 'new' types of operations in allegedly 'new' wars stand out, such as peace-enforcement, state-building, counter-insurgency, humanitarian aid, and not the least, counter-terrorism. The success rate of these missions has, however, been mixed, providing fuel for an increasingly loud debate on the utility of force in modern war. This edited volume poses as its central question: what *is* in fact the utility of force? Is force useful for anything other than a complete conventional defeat of a regular opponent who is confronted in the open field?

This book will be of much interest to students of strategic studies, war and conflict studies, counter-insurgency, security studies and IR.

**Jan Angstrom** is a researcher at the Swedish National Defence College.

**Isabelle Duyvesteyn** is Associate Professor at the Department of History of International Relations, Utrecht University in the Netherlands.

## Cass military studies

**Intelligence Activities in Ancient Rome**
Trust in the gods, but verify
*Rose Mary Sheldon*

**Clausewitz and African War**
Politics and strategy in Liberia and Somalia
*Isabelle Duyvesteyn*

**Strategy and Politics in the Middle East, 1954–60**
Defending the northern tier
*Michael Cohen*

**The Cuban Intervention in Angola, 1965–1991**
From Che Guevara to Cuito Cuanavale
*Edward George*

**Military Leadership in the British Civil Wars, 1642–1651**
'The genius of this age'
*Stanley Carpenter*

**Israel's Reprisal Policy, 1953–1956**
The dynamics of military retaliation
*Ze'ev Drory*

**Bosnia and Herzegovina in the Second World War**
*Enver Redzic*

**Leaders in War**
West Point remembers the 1991 Gulf War
*Edited by Frederick Kagan and Christian Kubik*

**Khedive Ismail's Army**
*John Dunn*

**Yugoslav Military Industry 1918–1991**
*Amadeo Watkins*

**Corporal Hitler and the Great War 1914–1918**
The list regiment
*John Williams*

**Rostóv in the Russian Civil War, 1917–1920**
The key to victory
*Brian Murphy*

**The Tet Effect, Intelligence and the Public Perception of War**
*Jake Blood*

**The US Military Profession into the 21st Century**
War, peace and politics
*Edited by Sam C. Sarkesian and Robert E. Connor, Jr.*

**Civil–Military Relations in Europe**
Learning from crisis and institutional change
*Edited by Hans Born, Marina Caparini, Karl Haltiner and Jürgen Kuhlmann*

**Strategic Culture and Ways of War**
*Lawrence Sondhaus*

**Military Unionism in the Post Cold War Era**
A future reality?
*Edited by Richard Bartle and Lindy Heinecken*

**Warriors and Politicians**
U.S. civil–military relations under stress
*Charles A. Stevenson*

**Military Honour and the Conduct of War**
From Ancient Greece to Iraq
*Paul Robinson*

**Military Industry and Regional Defense Policy**
India, Iraq and Israel
*Timothy D. Hoyt*

**Managing Defence in a Democracy**
*Edited by Laura R. Cleary and Teri McConville*

**Gender and the Military**
Women in the armed forces of western democracies
*Helena Carreiras*

**Social Sciences and the Military**
An interdisciplinary overview
*Edited by Giuseppe Caforio*

**Cultural Diversity in the Armed Forces**
An international comparison
*Edited by Joseph Soeters and Jan van der Meulen*

**Railways and the Russo-Japanese War**
Transporting war
*Felix Patrikeeff and Harold Shukman*

**War and Media Operations**
The US military and the press from Vietnam to Iraq
*Thomas Rid*

**Ancient China on Postmodern War**
Enduring ideas from the Chinese strategic tradition
*Thomas Kane*

**Special Forces, Terrorism and Strategy**
Warfare by other means
*Alasdair Finlan*

**Imperial Defence, 1856–1956**
The old world order
*Greg Kennedy*

**Civil–Military Cooperation in Post-Conflict Operations**
Emerging theory and practice
*Christopher Ankersen*

**Military Advising and Assistance**
From mercenaries to privatization, 1815–2007
*Donald Stoker*

**Private Military and Security Companies**
Ethics, policies and civil–military relations
*Edited by Andrew Alexandra, Deane-Peter Baker and Marina Caparini*

**Military Cooperation in Multinational Peace Operations**
Managing cultural diversity and crisis response
*Edited by Joseph Soeters and Philippe Manigart*

**The Military and Domestic Politics**
A concordance theory of civil–military relations
*Rebecca L. Schiff*

**Conscription in the Napoleonic Era**
A revolution in military affairs?
*Edited by Donald Stoker, Frederick C. Schneid and Harold D. Blanton*

**Modernity, the Media and the Military**
The creation of national mythologies on the Western Front 1914–1918
*John F. Williams*

**American Soldiers in Iraq**
McSoldiers or innovative professionals?
*Morten Ender*

**Complex Peace Operations and Civil–Military Relations**
Winning the peace
*Robert Egnell*

**Strategy and the American War of Independence**
A global approach
*Edited by Donald Stoker, Kenneth J. Hagan and Michael T. McMaster*

**Managing Military Organisations**
Theory and practice
*Edited by Joseph Soeters, Paul C. van Fenema and Robert Beeres*

**Modern War and the Utility of Force**
Challenges, methods and strategy
*Edited by Jan Angstrom and Isabelle Duyvesteyn*

# Modern War and the Utility of Force

## Challenges, methods and strategy

**Edited by Jan Angstrom and Isabelle Duyvesteyn**

Routledge
Taylor & Francis Group
LONDON AND NEW YORK

First published 2010
by Routledge
2 Park Square, Milton Park, Abingdon, Oxon OX14 4RN

Simultaneously published in the USA and Canada
by Routledge
270 Madison Avenue, New York, NY 10016

*Routledge is an imprint of the Taylor & Francis Group, an informa business*



Typeset in Baskerville by Wearset Ltd, Boldon, Tyne and Wear



*British Library Cataloguing in Publication Data*
A catalogue record for this book is available from the British Library

*Library of Congress Cataloging-in-Publication Data*
A catalog record has been requested for this book

ISBN10: 0-415-57595-8 (hbk)
ISBN10: 0-203-85146-3 (ebk)

ISBN13: 978-0-415-57595-9 (hbk)
ISBN13: 978-0-203-85146-3 (ebk)

# Contents

*List of illustrations* ix
*Notes on contributors* x
*Foreword* xiv
GENERAL SIR RUPERT SMITH

*Preface and acknowledgements* xvii

1 War, what is it good for? 1
JAN ANGSTROM AND ISABELLE DUYVESTEYN

2 From victory to success: the changing mission of Western armed forces 16
CHRISTOPHER DANDEKER

3 The international private security company: a unique and useful actor? 39
CHRISTOPHER SPEARIN

4 Great expectations: the use of armed force to combat terrorism 65
ISABELLE DUYVESTEYN

5 Why does peacekeeping succeed or fail? Peacekeeping in the Democratic Republic of Congo and Sierra Leone 90
SARAH KREPS

6 Exploring the utility of armed force in peace operations: German and British approaches in northern Afghanistan 119
KERSTI LARSDOTTER

7 Counterinsurgency doctrine: the US experience 142
THOMAS MOCKAITIS

8 No simple formula: the use of military force in counterinsurgency 159
JAMES S. CORUM

9 Adapting to counterinsurgency: the US Army in Iraq, July 2004–December 2006 181
BRIAN BURTON AND JOHN NAGL

10 Inviting the Leviathan: external forces, war and state-building in Afghanistan 208
JAN ANGSTROM

11 Between reluctance and necessity: the utility of military force in humanitarian and development operations 231
ROBERT EGNELL

12 War, what it is good for ... 258
ISABELLE DUYVESTEYN

*Index* 279

# Illustrations

## Figure

9.1 Chiarelli's concept of full spectrum operations in counterinsurgency 190

## Tables

5.1 Comparison of situational factors for DRC/MONUC and Sierra Leone/UNAMSIL 94
6.1 Attacks in Afghanistan and two provinces, 2004–2006 133
9.1 Iraqi civilians killed as a result of war-related violence, July 2004–December 2005 197
9.2 Iraqi civilians killed as a result of all violent causes, January 2006–December 2006 197
9.3 Average number of daily attacks by province, August 2005–February 2007 198
9.4 Multiple-fatality bombings in Iraq, July 2004–December 2006 199

# Contributors

**Jan Angstrom** is a researcher at the Swedish National Defence College. He holds a PhD from the Department of War Studies at King's College in London. His research interests include the nature and conduct of modern war as well as the development of strategic thought. He has published articles in several journals, including *Civil Wars, Security Studies, Studies of Conflict and Terrorism* and *European Political Science.* He is also the co-editor (with Isabelle Duyvesteyn) of *Rethinking the Nature of War* (Frank Cass, 2005) and *Understanding Victory and Defeat in Contemporary War* (Routledge, 2007). He has co-authored a textbook on military theory in Swedish, *Militärteorins grunder* (2005).

**Brian Burton** is a research intern at the Center for a New American Security and a master's degree candidate in Georgetown University's Security Studies Program. He graduated with a bachelor's degree in political science from Stanford University, where he also received Interdisciplinary Honors in International Security Studies and the Firestone Medal for Excellence in Undergraduate Research for his thesis entitled, 'Counterinsurgency Principles and US Military Effectiveness in Iraq'.

**James S. Corum** is Dean of the Baltic Defence College and formerly Professor of Military History in the Department of Joint and Multinational Operations at the US Army Command and General Staff College at Fort Leavenworth. He also taught comparative military studies at the School of Advanced Airpower Studies, at Maxwell Air Force Base. He is the author, together with Dennis Showalter, of *Bad Strategies: How Major Powers Fail in Counterinsurgency* (Zenith, 2008). He has also published, among others, *The Luftwaffe: Creating the Operational Air War, 1918–1940* (University Press of Kansas, 1997) and *The Roots of Blitzkrieg: Hans von Seeckt and German Military Reform* (University Press of Kansas, 1992). Dr Corum, a lieutenant colonel in the Army Reserve, is a graduate of Army Command and General Staff College and Air War College.

**Christopher Dandeker** is Professor of Military Sociology in the Department of War Studies at King's College London. He served as Head of the School of Social Science and Public Policy from 2005–2008, and from 1997–2001 he was Head of the Department of War Studies. He is a Fellow of the Inter-University Seminar on Armed Forces and Society (IUS) and a member of its Council. He is also an Associate Editor of *Armed Forces and Society*. In addition, he is a Vice-President of the Research Committee, Armed Forces and Conflict Resolution of the International Sociological Association. His most recent publications include (with D. Segal) 'Conflict, Competition and Cooperation in Twenty-First-Century Military Peacekeeping Operations', in Ann Denis and Devorah Kalekin-Fishman (eds) *The New ISA Handbook of Contemporary Sociology* (Sage, 2009).

**Isabelle Duyvesteyn** is Associate Professor at the Department of History of International Relations, Utrecht University in the Netherlands. She is currently heading a research project into the history and effectiveness of counter-terrorism and counter-insurgency. She holds a PhD from the Department of War Studies at King's College in London. She has published, among others, in *Civil Wars, Security Studies* and *Studies in Conflict and Terrorism.* She edited, together with Jan Angstrom, *Understanding Victory and Defeat in Contemporary War* (Routledge, 2007) and contributed chapters to Magnus Ranstorp (ed.) *Mapping Terrorism Research* (Routledge, 2007) and Gelijn Molier and Eva Nieuwenhuys (eds) *Peace, Security and Development in an Era of Globalization: A Multidisciplinary Approach to the Process of Peace Building After Armed Conflict* (Leiden Brill Academic Publishers, 2009).

**Robert Egnell** is Lecturer in War Studies at the Swedish National Defence College, working on peace-support operations, civil–military relations and African security. He received his doctorate from the Department of War Studies, King's College London, with a thesis titled, 'The Missing Link: Civil–Military Aspects of Effectiveness in Complex Irregular Warfare'. He has previously been a researcher at the Swedish Defence Research Agency, a lecturer at the Department of Political Science at the University of Dar es Salaam, and a guest researcher at the Swedish National Defence College. Dr Egnell is also a Captain in the Swedish Army Reserves with international experience from the first Swedish battalion in Kosovo.

**Sarah Kreps** is Associate Professor in international relations at Cornell University's Department of Government. Her research interests include international conflict and cooperation, alliance politics and proliferation of weapons of mass destruction. Her work has appeared in *The Journal of Strategic Studies, Security Studies, Political Science Quarterly, The African Security Review, Duke Journal of Comparative and International Law, The International Journal of Sustainable Development and World Ecology* and *Intelligence and National Security.* She has, furthermore, contributed to the

*Washington Post, International Herald Tribune* and *Christian Science Monitor.* She obtained her PhD from Georgetown University and was a postdoctoral fellow at Harvard University's Belfer Center for Science and International Affairs, and a senior fellow at Georgetown University's Institute for International Law and Politics.

**Kersti Larsdotter** is a Research Assistant at the Department of War Studies, Swedish National Defence College, and a research student at the School of Global Studies, University of Gothenburg. Her research interests include military conduct in peace operations and counterinsurgency operations as well as the impact of culture on military interventions. She has published, amongst others, on new wars, culture and military intervention. She is currently completing her PhD at the University of Gothenburg.

**Thomas Mockaitis** is Professor of History at DePaul University. As an adjunct faculty member of the Center for Civil Military Relations, he has co-taught several terrorism/counterterrorism courses and worked with a team of specialists to help the Romanian Army rewrite its civil–military cooperation doctrine. He has lectured at the US Marine Corps Command and Staff College and the Canadian Forces Staff College, and presented papers at the Pearson Peacekeeping Center (Canada), the Royal Military Academy Sandhurst (UK), and at conferences co-sponsored with the Military Science Department of the Austrian Ministry of Defense. A frequent media commentator on terrorism and security matters, Dr Mockaitis has appeared on Public Television, National Public Radio, and various Chicago radio and TV stations. He appears regularly as a terrorism expert for WGN TV News (Channel 9). He co-edited *Grand Strategy and the War on Terrorism* (Frank Cass, 2003) with Paul Rich. He is the author of *British Counterinsurgency: 1919–1960* (Macmillan, 1990), *British Counterinsurgency in the Post-Imperial Era* (University of Manchester Press, 1995) and *Peacekeeping and Intrastate Conflict: The Sword or the Olive Branch?* (Praeger, 1999), and *The Future of Peace Operations: Old Challenges for a New Century* (Frank Cass, 2004), co-edited with Erwin Schmidl. Dr Mockaitis earned a BA in European History from Allegheny College in Meadville, PA, and his MA and PhD in Modern British and Irish History from the University of Wisconsin-Madison.

**John Nagl** is Senior Fellow at the Center for a New American Security. A retired Army officer with combat service in Operations Desert Storm and Iraqi Freedom, he concluded his military career as the commander of the 1st Battalion 34th Armor, training transition teams for service in Iraq and Afghanistan. Nagl is a West Point graduate with a DPhil in international relations from Oxford University. He is the author of *Learning to Eat Soup with a Knife* and was on the writing team that produced *FM 3–24, Counterinsurgency.*

**Christopher Spearin** holds a BASc from McMaster University and an MA in International Affairs from Carleton University. He conducted his doctoral work in political science, with a speciality in security studies, at the University of British Columbia. Prior to joining the staff of the Canadian Forces College in the summer of 2003, Dr Spearin was a Department of National Defence R.B. Byers Post-Doctoral Fellow at York University's Centre for International and Security Studies. Dr Spearin contributes to the College's programmes, including the Joint Command and Staff Programme, the Advanced Military Studies Programme, the National Security Studies Programme, and the Canadian Security Studies Programme, through seminars and lectures related to security concepts, Canadian foreign and defence policy, aerospace studies, the nature and actors of contemporary conflict, and military strategy. In January 2007, Dr Spearin became the Chair for the Department of Security and International Affairs. Previously, he had held the posts of Deputy Director Research and Co-Chair of Aerospace Studies. Dr Spearin's research concerns change in militaries, global security governance, non-state actors, mercenaries, the privatisation of security, and Canadian foreign and defence policy. His work has been published in a variety of forums, including *Canadian Foreign Policy, Canadian Military Journal, International Peacekeeping, Journal of Conflict Studies, Civil Wars, World Defence Systems, Contemporary Security Policy, International Journal, International Politics* and *Parameters*.

# Foreword

When I was asked to write the foreword to *Modern War and the Utility of Force*, I was curious as to the interest: I had previously been impressed by the work of Dr Isabelle Duyvesteyn, but had little idea she had an interest in my own. To therefore find that the ideas in my book, *The Utility of Force, The Art of War in the Modern World*, had been part of the foundation of the inquiry set out in this book was both a revelation – and flattering indeed.

*The Utility of Force* was created from thoughts and events drawn from my experience of operations. One should expect each operation to be different: they occur in different places against different enemies. But these were all different for another reason: they were totally distinct from the operations we were trained, organised and equipped for – and each was seen as an aberration from a norm of war, usually referred to as Inter-State Conventional War. This paradigm of war had been the evolutionary crucible in which all our military and civil institutions that apply and control armed force had developed. The consequence of this institutional view and mindset was that each operation was labelled for its aberrational purpose, such as: peacekeeping, humanitarian, stabilisation, etc. As a result, there was no unifying theory as to how armed force was used in these new and real circumstances since it was difficult to carry the lessons from one operation to another. And yet, these operations were the norm, at least in my experience – and as I conducted operations I therefore searched for a unifying theory. It is very good to find in the chapters of this book so much support from academic inquiry for the theory I came to from practical experience.

With hindsight, and before being asked to write this foreword, I came to think there were two matters in particular that should have been given greater emphasis in writing *The Utility of Force*. The first is to understand that the nature of the battle or fight is not the same as the nature of the war in which it is being fought to win. Battle is, and will remain, as bloody and as frightening as it has ever been. Battles can be small and involve a few personnel, ships or aircraft exchanging fire for as long as they have ammunition, or big with formations of all arms being manoeuvred and supplied over considerable distances. For example, I have never found the ideas of High and Low Intensity Conflict helpful; for those who have been in a battle, big or small, the experience is very intense and it is not clear what else but conflict or fighting is being measured for intensity. What is important to understand is the purpose of the battle in winning the war. If this crucial linkage is not made and understood, then you have the all-too-observable phenomena of winning every fight and losing the war.

The second matter is related both to the first and to the objectives of the battle and the war. The two sides in a fight or a battle are in a bilateral relationship, otherwise there would not be a fight – but each is seeking a unilateral outcome, win or lose. If the strategic objective, the aim of the war, can be achieved directly by armed force then that objective can be expressed in such unilateral terms. But if the overall strategic objective cannot be achieved directly by force of arms, then it must be achieved by other means and understood in different terms to those of the battle. In essence, these objectives are to do with establishing a condition of security.

As opposed to defence, security exists in bilateral or multilateral relationships which are joined as a result of unilateral decisions – and there can be a multilateral outcome. They have to be unilateral because there is no absolute definition of security: each party measures it in terms of its own perception of its vulnerabilities. That is why a multilateral resolution is also possible, if a way can be found to address the vulnerabilities of each party.

A central theme of *The Utility of Force* is that we no longer live in a world of war and peace, but rather one of confrontation and conflict. You and the armed opponent have directly opposed objectives in the conflict; they can be understood as the opposite sides of the same coin that you are fighting for. But this is not the case in the confrontation: in this case we and our opponent operate amongst the people, both literally and through the media. Each group – us, the opponent and the people – has a different objective, and these objectives need to be aligned in order to resolve the confrontation.

The conflict, fight or battle, nests within the confrontation. Each has a different objective. The objective of the conflict or fight, of whatever level or scale, is achieved by military force. The objective of the confrontation is achieved by a combination of means – political, economic, humanitarian and possibly military. The achievement of both objectives is closely related and occurs in parallel rather than in sequence. The actions in the conflict and those in the confrontation must be linked and coherent with a single driving logic. The art is in the choice of objectives and the pattern of their achievement, so that those of the conflict reinforce and complement those of the confrontation to the disadvantage of the opponent. The skill is in the way armed force is applied in conjunction with the other means available. Together they work in building the narrative that exposes and isolates the armed opponent so as to allow his defeat in battle. The desired result of this activity is to lead the people – amongst whom you fight – to make their unilateral decisions to your advantage in the confrontation.

These ideas, like those in my book, evolved over time. The practice of war, like that of all arts, must begin with thoughts, which become honed through experience. This book *Modern War and the Utility of Force* is a crucial part of the honing.

General Sir Rupert Smith KCB DSO OBE QGM

# Preface and acknowledgements

This book is about the usefulness and utility of force. General Sir Rupert Smith published a book in 2005 entitled *The Utility of Force*. According to the book reviews, Smith's ideas and provocative thinking made him the 'new Clausewitz'. Considering that Clausewitz's texts, written almost 200 years ago, still form a benchmark in the field of war and peace studies, this is quite a large claim. Even though the book was well received, the main point Smith makes – that military force today is mainly of tactical utility – we argue, has not received the attention and debate it deserves. Smith notes that we have entered a new paradigm of war among the people. In this new paradigm, it has become more difficult to translate the use of the military instrument into political success, as achieving one's aims depends upon changing the will of the opponents rather than breaking their strength. For all its strengths – and they are considerable – General Smith's book and theory of modern war is in need of further empirical examination, and his thoughts should be contrasted with other reasoning on the utility of force. This necessity is most acute for types of operations that Western states' organized militaries were not designed to perform. These operations include peacekeeping and peace operations, humanitarian interventions, state-building operations, counter-insurgency operations and counter-terrorism operations. How force has been and can be useful in these circumstances is the subject of this volume.

## Acknowledgements

This book has been long in the making. Its drawn-out gestation process, however, contributed to its quality. A first test run of the arguments was presented during the British International Studies Annual Conference in Cambridge in December 2007. We would like to thank all the participants in the panel discussion for their constructive feedback. The arguments benefited from this debate. Subsequently, a special issue of the journal *Small Wars and Insurgencies* was published on the topic of the utility of force in 2008. The contributions to this issue have been included in this collection. We would like to thank Taylor and Francis for the permission to reprint these articles here. After the publication of the special issue, the comments and feedback were such that we contemplated an extension in the shape of this edited volume. Much to our pleasure, we attracted the attention of a series of scholars to deepen and broaden the debate. The end result is here before you.

The editors would like to thank Andrew Humphrys and Rebecca Brennan at Routledge for their encouragement and faith that the volume would materialise, even when things were not going smoothly. We would like to thank Allie Waite and Carl Gillingham for their professionalism and assistance towards the production of the final manuscript. Without the input and hard work of the contributors, however, this volume would not have seen the light of day. A large part of our gratitude goes to them and their insightful comments on the project along the way. We would also like to thank a number of our colleagues and friends – no one named, but no one forgotten – who have put up with reading early drafts and hearing near endless proclamations on how interesting the utility of force discussion is. Finally, our families deserve a world of thanks. Isabelle's tradition of 'one book, one baby' was continued with the arrival of little one number four and the delivery of the manuscript in the autumn of 2009. Jan, meanwhile, is only half as effective in the first regard, but has proudly progressed to using a snow shovel to make way through heaps of Barbie dolls and Star Wars Lego every night.

The utility and effectiveness of military force in modern conflicts is a topic that engages many. It goes to the heart of many questions on what societies currently value and the extent to which they are willing to pay dearly for these values. As such, it stirs fierce debates among politicians, professional militaries, peace activists, academics, and among the electorates of many states, whose men and women are engaged in conducting military operations. It is impossible to describe these categories collectively in any other way, but as: students of the utility of force. To those, our fellow strugglers in understanding if and how military force can be used to achieve political ends, we dedicate this book.

Jan Angstrom and Isabelle Duyvesteyn
October 2009

# 1 War, what is it good for?

*Jan Angstrom and Isabelle Duyvesteyn*[1]

War is among the most destructive of humanity's inventions. It takes lives, destroys homes, property and livelihood, it deprives people of their dreams, and scars survivors for life. Still, the use of military power is not seldom turned to by decision-makers and individuals. What is war good for, to paraphrase American soul artist Edward Starr? What is the role and utility of the use of armed force in modern conflicts? After the Cold War, Western armed forces have increasingly been called upon to intervene in internal conflicts in the Developing World. Wars have ravaged Central Asia, Western and Central Africa, the Balkans, South East Asia and the Middle East, spurring governments in the West to action. Western armed forces, in particular, have been called upon to carry out missions that they traditionally have not been trained and equipped for, in environments that they traditionally have not been prepared for. A number of these 'new' types of operations in allegedly 'new' wars stand out, such as peace-enforcement, state-building, counter-insurgency, humanitarian aid, and not least, counter-terrorism. The success rate of these missions has, however, been mixed, providing fuel for an increasingly loud debate on the utility of force in modern war.

One of the most outspoken of these voices belongs to former British General Sir Rupert Smith, who, upon retirement, published his thoughts on the use of force in a seminal 2005 book entitled *The Utility of Force: The Art of War in the Modern World.*[2] The book has a revolutionary aura around it, with the author being hailed as the new Clausewitz;[3] Smith himself has called for nothing less than a complete paradigm shift from what the military establishment sees as regular war to 'war amongst the people'.[4] Smith fits into a very long tradition of former military commanders sometimes referred to as 'soldier-scholars', who after their active service reflect upon their experiences. In the present wave of writings by former military commanders, there is a strong critical undercurrent. Commanders such as Sir Michael Rose, Wesley Clark and Sir Rupert Smith have taken up their pens to share their points of view.[5] Their books and articles criticise Western military performance, including their own. Smith's book stands out because of its scope and penetrating analysis of contemporary strategic issues. Essentially, Smith pointed out that a lack of coherent strategy for the use of force and a failure to grasp the new strategic environment had hampered Western intervening forces from reaching their desired political aims in the wars in the Balkans and elsewhere during the 1990s. Smith argues that we have entered a new paradigm of war among the people. In this new paradigm it has become more difficult to translate the use of the military instrument into grand strategy and political success. Military force, Smith suggested, 'is no longer used to decide the matter but to create the conditions in which the strategic result is achieved'.[6] Military force had been reduced to tactical utility. The role and utility of force, following Smith, had changed after the Cold War, but Western military and political institutions had at large failed to appreciate the depth of this change and its consequences.

This volume takes up the central theme of Smith's book. As James Gow has lamented, 'no one has grasped the essential questions the General addresses, which concern the utility of force – that it is, how to use armed force to achieve a desired and stable political outcome.'[7] This present volume poses as its central question: what *is* in fact the utility of force? Is force useful for anything other than a complete conventional defeat of a regular opponent, who is confronted in the open field? Smith himself has proposed that the utility of force is not clearly understood; neither by the military establishment, nor by the political leadership. He writes:

> Various terms are now used to apparently define what it is the military are expected to achieve – terms such as 'humanitarian operation', 'peacekeeping', 'peace enforcement', 'stabilization operations', 'achieving a stable and secure environment' – yet these are in reality more of a description of the activity rather than the outcome. None the less, many people, including senior decision makers and policy makers, use and understand them as descriptive of a good outcome, and this can lead to confusion of purpose.[8]

This confusion of means and ends is painfully clear in many Western-run operations today. General Smith's book is certainly well-argued and an important contribution to the debate on the utility of force, but it still has some shortcomings. James Gow has, for example, argued that its conceptualisation of war 'is wrong – or at least open to question, or possibly confused'.[9] The notion of 'war amongst the people' as a fruitful category of war is therefore contentious. Furthermore, the type of war Smith focuses on is in fact not new and therefore one can question the concept of a new paradigm. Irregular wars have been an important type of war since the nineteenth century, and have been the dominant form of war – at least from a quantitative perspective – since 1945. Finally, one could add that the book and its conclusions have not been subject to more careful empirical scrutiny. Although based upon four decades of personal experiences in the British armed forces, the book was arguably not intended to be a work of social science in which selection of cases and sources is critical, and an – often theoretically grounded – analytical instrument is carefully employed.

In this volume, the overall aim is to contribute to filling this void. In particular, we aim to analyse the utility of force in missions for which military force, as it has been institutionalised in the West, was not originally designed or intended. We know that military force can have a distinct effect and result when employed in conventional war of state pitted against state. On the other side of the spectrum, we know from other studies that deterrence, the threat of the use of force, has a proven track record. Everything in-between – ranging from (counter-)insurgency, peacekeeping and peace operations, state-building operations, stabilisation operations, humanitarian intervention, to (counter-)terrorism – seems a lot more complicated. The usefulness of military force in these instances is less clear-cut, which is the main reason for focusing on these missions. We will jointly deal with each of these types of operations below. Moreover, collectively the volume also contributes to the burgeoning debate on the changing conditions for the creation of military force. The link to utility of force is of course obvious. The types of forces created, their doctrine, and their relationship with other actors, are important aspects to consider when analysing the utility of force in modern conflict.

The amount and quality of research on the utility of force in these operations vary to a high degree, and will therefore be dealt with in the various chapters respectively. In spite of the fact that using a conceptual framework is unavoidable when dealing with the utility of force in these missions, we have not tried to impose a conceptual straitjacket on these partly new tasks of the military. The reason of course is that deciding upon such a conceptual framework prior to empirical analysis can impede our understanding of the conditions under which force has utility. Instead, an underlying aim of the volume is also to deal with how we should relate these missions to one another when it comes to the utility of force. What differences and similarities are there between types of operations? Is it possible, for example, to make valid generalisations on counter-terrorism based on research on counter-insurgency? Or to what extent do humanitarian operations, peace operations and state-building operations overlap? How can modern Western armed forces best be utilised for effect? How is military power created? Are Western armed forces supposed to form the bulk of the 'boots on the ground', or should they operate as trainers and advisors for local forces or private security companies? And how does the conceptual framework of the armed forces develop?

Specifically, we engage with Smith's book and the broader topic of utility of military force in several ways. First, by focusing on the available empirical evidence, the contributors to this volume have tried to test to what extent force serves a purpose. While Smith's study contained a theoretical argument infused with his first-hand experiences in recent conflicts, this volume tries to ground the discussion more strongly within the context of available empirical data. Second, although Smith's work is impressive, it did not cover all aspects of the utility of force in modern conflict. Hence, another way of engaging with his ideas is to broaden the discussion to include more facets of force. One example of this approach is Christopher Spearin's chapter on private security companies and how and to what extent they can supplement or compete with regular state-run forces when using military power. Third, while Smith's book certainly can be said to include a theory of modern war, collectively the authors of the present volume also consider alternative theoretical frameworks, while actively engaging with Smith's thoughts. Hence, it is along these three avenues that the contributions of the volume challenge, extend and further elaborate on the utility of force in modern conflict.

## Discussing the utility of force

The current wave of interest in the relevance and utility of armed force has developed in a specific context. Over the years, the utility of force has been widely researched and debated. However, the overwhelming majority of this research has been focused on traditional interstate conventional confrontations. It was within this context that concepts such as coercion, compellance, brute force and deterrence – to name a few – emerged. With the development of nuclear weapons during the height of the Cold War, Thomas Schelling famously argued that 'strategy can no longer be thought of, as it could for some countries in some eras, as the science of military victory'. Instead, he suggested, it 'has become the diplomacy of violence'.[10] Although concepts derived from the study of interstate war proved partly valuable even in the study of irregular warfare during the Cold War, the literature on the utility of force in irregular contests seemed to have lived a separate life from the mainstream literature on the utility of force for too long. And due to that most Western strategic thought during the Cold War was undoubtedly directed at the danger of a nuclear confrontation with the Soviet Union, peacekeeping, counter-insurgency and other non-interstate phenomena were downgraded.[11]

The end of the Cold War and the lull of the Western strategic interlude during the 1990s were interrupted unexpectedly by the terrorist attacks in New York and Washington, DC on September 11, 2001. The attacks made it apparent that Western patterns of thought were partly outdated. Since then, the development of strategic thought has followed a logical path. At first, there was an upsurge of interest in changes in the conduct of war. Martin van Creveld, even preceding this discussion, proclaimed that war had changed its face and the low-intensity conflicts in the Third World had surpassed Great Power interstate wars as the main security problem globally.[12] Following the predominantly intrastate wars of the early 1990s, Kalevi Holsti,

Herfried Münkler and Mary Kaldor each authored influential books on new wars, describing changes in the conduct and purpose of war.[13]

The second development of post-Cold War strategic thought was an intensified debate on how war should be conceptualised. Had war become something else than Prussian strategist Major General Carl von Clausewitz had envisaged in his understanding of war as politically instrumental and a tool of statecraft? In this debate, a number of experts pronounced war in the post-Cold War world as having become 'post-modern', a 'spectator-sport' or indeed not driven by political motives but greed and ethnicity.[14] An equally large – or larger – contingent of scholars soon emerged to criticise these perspectives and maintained that war still could be viewed as instrumental and politically rational.[15]

The third major debate in post-Cold War strategic thought raised the question of whether or not the meaning of victory and defeat in modern war had changed. This question was more or less unavoidable, as the US president George W. Bush's declaration in 2003 that the war in Iraq had been won turned out to be somewhat premature. In this debate, it became clear that not only was there a distinct possibility that the game had changed, but it was also possible that the way one kept the score of the game had changed.[16] The findings of this literature partly coincided with, and were eventually reinforced by, new developments in conflict research aiming to investigate the effectiveness of peacekeeping missions.[17] Yet another strand of research emerging from this problem complex dealt with military effectiveness in terms of how effective military power can be created and sustained.[18]

With even the contents and conditions of victory and defeat in flux, it did not take long until questions were raised surrounding the utility of force. Among the sceptics towards the utility of force we count, in perhaps its most extreme form, John Mueller. He wrote in his book *The Remnants of War* that 'even if we consider criminal civil conflicts to be warfare, we are dealing increasingly not with a new kind of war but with the residue of warfare'; it is 'the only kind still going on … war has increasingly been reduced to its pathetic, if often highly destructive, remnants'.[19] Barry Posen, another sceptic, has argued that force in relation to refugee crises has limited utility:

> All except full-scale war are temporary expedients. They reduce hardship and save lives but they do not solve the original political problems that produced the refugees. It is probably true that a full-scale war is the best military answer to refugees produced by cruel occupations or highly repressive indigenous regimes.[20]

Condoleezza Rice warned in an article in *Foreign Affairs* in 2000, before the outbreak of the 'war on terror', that force has a circumscribed applicability:

> The president must remember that the military is a special instrument. It is lethal, and it is meant to be. It is not a civilian peace force. It is not a political referee. And it is most certainly not designed to build a civilian society. Military force is best used to support clear political goals, whether limited, such as expelling Saddam from Kuwait, or comprehensive, such as demanding the unconditional surrender of Japan and Germany during World War II.[21]

Many others are thoroughly convinced that military force still has utility beyond this industrial war paradigm. On the one hand, there are those who argue that 'the morality and legality of the use of force automatically defined its utility: if it was moral and legal it would be successful and deliver the desired result.'[22] This current is strongly visible in the many peace operations and its concomitant peacekeeping ideology established since the 1990s. On the other hand, there are those who frame it in terms of necessity, because we have the means available and there is a (security) space to fill, the logical choice is the use of the military. Madeleine Albright, as American Ambassador to the United Nations, exhorted to Colin Powell: 'What's the point of having this superb military that you're always talking about if we can't use it?'[23] Sam Sarkesian and Robert Connor claim that: 'This "Albright Syndrome" seems to pervade in civilian circles, particularly those with little or any military experience.'[24]

Morality and necessity aside, what available evidence is there that military force is indeed useful – in particular in refugee crises, but also in other emergencies, such as in dealing with terrorism, peace operations, counter-insurgency and nation building? Force, writes Smith, 'does not have an absolute utility, other than its basic purpose of killing and destroying'.[25] To kill or to destroy must have a direct link with the larger aims of the clash of wills, i.e. to make the opponent do as you desire. In Smith's words:

> Since the overall objective we seek by employing force is to win the clash of wills, it follows that every trial of strength must be won in such a way that each success complements and supports the measures to win the clash of wills. Only then will the forces we send have the utility and deliver the political result desired.[26]

When force is employed, this leads to death and destruction. However, these outcomes need to deliver the political goals for which force was ultimately employed in the first place. Smith calls many of the contemporary uses of force 'sub-strategic'. While the strategic use of force involves deployment and employment of force towards political goals, the sub-strategic use merely involves deployment towards creating certain conditions. In the sub-strategic use of force, the political goal – capturing the will of the people – should be reached through other means:

> we intervene in, or even decide to escalate to, a conflict in order to establish a condition in which the political objective can be achieved by other means and in other ways. We seek to create a conceptual space for diplomacy, economic incentives, political pressure and other measures to create a desired political outcome of stability and if possible democracy.[27]

The link between the military goal and the political aim has become less direct. In the past, the military aim, such as occupation and the removal of regimes, was directly linked to the political aims, including territorial enlargement or removing undesirable political opponents. As Christopher Dandeker writes in his contribution to this volume:

> In essence, in wars amongst the people, the objective for an intervening state or coalition of states, [is] to change the intentions of insurgents and, ultimately, win over the people in an ideological battle or 'war of ideas' for 'hearts and minds'.

At present, however, the political leadership confuses the deployment of force with the employment of force.[28]

Smith makes a crucial distinction between confrontation and conflict. They differ in regards to their goals.

> In confrontations the aim is to influence the opponent, to change or form an intention, to establish a condition and, above all, to win the clash of wills. In conflicts the purpose is to destroy, take, hold: to forcibly attain a decisive outcome by the direct application of military force.[29]

The essence of the new paradigm, he argues, is establishing conditions among the people. As Christopher Dandeker points out, there is an 'inherent plasticity of the idea of condition', and 'strategic patience' is required for its achievement.

Interestingly, this idea of condition rather than victory defining war has been substantiated by several empirical studies; one recent study has pointed out that the use of force can lead directly to the desired effect in cases of occupation and regime change, i.e. Smith's idea of destroy, take and hold.[30] When it comes to changing behaviour and attitudes of the opponent and opposing populations, the utility of force seems to be more limited. In Iraq, for example, a change in behaviour was a main issue at stake, namely supporting and trusting the newly established political regime before a withdrawal could take place. Another study concludes that, in general, when political guidance in military undertakings is strong, success is more likely.[31] The key to these uses of force is, according to Smith, formed by attacks against 'a target which is important to the opponent in order to focus his attention and change his intentions'.[32] In the end,

> in war amongst the people no act of force will ever be decisive: winning the trial of strength will not deliver the will of the people, and at base that is the only true aim of any use of force in our modern conflicts.[33]

At most, 'military force can only achieve tactical results and to have more than passing value these must be stitched into a greater plan'.[34]

Fittingly, General Smith's thoughts also resonated with other findings in the field. Although not using the concepts of integration, responsiveness, skill and supply in the way they are used by, for example, Risa Brooks, one could easily fit his ideas in the interpretative framework of 'military effectiveness'.[35] The focus of the latter literature has been to develop a framework for assessing when military forces might be exhibiting qualities that allow them to utilise their resources with maximum gain to attain political goals. 'Effective' militaries are those that are able to take basic resources in population, wealth and technology and translate those into actual military power in a conflict situation. As such, effective militaries exhibit four key properties.

First, they exhibit high levels of integration, or the capacity to ensure consistency both within and across levels of political–military activity (i.e. integrating military activity at the strategic, operational and tactical levels with the force development activities, such as training, doctrinal formulation, etc., that intersect those levels of activity). Although Smith does not use the concept of integration, the concept that tactical military activity must be tailored with a state's political goals and military strategy is central to his thesis in the book. Without such integration, even large commitments of resources and uses of force are, at best, likely to be wasted, and at worst counter-productive to larger political goals in contemporary conflicts.

Second, effective militaries exhibit high degrees of responsiveness. That is, they can tailor their military activity both to their own internal and environmental constraints, as well as to their adversaries' military activities and strengths and weaknesses. Arguably, for any military to adapt and respond to the type of conflict scenarios to which Smith points, they will need to exhibit high levels of responsiveness.

Third, militaries must be able to produce high levels of skill, which captures both the aptitude of personnel to undertake complex tasks, as well as their motivation to do so.

Finally, the capacity to supply an armed force with quality weapons and equipment (in addition to appropriate systems, which is a question of integration and responsiveness) is essential to realizing the potential of basic resources and investments in military forces. The more a military exhibits these four properties, the more effective it is; that is, the better able it is to realise the intrinsic contribution military force can make in an armed conflict. More effective militaries – those exhibiting these four attributes and organisational competencies – will be better able to internalise the lessons Smith articulates and adapt to the form of warfare and conflict scenarios he highlights.

Questions, however, have also been raised even among those who insisted that war is still a relevant tool of statecraft. For example, Gow suggests that 'War retains its relevance. But the balance between destructive force and politics needs to be calibrated by a greater admixture of politics and other means alongside the use of the military.'[36] And it is in this context this volume should be seen. Provided that war may not be an ideal form of communication or bargaining between at least two actors, what is the utility of force in modern war when the aims of war, the means of war and the methods of war seemingly have changed? Indeed, what is the utility of force when our understanding of victory and defeat are in flux? And what implications does this have for the use of force and creation of military power?

## The importance of studying utility of force

At least four sets of interrelated paradoxes make studying the utility of force in modern conflict important. First, of late the state's monopoly of legitimate violence has been challenged by an array of different actors using military means. This has been a common trait of modern armed conflict. Rebels, terrorists and private security companies, employed both by intervening states, international organizations and the parties in the conflict themselves, combine to make the use of force dispersed. Considering that the wars are seemingly not solved by military means, why do actors challenging the state with military means emerge in these conflicts? What is the use of armed forces in modern warfare?

This paradox is further compounded by a second one. The concentration of force at the decisive point as the surest way to military victory was a tenet formulated by Carl von Clausewitz almost two centuries ago. In regular interstate conventional warfare, this claim has had some merit. In irregular warfare, however, this principle is hardly applicable because it has been often pointed out that the decisive point is intangible, i.e. the political programme of the insurgents or the population at large. Still, the overwhelming majority of Western armed forces prepare and develop tactics primarily for conventional warfare. What is the utility of armed forces in light of this extreme paradox in modern warfare?

Third, a number of scholars have found that wars concluded by a military victory provide a more stable peace compared to wars concluded by negotiations (see also the more elaborate discussion in the conclusion, pp. 258–278). This finding stands in stark contrast to the feasibility of a military solution in contemporary conflict. It is well know that the majority of wars are civil wars, fought in an irregular manner, and a military solution in irregular war is extremely rare. What is the role of armed forces in a situation where its use is extremely problematic, yet their positive effects are strongly desired, not least by the need for humanitarian aid?

Finally, lately there have been several major changes for Western armed forces. In the jargon of the Pentagon, these changes have been coined as transformation of the armed forces from deterrence to expeditionary warfare, from attrition to manoeuvre, and from conscript to professional. It thus appears that Western armed forces are in the process of redefining their objectives, methods and available means. Still, the overwhelming majority of Western armed forces are developed, equipped and trained to fight interstate conventional wars, rather than what arguably is demanded of them and what they will be challenged by in the near future.[37] What is the utility of armed force and how should they be shaped?

## Structure of this volume

The volume continues as follows. In the second chapter, Christopher Dandeker argues that analysts of the utility of force in modern conflicts need to recognise that the political aims have changed in the current strategic context. Moving from victory to an acceptable condition, as Dandeker puts it, has several consequences for the use of force. Based on an analysis of the Afghanistan case, he discusses how the notion of a strategic narrative has changed due to the shift from victory to condition and raises important questions of civil–military relations and force generation for current conflicts. Dandeker addresses several of the main points of Smith's book and also shows the degree to which Smith relies on a particular approach to military sociology.

In the following chapter, Christopher Spearin analyses the emergence of so-called private security companies. Spearin evaluates how private actors may or may not be effective replacements for regular forces in modern conflicts, and whether private actors can train indigenous forces and serve as advisors. As such Spearin demonstrates how private actors can contribute to reaching political goals in the modern strategic context, thereby pointing to how other actors than state-organised forces can reach their goals. Of particular relevance for the utility of force, Spearin identifies a further area of utility of force, namely how Western forces (and even private ones) can be used effectively to train indigenous units. Spearin argues that, although private security companies can serve as an alternative way to solve military problems, it is unlikely that they can solve problems associated with force generation, not the least because their effectiveness tends to rely on the already trained and experienced soldiers and officers trained by the regular army.

In Chapter 4, Isabelle Duyvesteyn examines the use of force in counterterrorism operations. Duyvesteyn argues that the military option in response to terrorism seemingly is not used as often as its detractors suggest. She finds the use of force against terrorism of limited utility. In particular, it seems that using military power against terrorists plays into the hands of terrorist strategy aiming to provoke a military overreaction. In turn, this makes it easier for terrorists to recruit and gain legitimacy. Duyvesteyn thus engages with Smith's conclusions from an empirical standpoint.

In the following chapter, Sarah Kreps analyses the utility of force in peacekeeping missions. Based on an analysis of the contrasting outcomes of peacekeeping missions in Sierra Leone and the Democratic Republic of Congo, Kreps identifies two seemingly critical factors to explain success in peacekeeping missions. She argues that the skill and capacity of the peacekeeping units, including a dense force–space ratio, as well as ambitious or ambiguous political aims, can explain the outcome of peacekeeping missions. Both these factors are intimately linked to the overall puzzle of how and under what conditions force is effective in modern conflicts.

While Kreps can be said to examine utility of force on the *strategic* level of peace support operations, in Chapter 6, Kersti Larsdotter analyses the use and utility of military force in peace-support operations in Afghanistan on the *tactical* level. Larsdotter differentiates between two distinct tactics in peace operations: the more robust 'show of force tactics' used by, among others, US and German forces and the less-threatening force posture of 'minimum force tactics' used by, among others, British forces. Based on a case study of two provincial reconstruction teams in Northern Afghanistan, she argues that the commonly held superior minimum force approach in these operations did not correlate with a better outcome. This raises questions about the degree to which we can generalise among peace operations, and it also raises questions on the methods of measuring success in peace operations. In the context of the utility of force, Larsdotter's results indicate that there are contexts in which more force may be better. Her results can, therefore, in the long run, contribute to a better understanding of under which conditions force can be used efficiently in peace operations.

In Chapter 7, Thomas Mockaitis provides an analysis of the US armed forces' ambivalence towards counter-insurgency. Despite being involved in what could be termed counter-insurgency throughout its history, Mockaitis shows that the US armed forces have done so reluctantly. He argues that the armed forces do not develop in isolation from the rest of the society. The perceived apparent US success in conventional war and failure in counter-insurgency has created an aversion towards the latter. A combination of external threats, perceived historical experience and self-identity of the armed forces have relegated counter-insurgency as a second-hand object of the armed forces, and therefore there has been a reluctance to deal with it explicitly in doctrines.

In Chapter 8, James Corum analyses the use of military force at a strategic level in counter-insurgency operations. Corum engages with the most critical questions of modern conflicts: how much force should be used to reach one's strategic goals and when does the use of force become its own enemy? By surveying cases ranging from the Philippines to modern-day Iraq, Corum critically evaluates a number of possible explanations, conditions and scenarios where force has become less productive. Corum concludes that, in some cases, it seems that the counter-insurgency effort succeeds by default, rather than by design. Because the conflict is not deemed important enough to warrant a large troop build-up, it is forced to solve the task in such a way that incorporates the local elites in the decision-making, thereby strengthening the legitimacy of the operations.

In the following chapter, Brian Burton and John Nagl analyse the utility of force at an operational and tactical level in counter-insurgency operations. They focus in particular on the virtues of transition as a counter-insurgency strategy. Based in no small part upon first-hand sources and experience, they argue that the US conduct of operations in Iraq was a strategic failure up to the end of 2006. Although the situation began to show signs of progress gradually during 2004 on the local level in some regions, they maintain that it was dependent upon the awareness of individual commanders of counter-insurgency, and not upon a thought-through consciously implemented strategy. The use of force, they conclude, is necessary, but needs careful constraints in counter-insurgency.

In Chapter 10, Jan Angstrom examines the utility of armed force in state-building operations. Analysing the four attempts of externally supported state-building during the past 30 years of armed conflict in Afghanistan, Angstrom illustrates how they have failed due to a flawed conception of legitimacy. The explanation suggests that international state-building attempts condition and shape a new strategic environment for the local actors. In doing so, the mode of state-building can create incentives for continuing the war or, carried out differently, create incentives for the parties to contribute in building a legitimate state from the rubble of the old state. So far, however, the different ways of using force in Afghanistan have created situations where continued war has been the rational solution on behalf of the local actors. Angstrom thus engages with Smith's thoughts by discussing the conditioning effect of military force.

In the final substantive chapter, Robert Egnell analyses the utility and trade-offs of military force in humanitarian aid operations. Using the opposing pair of concepts 'militarisation of humanitarian aid' and 'humanitarianisation of military operations', Egnell discusses the trade-offs and shifting interface between politics and force. Egnell argues that the effects of integrating these tasks in an overall approach are not only positive. Instead, they may have drawbacks both for the military and the civilian purposes, including the non-governmental-organisations community. Military force in these operations, Egnell suggests, can be both productive and counter-productive.

In the concluding chapter, Isabelle Duyvesteyn summarises the findings and relates the discussion to the analytical categories of the use of force advanced by General Sir Rupert Smith. She argues that the chapters collectively advance our knowledge on the utility of force in modern conflict in terms of how armed forces can work as coercion and deterrence, containment, and amelioration. By placing the results of the substantive chapter into a wider scholarly context, she is also able to suggest avenues for further research.

## Notes

1 An earlier version of this revised chapter was published as Jan Angstrom, 'Introduction: Exploring the Utility of Armed Force in Modern Conflict', *Small Wars and Insurgencies*, 19, 3 (2008), 297–302. The first part has been adapted from the introductory paragraph of Isabelle Duyvesteyn, 'Exploring the Utility of Force: Some Conclusions', *Small Wars and Insurgencies*, 19, 3 (2008), 423–443. We thank Taylor and Francis for kind permission to reprint this material.
2 Smith, *The Utility of Force*.
3 Gow, 'The New Clausewitz?'. Indeed Smith is a true Clausewitzean, e.g. chapter 1 and passim.
4 Taking up this argument is Betz, 'Redesigning Land Forces for Wars Amongst the People'.
5 Rose, *Fighting for Peace*; Rose, *Washington's War*; Clark, *Waging Modern War*; Clark, *Winning Modern Wars*.
6 Smith, 'Thinking About the Utility of Force in War Amongst the People', 28.
7 Gow, 'The New Clausewitz?', 1161.
8 Smith, *The Utility of Force*, 375.
9 Gow, 'The New Clausewitz?', 1160.
10 Shelling, *Arms and Influence*, 34.
11 Mahnken, 'The Future of Strategic Studies'.
12 Van Creveld, *The Transformation of War*.
13 Holsti, *The State, War, and the State of War*; Kaldor, *New and Old Wars*; Münkler, *Die Neuen Kriege*.
14 See, for example, McInnes, *Spectator Sport War*; Coker, *Humane Warfare*; Hables Gray, *Postmodern War*.
15 See, for example, Duyvesteyn and Angstrom, *Rethinking the Nature of War*.
16 See, for example, Angstrom and Duyvesteyn, *Understanding Victory and Defeat in Contemporary War*.
17 See, for example, Page Fortna, *Does Peacekeeping Work?* and Morje Howard, *UN Peacekeeping in Civil Wars*.
18 See, for example, Brooks and Stanley, *Creating Military Power*.
19 Mueller, *The Remnants of War*.
20 Posen, 'Military Responses to Refugee Disasters', 108.
21 Rice, 'Campaign 2000', 53.
22 Smith, *The Utility of Force*, 9. See also Egnell: 'Peace and support operations are inherently political and value laden, often based on Just War theory and international legal frameworks' (Egnell, 'Between Reluctance and Necessity', in this volume).
23 Quoted in Powell, *My American Journey*, 576.
24 Sarkesian and Connor Jr, *The U.S. Military Profession into the Twenty-First Century*, 195.
25 Smith, *The Utility of Force*, 18.

26 Smith, *The Utility of Force*, 278.
27 Smith, *The Utility of Force*, 270.
28 Smith, *The Utility of Force*, 4.
29 Smith, *The Utility of Force*, 182.
30 Sullivan, 'War Aims and War Outcomes'.
31 Fishel, *Civil Military Operations in the New World.*
32 Smith, *The Utility of Force*, 182.
33 Smith, *The Utility of Force*, 331.
34 Smith, *The Utility of Force*, 372.
35 Brooks and Stanley, *Creating Military Power.*
36 Gow, 'The New Clausewitz?', 1161.
37 See, for example, Thornton, *Asymmetric Warfare.*

## Bibliography


Angstrom, Jan and Isabelle Duyvesteyn (eds), *Understanding Victory and Defeat in Contemporary War.* London: Routledge, 2007.

Betz, David, 'Redesigning Land Forces for Wars Amongst the People', *Contemporary Security Policy*, 28, 2 (2007), 221–243.

Brooks, Risa A. and Elizabeth A. Stanley (eds), *Creating Military Power: The Sources of Military Effectiveness.* Stanford, CA: Stanford University Press, 2007.

Clark, Wesley K., *Waging Modern War: Bosnia, Kosovo and the Future of Combat.* New York: Public Affairs, 2001.

Clark, Wesley K., *Winning Modern Wars: Iraq, Terrorism and the American Empire.* New York: Public Affairs, 2003.

Coker, Christopher, *Humane Warfare: The New Ethics of Postmodern War.* London: Routledge, 2001.

van Creveld, Martin, *The Transformation of War*, New York: The Free Press, 1991.

Duyvesteyn, Isabelle and Jan Angstrom (eds), *Rethinking the Nature of War.* London: Frank Cass, 2005.

Fishel, John, *Civil Military Operations in the New World.* Westport, CN: Praeger, 1997.

Gow, James, 'The New Clausewitz? War Force, Art and Utility – Rupert Smith on 21st Century Strategic Operations, Tactics in a Comprehensive Context', *Journal of Strategic Studies*, 29, 6 (2006), 1151–1170.

Hables Gray, C., *Postmodern War: The New Politics of Conflict.* London: Routledge, 1997.

Holsti, Kalevi, *The State, War, and the State of War.* Cambridge: Cambridge University Press, 1996.

Kaldor, Mary, *New and Old Wars: Organized Violence in a Global Era.* Cambridge: Polity, 1999.

McInnes, Colin, *Spectator Sport War: The West and Contemporary Conflict.* Boulder, CO: Lynne Rienner, 2002.

Mahnken, Thomas, 'The Future of Strategic Studies', *Journal of Strategic Studies* 26, 1 (2003), x–xviii.

Morje Howard, Lise, *UN Peacekeeping in Civil Wars.* Cambridge: Cambridge University Press, 2008.

Mueller, John, *The Remnants of War.* New York: Cornell University Press, 2004.

Münkler, H., *Die Neuen Kriege.* Reinbeck bei Hamburg: Rowohlt, 2002.

Page Fortna, Virginia, *Does Peacekeeping Work?* Princeton, NJ: Princeton University Press, 2008.

Posen, Barry, 'Military Responses to Refugee Disasters', *International Security*, 21, 1 (1996).

Powell, Colin, *My American Journey*. New York: Random House, 1995.

Rice, Condoleezza, 'Campaign 2000: Promoting the National Interest', *Foreign Affairs*, 79, 1 (2000).

Rose, Michael, *Fighting for Peace: Bosnia 1994*. London: Harvill, 1998.

Rose, Michael, *Washington's War: From Independence to Iraq*. London: Weidenfeld and Nicholson, 2007.

Sarkesian, Sam C. and Robert E. Connor Jr., *The U.S. Military Profession into the Twenty-First Century: War, Peace and Politics*. London: Routledge, 2006.

Shelling, Thomas C., *Arms and Influence*. New Haven: Yale University Press, 1966.

Smith, R., *The Utility of Force: The Art of War in the Modern World*. London: Allen Lane, 2005.

Smith, R., 'Thinking About the Utility of Force in War Amongst the People', in J.A. Olsen (ed.), *On New Wars*, Oslo Files 4/2007. Oslo: Norwegian Institute for Defense Studies, 2007.

Sullivan, Patricia, 'War Aims and War Outcomes: Why Powerful States Lose Limited Wars', *Journal of Conflict Resolution*, 51, 3 (2007), 496–524.

Thornton, Rod, *Asymmetric Warfare: Threat and Response*. Cambridge: Polity, 2007.

# 2 From victory to success

## The changing mission of Western armed forces

*Christopher Dandeker*[1]

From time to time, we are reminded of the relevance of Clausewitz's advice to any political leader who was considering launching the inherently escalatory, violent and terrible act of war: be sure of the objective(s) one is seeking to achieve and the method by which such an act is to be carried out. Just such an occasion happened recently as scholars sought to evaluate the costs and benefits of 'Operation Cast Lead' conducted by the Israel Defence Forces (IDF) in Gaza during the weeks that fell between the result of the US Presidential election and the inauguration of President Barack Obama in January 2009.[2] This operation was, for some commentators at least, by no means an unqualified and unambiguous success – let alone a 'victory' – when measured against Clausewitz's strictures. Although the Israeli government engaged in a punitive mission against Hamas and also sought to reinstate the power of deterrence, some have doubted whether Hamas has been deterred from attacking Israel in future or given sufficient cause to reach an accommodation with the state of Israel.[3] As Hamas could and did say in effect: 'we are still here and have no intention of submitting to or accepting the legitimacy of the state of Israel.'

Was the seemingly inconclusive outcome of 'Operation Cast Lead' – the elusiveness of victory, or the lack of a clear-cut success – not an isolated phenomenon but rather confirmation of a deeper trend; a shift away from the possibility of 'victory' to the less ambitious (and malleable) idea of success? Perhaps not, when one considers the most important case of current Western intervention in the affairs of other states – Afghanistan – which has grown in significance as the major withdrawal of coalition forces (including the end of the UK's if not the US's military involvement) from Iraq in mid-2009 is implemented. One significant feature of recent political and military debates about coalition operations in Iraq is that arguments have tended to work on the assumption that 'we may not be winning but at least we are not losing'. The sense of not losing, and some grounds for optimism, have been encouraged by the apparent successes of the surge of US military forces in containing violence and providing (with indigenous Iraqi police and military organisations) the basis of security and thus a framework for economic development. Meanwhile, so far as progress in Afghanistan is concerned, arguments tend to be based on the idea that 'we are not losing yet, so let us make very sure that we do not lose even if we are facing serious challenges'.[4] Some have claimed that operations should be judged not in terms of 'winning or losing' but in terms of achieving 'success'.[5]

If there is an ongoing shift from the mission of victory to success, why has it occurred? What are the implications for how military and other actors, engaged in achieving success, persuade the audiences observing and/or engaged with their activities that these military operations are worth their continued support: that they have legitimacy; that a persuasive narrative can be provided to generate their support and commitment to see them through to a conclusion? These questions provide the focus of this chapter, which in answering them, analyses the connections between two recent developments: one is in the field of war and military affairs; the other concerns the area of communications and media coverage of the conduct of armed forces in contemporary operations, with particular attention paid to what have become known, since the US-led coalition operations in Afghanistan and Iraq, as intervention and stabilisation operations.

## From victory to success – themes in classic civil–military relations theory

The first questions to be addressed are: has there been a shift away from the idea of the *pursuit of victory*[6] to that of *success* in thinking about the use of Western armed forces? And, if so, why? Is this shift in language any more than a relatively trivial redrawing of the idea of victory in less-ambitious terms and tone?[7] Answering these questions involves locating the idea of 'success' in wider debates about the changes in the nature of contemporary warfare. In particular, attention needs to be paid to General Sir Rupert Smith's influential notion of the pursuit of a 'condition' in what he has called 'wars amongst the people', and how armed forces should be configured to conduct operations of this kind.[8] However, before considering this idea we need to understand its deeper roots; it echoes much earlier discussions in sociology and politics about the use of force in international relations – specifically the absolutist and pragmatic perspectives of Huntington and Janowitz, respectively, half-a-century ago.[9]

Janowitz's influential and prescient remarks on the 'constabulary force' served as both a description of overarching military trends as well as a prescription; they comprise a recommendation for how best to organise the military in the context of (nuclear) Cold War international affairs. His remarks anticipate some of Rupert Smith's formulations that will be analysed later. As Gray has argued:

> In his sociological study of military professionalism, Morris Janowitz distinguished between absolutists and pragmatists. Absolutists are those officers *who prefer total victory* and minimal civilian interference in matters relating to war. Pragmatists are more concerned with [the] *measured application of force* and its political outcomes [emphasis added].[10]

In a famous passage, Janowitz stated that,

> [t]he military establishment becomes a constabulary force when it is continuously prepared to act, committed to the minimum use of force, and seeks viable international relations, rather than victory, because it has incorporated a protective military posture. The constabulary concept is grounded in, and extends, pragmatic doctrine.[11]

These three components of the constabulary concept are worth disentangling. First, the military should be continuously prepared to act; that is to say, ready to deter or to be used in circumstances below the threshold of nuclear confrontation and designed to minimise the logic of escalation. Consequently, and second, it was configured for the minimum (a better term in this connection, as Gray suggests, is the 'measured') use of force.[12] This idea of a measured use of force is more consistent with Janowitz's formulation that the

> constabulary force concept encompasses the entire range of military power and organization. At the upper end are the weapons of mass destruction; those of flexible and specialized capacity are at the lower end, including the specialists in military aid programs, in para-military operations, in guerrilla and counter-guerrilla warfare.
>
> (Janowitz 1971: 418)

With regard to the third component of the constabulary concept, Janowitz was arguing that the military should be focused not on the absolutist goal of victory but on the pursuit of viable international relations. This theme reflected the way in which he was influenced by Cold War fears of nuclear escalation; it was the expression of a pragmatic alternative to absolutism (as found in Samuel Huntington's *Soldier and the State*), to 'victory' or decisive victory, as well as an alternative view of the relationship between the military profession and politics in democratic states.[13]

Huntington, whose views on the military profession (as set out in 1957) were not that much influenced by the impact of nuclear weapons, was committed to the idea that its effectiveness in providing security for the state was best served by preserving its professional autonomy. This meant keeping liberal civilian values of wider society at bay; and, ideally, the military would be supported by a society whose value system, he advocated, should be conservative and sympathetic to its way of life and ideological assumptions. Together with a sharp division between the political objectives of war and the means of fighting it, these arrangements would provide a framework (what he called 'objective civilian control') within which the military could focus effectively on its key task: fighting and winning the nation's wars.[14]

For Huntington, the core of military expertise was focused on the 'science of war' in contrast to the world of setting political objectives, which he referred to as the 'art of war' – a distinction that is similar to that made by Max Weber between the means and ends in his theory of bureaucracy and the state.[15] Huntington's account is rooted in the idea of what might be termed a 'pure military space' occupied by a military profession using legitimate violence to achieve victory. At the heart of this space are the phenomena of combat between uniformed armed forces pursuing goals set by sovereign states, battle as the exchange of lethal force, the use of fire, manoeuvre and shock effect, and values of courage, obedience, loyalty and masculinity, together with other features of conservative realist military culture.

In addition, although the military should not 'second guess' the political guidance (deriving from the realm of the art of war) that it is given on the objectives set for its operations, the military profession *is entitled to resist* the intrusion of unwarranted, extraneous political interventions in its field of professional judgement. As Huntington argued: '[t]he statesman has no business deciding, as Hitler did in the later phases of World War II, whether battalions in combat should advance or retreat.'[16]

Huntington's analysis leaves two important issues in its wake: first, how our understanding of professional space needs to be altered in the light of the increasingly 'crowded battlefield' in which a variety of military and non-military actors are engaged in the field of operations that encompasses a far broader range of security objectives. And that space is increasingly subject to monitoring by other actors outside the immediate field of operations with access to a variety of electronic means of communications. Second, there is the matter of whether, then or now, it is quite so easy to circumscribe a field of military judgement immune from political scrutiny and questioning. Indeed, not just in the field of operations but right across the spectrum of professional space, a range of social and political considerations render the claims of professional military judgement conditional upon the armed forces being able to sustain their claims in a variety of settings, not least in courts of law, as current debates on the integration of women and gays into the military illustrate.[17]

Janowitz, by contrast, *was* influenced by the strategic impact of nuclear weapons when he formed his views on the military profession as part of his development of the constabulary concept. The sharp, Huntingtonian, distinctions between the means (science) and ends (art) of war, the respective spheres of political and military professional activity and competence, as well as the conditions of peace and war, were all seen as interpenetrating, and necessarily so, by Janowitz. This process had been driven by the development of the military and political bureaucracy during the Second World War, and had continued with the development of a permanent war machine to counter the Soviet threat in the post-1945 period.[18] The tensions of the Cold War, with the risks of small-scale conflicts escalating into serious, possibly nuclear, confrontations between the superpowers, made it difficult to maintain a sharp distinction between peace and war. As a result, with the military increasingly drawn into the political decision-making process about how to handle Cold War tensions and contribute to 'viable international relations', it became evident that the military profession required political sensitivity about the context of its missions; and this involved a major constraint on any untrammelled definition of war-fighting and winning as its core role.[19] The risks of a nuclear Armageddon meant that, while the military profession acquired political influence and engagement, its autonomy had to be limited and, so far as was possible, its values should be in tune with the civilian and liberal sensibilities of contemporary capitalism.

Critical to understanding these differences of view between Huntington and Janowitz is a clarification of the concept of 'victory'. Victory has two intertwined connotations: first, militarily, it involves a decisive encounter between the armed forces of states in which one party defeats the other so that it can subject it to its will, and the losing party is unable to muster enough collective power to resist and to challenge for supremacy on a future occasion. Second, politically, the military contest leads to the central authority of a state that has been militarily defeated to recognise that fact and to seek terms that will lead not only to an end of hostilities but to a political settlement that creates the conditions for a lasting peace (that is, not a truce or temporary cessation of hostilities). This political settlement might involve a range of scenarios, depending upon what objectives both states sought at the commencement of an armed conflict – from territorial or other concessions through to unconditional surrender, in which latter case the initiative for a political settlement lies overwhelmingly with only one of the parties to the conflict and is normally enforced through occupation (as in the cases of Germany and Japan in 1945).

Huntington, as mentioned earlier, adopted an absolutist position with a preference for an ideal that saw the military profession securing 'total victory'– to be sure, within a political framework established by civilian political authority – but with minimal civilian interference in matters relating to war. Janowitz saw the war-fighting (and winning) mission of the military as necessarily constrained by the overarching goal of securing viable international relations in which military professionals and politicians were closely engaged as partners in a relationship of subordination.

Although Smith says nothing about Janowitz in his work on the nature of contemporary warfare, his views are, in fact, very much in that tradition of thought, as we shall now see.

## Smith's 'end of war' and of 'winning': the pursuit of a condition not victory

The key idea in Smith's work is what he calls the 'end of war': the post-1945 shift from industrial war to 'war amongst the people'. In industrial war, armed forces are used to resolve conflicts through decisive strategic encounters with their opposite numbers on the battlefield: war is a rational, strategic use of force to win a political contest, and this remained the case from the early nineteenth century through the wars of German unification, the American Civil War through to the two world wars of the twentieth century. (The Falklands Conflict in 1982 was a rare example of post-1945 'industrial warfare'.) The application of science and technology in industrial war led to the creation of nuclear weapons, which also, ironically, led to its decline as states could no longer use war as a rational instrument of policy to achieve military victory (an echo of Janowitz's concerns as discussed above, p. 20).

The so-called 'end of war' (that is, industrial war) does *not* mean the end of armed conflict between states or between states and other political communities – but, rather, an alteration in the calibration of military means and political objectives as well as the kind of actors involved. A new political landscape appeared, which predated but became more obvious after the end of the Cold War: a world of 'wars amongst the people'. These are not interstate, strategic military encounters between one or more states, but engagements between states, and non-state or proto-state actors, which comprise the majority of the parties to these conflicts.[20] The objective of these actors in the post-1945 period was to avoid a contest with an industrial state where the latter was strong but, from a position within the (more or less supportive) surrounding civilian population, to inflict a series of smaller tactical attacks on it. As a result, so it was hoped, the state would become tired of its involvement in the guerrilla and revolutionary wars during the period after 1945: in Korea, Malaya, Vietnam, Bosnia; the recent insurgency in Iraq has provided a further example, with the 'Petraeus' military surge' lessening its impact, although the Taliban insurgency continues apace in Afghanistan.

The complex armed conflicts in which Western states have become involved, inspired largely by forms of conservative as well as liberal interventionism during the post-Cold War decades,[21] have encouraged some writers to think of them as Smith has done, if not with the same language. In the context of the war in Iraq since 2003, Richard Holmes's remarks, for example, are of interest. He speaks of 'postmodern conflict':

> What happened in Iraq in mid-2004 was neither a conventional battle between two symmetrical enemies nor a peace-keeping operation, for the very phrase implies there is a peace to be kept. It was instead a postmodern conflict comprising extreme violence and near normality,

> formally structured military operations and sheer terrorism, diplomatic negotiations and Mafia-style power broking, all intertwined like the skeins of a rope.[22]

Later in his text, Holmes concedes that this operation was an example of Smith's 'war amongst the people'.[23] In essence, in wars amongst the people, the objective for an intervening state or coalition of states cannot be to defeat an enemy in a trial of strength on the battlefield but, through a contest of wills, to change the intentions of insurgents and, ultimately, win over the people in an ideological battle or 'war of ideas' for 'hearts and minds'. By contrast with industrial war, where decisive outcomes are achieved via strategic military encounters, force can only be effectively applied at sub-strategic levels, with political and diplomatic agencies in the lead.

The context in which Smith first started to develop these ideas – 1992–1994 as Assistant Chief of Defence Staff responsible for overseeing UK operations in Bosnia, as Force Commander in 1995, and then 1999 as Deputy Supreme Commander Allied Forces Europe (DSACEUR) in the preparation of operations against Serbia over Kosovo – formed the thinking of other analysts in a similar way during that period.[24] One of these, Michael Ignatieff, wrote 'Virtual War' in which both of the developments focused on in this chapter were also addressed. So far as the idea of victory was concerned, he argued,

> Virtual war proceeds to virtual victory. Since the means employed are limited, the ends achieved are equally constrained: not unconditional surrender, regime change or destruction of the war-making capacity of the other side, only an ambiguous 'end state'. Instead of Serb surrender in Kosovo the NATO alliance contented itself with a 'military technical agreement' which, as its name implies, specified the terms and timing of Serbian withdrawal and the entry of NATO troops, but left entirely undefined the juridical territory over which the war was fought.[25]

The uncertain 'end game' of which he spoke was accompanied by the fact that, in this case, the conduct of operations was as much about competing images and media narratives of the objectives, costs and achievements of the war as the realities on the ground; this was magnified by the extent to which the campaign was fought from the air and, perforce, reliant upon the images and information that could be generated from aircraft and the electronic systems that supported them. In the Kosovo campaign, for example, Ignatieff, amongst others, was concerned to highlight one aspect of virtual war that had important narrative implications: 'NATO's unease about its own impunity', having not suffered a single combat fatality, and thus (citing Paul Kahn), whether '"riskless warfare in pursuit of human rights" is a moral contradiction'.[26]

So far as Iraq and Afghanistan are concerned, the issue has shifted somewhat: is the sacrifice of casualties of Western states worth it in exchange for the purported condition being sought? That having been said, there is also a degree of continuity between 1999 and more recent events in that some scholars contend that the number of Western casualties is kept low at the unacceptable moral price of non-combatant dead and injured, while enemy forces are killed as efficiently and silently as possible, virtually hidden from the public eye. These ideas were taken up in Shaw's analysis of 'risk-transfer' warfare.[27] He contends that controlling the narrative through the management of the media is a critical element in this form of war.[28] Whether or not contemporary military operations are fought largely from the air, the problem of managing the narrative when states pursue a condition or 'success' and not decisive victory looms as large now as it did then. However, this is not the only problem facing those who wish to achieve success by intervening in wars amongst the people, as we shall now discuss, before considering the particular difficulties associated with managing the narratives of war.

## Conducting intervention–stabilisation operations: problems encountered in the pursuit of success

From the point of view of states engaged in wars amongst the people, a number of problems need to be addressed as a result of this change in mission. For example, while Smith was most concerned with arguing for a change in *thinking* about how the armed forces and force might best be used in such conflicts, he was also aware of the importance of matters of organisational design and technology. In this connection, the first key issue is whether states should continue to be prepared for industrial war, and whether their commitment to it is based on conservatism (as Smith thought) rather than on a rational calculation of the strategic probabilities.

Recently, there has been further thinking on this issue in the UK context. US and UK Army doctrine is based on the idea that maintaining a high-intensity war machine serves as a good basis for responding to the variety of risks and threats in international affairs. It provides a means of deterring and, if necessary, fighting an interstate conflict while also allowing for adjusting 'down and across', as it were, to other kinds of mission, such as complex peace support and humanitarian relief operations.[29] Another way of putting this point is to argue that one needs to retain an ethos that involves being prepared to 'train heavy' but, if necessary, 'fight light'.[30] This preference is an organisational and technological expression of the concept of victory within the military.

In reflecting on a decade of foreign-policy activism and a plethora of military operations, especially the major interventions in Iraq and Afghanistan, the UK's Army Chief of General Staff (CGS) asked, early in 2009, whether this doctrine now needed to be amended, with intervention and stabilisation operations being made what he referred to as the 'default setting' rather than asking the Army to adapt to intervention/stabilisation operations on each occasion as required. On the other hand, this is on closer inspection a subtle shift of emphasis. CGS (in what is, in essence, a critique of Smith's view that 'industrial war' has ended) also called for a *continuation* in the UK's capacity to deliver a division of 25,000 personnel for high-intensity interstate-type warfare; and, indeed, he argued for keeping alive the 'conceptual flame' of manoeuvre warfare (as was once routinely prepared for during the late Cold War). In any event, CGS argued forcefully for the investment in the means to deliver this capability, broaching a central political issue: does the UK have the political will as well as the economic means to provide a military basis to underpin its commitment to the US alliance, which it sees as fundamental to its security interests but which, as the US has made plain, requires a commitment from the UK proportionate to the support it expects to receive from the US in return.[31]

A second problem for states intervening in 'wars amongst the people' is the inherent *complexity* of the mission, which is consequent upon the fact that the objective is not 'victory', and that the military has to work with a number of non-military organisations as well as provide a mix of kinetic and non-kinetic activities to achieve an acceptable 'condition'. As CGS put it: 'we are not talking about "winning", we are talking about *succeeding* in a multi-agency and cross-Government operation.'[32] Apart from establishing organisational coherence amongst overlapping, sometimes cooperating, sometimes competing military, development, humanitarian state and non-state actors,[33] and dealing with professional challenges to how the military conducts itself,[34] to whom is one talking and how might one tell one's interlocutors a convincing story? This is the third problem to be faced in conducting intervention–stabilisation operations, to which we now turn.

## Success – not winning: the implications for the narratives of war

What are the effects of a shift from the pursuit of victory to that of 'success' on the political narratives of war, and attempts by government to gain legitimacy from a variety of civilian as well as their own military audiences for the missions they ask their armed forces to undertake, missions that are undertaken in conditions of 24/7 media coverage or 'global surveillance'?[35]

A number of strands in the discussion need to be disentangled. First, as mentioned above, intervention and stabilisation operations are characterised by a *complexity of organisational involvement*; different government departments (national and those of other partners/allies, including their armed forces) as well as a variety of non-governmental agencies are involved. Coherence amongst these different organisational actors, including both the messages they intentionally seek to convey and those they 'give off' unintentionally, is a key ingredient in achieving useful effects.[36] The military component is a critical instrument, albeit only one of many, used in intervention and stabilisation operations. There are two aspects to this: (a) the military needs to link its efforts effectively with non-military components – for example, with regard to governance, reconstruction, development and other functions, including the issue of whether it is a lead partner in such activities and whether the relationship is sequential or simultaneous;[37] (b) the military itself has to blend kinetic (application of force) and non-kinetic (hearts and minds activities) functions in order to provide the means of advancing security and development in the political community in which it is operating.[38]

A second strand involves an important change in language and conceptual thinking: from victory or 'winning' to success or achievement of a condition. As discussed earlier, victory and success are two different ways of thinking about the Clausewitzean idea that war involves the pursuit of a strategic political objective. In pursuing success, or a condition, intervening states combine kinetic (violence) and non-kinetic (hearts and minds) activities; military and non-military instruments to achieve a condition (success), that is to say, a political situation that is satisfactory from the standpoint of the intervening powers.

The idea of pursuing a condition raises a number of difficulties so far as conveying messages about success is concerned. Consider, first of all, the matter of success in controlling the violence and casualties caused by an insurgency. It may not, however, mean success in terms of reaching a 'condition' if that is defined too unrealistically. Take the examples of Iraq and Afghanistan. In the former case, the 'Petraeus surge' may be considered a success or regarded as a turn around of what looked like a disaster two or three years ago.[39] However, we cannot feel confident that Iraq will become a democracy in the Middle East (the original objective of the Bush administration) or a stable united state either, with the possibility of break up on Kurdish/Sunni/Shiite lines. As Max Hastings argued in a recent review of Ricks' *The Gamble*:

> he concludes his story of Petraeus's extraordinary tactical success [which is based on the trinity of concepts: 'clear, hold, build'] by raising doubts about whether it means a thing amid wider political failure. It is a remarkable achievement to have reduced the violence. But is Iraq viable? Will the Kurds and Sunnis eventually fall into civil war with the Shi'ites, and the country suffer partition? The author seems to believe that the country is most likely to end up as a military dictatorship because the army is the only institution that sort of works. In Iraq as in Afghanistan, widespread belief persists that American and allied forces are conducting military operations in support of political constructs that are unsustainable.[40]

Yet, as with earlier conflicts – for example, in Bosnia during the mid-1990s – there is a key point to be made here: in these conflicts, the intervening states hold the strategic initiative but the resolution of a conflict and the direction of travel for the state being intervened in is, and must remain, the responsibility of the local political actors.[41] Indeed, General David Petraeus, when giving evidence to Congress on the potential impact of the surge, suggested an objective that was akin to a condition not victory. He said: '"We're not after the holy grail in Iraq; we're not after Jeffersonian democracy," Petraeus later told the House Foreign Affairs Committee. "We're after conditions that would allow our soldiers to disengage."'[42]

To turn to the example of the intervention in Afghanistan, the pursuit of success has been hindered by the number of strategic goals set for the mission, as well as shifts in priorities:[43] defeating the Taliban and the eradication of their ability to threaten the West as well as regional neighbours (2001); the development of the Afghan economy – the security and development agenda (2002 onwards); the building of good governance (2005); the creation of stability, possibly via 'deals' or negotiation with the Taliban (2008), leading now to a reconsideration of strategic priorities. One likely outcome is a return to less-ambitious goals: to eradicate Taliban terror groups, their bases and capacity to mount a serious threat, accompanied by the creation of a robust Afghanistan military and police security force able to handle the Taliban threat, and then for the intervening forces to leave.[44] Possibly, this scenario would be linked with the political timetable of electoral politics: the possible re-election of President Obama to a second term of office in 2013. One key implication of this discussion is that the pursuit of success needs to be linked to a well-thought-through analysis of the condition that one is trying to achieve; that is, a set of realistic goals that can be understood and accepted by the audiences whose support is needed for an international mission.

The pursuit of success or a satisfactory condition leads on to a number of other issues connected with the narratives of war. One of these is the idea of 'strategic patience' in what Smith perceptively termed 'timeless missions'. By 'timeless' missions, Smith is reflecting on the fact that it is difficult to be exact about an 'end game' (to use a phrase that has circulated around Obama's team as it considers how best to approach coalition strategy in Afghanistan) in an intervention. This is because of the inherent plasticity of the idea of condition, including the key point that a resolution of the conflict in which an intervening state (or coalition state) has intervened is up to the conflicting parties themselves.

Timelessness, in this sense, is applicable to the *strategic* level of war and military operations. From a UK point of view, in terms of doctrine, the levels of war can be considered as follows. The 'strategic' level of war is that at which a nation or group of nations determines national or multinational security objectives and deploys national, including military, resources to achieve them. The 'operational' level is that at which campaigns and major operations are planned, conducted and sustained to accomplish strategic objectives within theatres or areas of operations. The 'tactical' level is that at which battles and engagements are planned and executed to accomplish military objectives assigned to tactical formations and units. So far as the issue of timeless operations in pursuit of a condition is concerned, it is important to realise that the concept of time operates differently at each of the levels of war. At the strategic level, one plans for activities months, sometimes years, in advance. At the operational level, military HQs may plan operations that run for several months ahead but also broad plans for up to seven days and detailed plans for 48–72 hours, while also dealing with immediate conduct of operations. Meanwhile, tactical commanders tend to be dealing with immediate operations which might extend out to 48 hours and not much more.

'Strategic patience', therefore, is required of elites and the wider publics when missions are 'timeless' at the strategic level, and thus difficult to garner legitimacy from publics. This is because they are not only long-term operations, but the condition which is the object of a campaign cannot be defined in a clear-cut fashion, even when coherence of objectives has been established. Timeless missions are thus chronically fragile in terms of legitimacy.

Consequently, in timeless missions it is critical to make the most of opportunities arising on the ground that can allow the momentum of an operation to be maintained; these opportunities have been called 'defining moments', and seizing them is all the more critical in the absence of a clear end-point, as in circumstances of unconditional surrender and/or suing for peace by a single political authority.[45] Defining moments need to be situated in a context where 'metrics of success' have been thought through.[46] In Afghanistan, such defining moments – or attempts at constructing them – included the courage and determination involved in ensuring the successful transport of a generator to the Kajaki dam in 2008. In addition to illustrating the technical competence and courage of the troops involved, it also displayed a commitment to using the military to provide not just security but the means of development via the provision of energy in Helmand province.[47]

The fact that the energy can be tapped by the Taliban and used as a de facto means of taxing local populations, while the timing of the operation was influenced by domestic US electoral politics (to achieve a success in order to garner Congressional support for funding for continued operations in Afghanistan), does not diminish the achievement. However, there are further complications. For example, it is misleading to talk about 'the Taliban': there are different Taliban political foci in Afghanistan and Pakistan; some groups are reconcilable with the current Afghan government, others less so; and – a key point – 'deals with the Taliban', so to speak, are not for the intervening forces to make but for the Afghan sovereign government. Yet the current government apparatus is riven chronically by corruption, making it very difficult to construct an effective, orderly, let alone democratic, polity. The elections have offered little prospect for significant improvements in this regard. Such are the constraints of a mission akin in political logic to what was referred to in the late 1990s with regard to Bosnia as 'strategic peacekeeping'.[48] However that may be, such events only reinforce the need for further momentum, and for them to be located in a convincing narrative about the ISAF mission.[49]

Defining moments need to be located within a broader narrative that can operate at the different levels of operations – from the strategic to the micro-tactical – that make sense and are persuasive.[50] They also need to underpin a sense of momentum that can gain support from audiences that ask questions about the nature and conditions of success and the sacrifices that need to be made to achieve it. Here, there are multiple audiences in a variety of settings, each with varying access to the manifold means of communication that are available from television, radio, the Internet and mobile phones, cameras and other information devices. Within this complex web of audiences, three can be highlighted as critical (and connected): (a) the military involved in conducting operations: for them, defining moments have to be defined, in part, as 'tactical victories' in order for their own sense of self-worth and professional identity to be sustained;[51] (b) the people who are the object of hearts and minds campaigns within the state that is the target of an intervention by other states; and (c) other audiences in the region and globally, whose involvement, depending on the nature and level of their interest, is made possible by the dramatic changes in information/media technology over the last two decades. And, of course, states seek, when and where they can, to limit the media's capacity to scrutinise what is going on in operational spaces – as in the control of journalists through the practice of embedding them with military units and their suspicions of civilians with mobile phones in operational areas.[52]

Meanwhile, those involved in terrorist campaigns seek to overcome their asymmetrical military disadvantages by magnifying their political effects through the use of violent spectacle. This was revealed, once again, in Mumbai as an echo of 9/11. The militant Islamic, terrorist outrage in Mumbai in late November 2008 was immediately referred to by some commentators as 'India's 9/11'. This was because, inter alia, the attacks were on symbols of Indian, capitalist cosmopolitanism, such as the Taj Mahal hotel; there was targeting of Westerners and Jews; a ruthless, violent pursuit of objectives with no moral or humanitarian scruple; and the attempt to leverage regional and global political effects (including a conflict between Pakistan and India) through these acts of 'symbolic violence'. Apparently, Blackberries were used by some of the terrorists to monitor how successfully their efforts had been in producing spectacular effects via 'global surveillance'.[53] The speed and agility of terrorist insurgents so far as their capacity to use kinetic means to deliver information effects can be impressive.

If the pursuit of a condition or success poses questions about the persuasiveness and coherence of the narratives of war, then who should play what role in the process of constructing and maintaining narratives to underpin complex intervention and stabilisation missions? An effective answer to this question requires activity at each of the levels of war discussed earlier, together with an awareness that the changing character of war and the technology of communications have led to an interpenetration of those levels such that low-level actions can have large-scale consequences, leading in turn to incentives for political interventions from the strategic level in lower-level conduct. This is known as the phenomenon of 'compression in the levels of war'.[54]

The provision of overarching, strategic coherence is the responsibility of political leadership of a coalition of states seeking to achieve success in an intervention and stabilisation mission. Whether, in this connection, the UK, for example, needs a national security advisor to provide such coherence is a moot point. It is at this level that key questions need to be addressed, and about which we already know a good deal from the 'casualties literature'.[55] For example, it is a well-established research finding that Western publics are not casualty-averse per se (although some members of national political elites persist in thinking they are), but are able to accept considerable sacrifices if the following conditions are present: the mission is regarded as of national importance and the aims and objectives of the mission are clear; that sacrifices are proportionate to the value of the mission; that momentum is being achieved; that those who are politically responsible for the mission seem to be competent leaders; and that, critically, *futile* loss of life – 'home', 'away' sides *and* civilian non-combatants – is avoided. The sensitivity to futile loss of life is magnified in contemporary intervention and stabilisation missions because liberal interventionism involves wars of choice which, increasingly and as a result of a decade of 'spin' (not confined to the UK), especially in relation to the build up to the invasion of Iraq in 2003, have become wars of *contested* choice.[56] Public scepticism and suspicion is quite entrenched as a result, with the problem of building national consensus or a coalition of public support for what are inherently long-term or timeless missions being a key challenge for those responsible for building narrative coherence at the strategic level.[57]

Activities at the strategic level need to be supplemented by accounts and actions on the operational and tactical levels – that is, on the ground – that make sense and establish through 'defining moments' a sense of success or progress on the ground – especially in the interlinked fields of security and development. As mentioned above, with reference to the idea of 'compression in the levels of war' (and this has been well-known since the publication of General Krulak's paper on the 'strategic corporal'), individual soldiers need to be aware that their actions can have neutral, positive and negative effects on the key audiences whose intentions they are trying to shape by what they do.[58] This is because technologies of military power and communications facilitate the compression of the levels of war – that is to say, allowing small-scale tactical events to have major strategic consequences. Cases of disproportionate use of force causing the death of non-combatants, the deliberate abuse of prisoners or unlawful killing, for example, can cause reputational damage amongst all key populations – local, regional and global audiences – and can also sap at the support required from the domestic population of the state contributing soldiers to a mission.[59]

## Conclusion

The changing character of war and the ways in which audiences observe and understand intervention and stabilisation operations highlight the need for narrative work at each of the levels of war to underpin the pursuit of success. These developments indicate how right Smith was to emphasise that the military commander is an actor in the theatre of war, a role that complements sociological ideas of the soldier–diplomat and the soldier–scholar.[60]

The narrative work to be done is made that much harder by the potential for confusion that is increased by the move way from victory. Narratives must be coherent and stand the test of challenges on empirical, logical, conceptual and theoretical grounds. With an absence of defining moments or, instead, ones that, as a result of ambiguity, can be turned to the advantage of opponents, conditions are likely to arise that will lessen the power of narratives to convince and generate political support for a mission. In order to persuade, narratives must be supported by credible evidence, and this is why defining moments can stand out as milestones to success or indicators of setbacks, and thus stimulate fears of failure. It is also why the power of defining moments can be dissipated by a lack of overall coherence in the strategic narrative – as, for example, in the tensions associated with attempts to win hearts and minds in Afghanistan, which have, until recently, been undermined by an ill-thought-through attempt to use force to interdict the poppy crop that has formed Afghanistan's major contribution to the global heroin supply.[61] This measure helps to drive the local population into the hands of the Taliban; far better to use softer techniques, such as providing economically more attractive means of making a living.[62]

Narratives come and go as the strategic landscape changes and leaders seek to reshape their objectives. For now, the global war on terror, the long war and now the 'long struggle' seem to indicate an underlying continuity in US strategic narratives, as the coalition in Iraq and Afghanistan grapples with the problem of how best to use its instruments of power – soft and hard – to defend the Western way of life against Jihadist extremism. One question for another occasion is whether grand strategic narratives themselves have much chance of gaining the support and commitment of public opinion in the context of a postmodern culture in which the default response may increasingly be a suspicion that what is being offered is 'spin', or that there is a deep uncertainty about which values are held to be enduring, sacred even, and so worth defending.

## Notes

1 I am grateful for comments on earlier drafts of this chapter from Eyal Ben-Ari, Stuart Cohen, Peter Feaver Richard Kohn, David Last and Kobi Michael. The roots of the ideas presented here lie in far earlier discussions during the mid-1990s with my colleague James Gow and General Sir Rupert Smith.

2 Cohen, 'The Futility of Operation Cast Lead'.

3 Cohen, 'The Futility of Operation Cast Lead'.

4 'President Obama conceded yesterday that America was not winning the war in Afghanistan and opened the way for negotiations with moderate elements of the Taleban, much as the US did with Sunni tribes in Iraq.' See Reid and Evans, 'Obama Says US is Losing War in Afghanistan and Hints at Taleban Talks'. I am indebted for this line of argument to Professor Peter D. Feaver, Duke University.

5 An emphasis on success was a key message in a recent speech by the British Army's Chief of General Staff. See General Sir Richard Dannatt, Chief of General Staff, British Army, *Transformation in Contact*.

6 The 'pursuit of victory' formed the title of Brian Bond's book published over a decade ago. See Bond, *The Pursuit of Victory*.

7 This was a question put to the author by Peter Feaver at the War, Peace and Defence Programme Seminar, University of North Carolina, April 2009.

8 Smith, *The Utility of Force*; Dandeker, 'The End of War?'. Smith's ideas have been influential not only in wider public debates about the use of force but also in UK Staff College discussions, that is to say professional military officers' own reflections on their profession. So far as this is the case, then the key objectives for Smith in writing the book will have been achieved. How far such Staff College discussions have been genuinely internationalised beyond the UK to the US and elsewhere is a moot point.

9 Janowitz, *The Professional Soldier*; Huntington, *The Soldier and the State*.

10 Gray, 'New Age Military Progressives', 65. Gray's analysis draws on Janowitz, 'Military Elites and the Study of War'; and *The Professional Soldier*, 264–279, 305–320; see Gray, 'New Age Military Progressives', 65, note 13.

11 Janowitz, *The Professional Soldier*, 418.

12 Gray argues, '[t]he pragmatic code primarily emphasizes the *measured* application of force and its political consequences.' Gray, 'New Age Military Progressives', 44 (emphasis added). This distinction between measured and minimum applications of force was developed in earlier work by Dandeker and Gow, 'Military Culture and Strategic Peacekeeping'.

13 Huntington, *The Soldier and the State*. On the differences and similarities between Huntington and Janowitz, see Feaver, 'The Civil–Military Problematique'. For an earlier, perceptive and influential analysis, see Larson, 'Military Professionalism and Civilian Control'. For an analysis of the differences as these relate to the USA and USSR during the Cold War, see Segal and Schwartz, 'Professional Autonomy of the Military in the United States and Soviet Union'. For an acute analysis of contemporary military professionalism, see Burk, 'Expertise, Jurisdiction and Legitimacy of the Military Profession'.

14 For the argument (from one of Huntington's former students) that the US won the Cold War *not* by following Huntington's prescriptions for objective civilian control but, rather, through a system based on a sharp difference of views and preferences between the military profession and civilian politicians, together with a system of close monitoring of the military and sanctions, as appropriate, in order to obtain military compliance, see Feaver, *Armed Servants.*
15 For Weber's application of the distinction to his analysis of the prospects for Parliamentary Democracy in the defeated Germany of 1918, see Weber, 'Parliament and Government in a Reconstructed Germany'. For an analysis of the contrasting views of Huntington and Janowitz, including the nuclear weapons dimension, see the perceptive essay by Boene, 'How "Unique" Should the Military Be?'.
16 Huntington, *The Soldier and the State*, 77.
17 On this key issue of 'professional space' and jurisdiction, see the seminal essay by Burk, 'Expertise, Jurisdiction and Legitimacy of the Military Profession'. On military professionalism and legal contestation or 'juridification', see the important essay by Rubin, 'United Kingdom Military Law'. On contemporary challenges to professional military space, see Forster, 'Breaking the Covenant'.
18 See Janowitz, 'Military Elites and the Study of War'; and Ulrich, 'Infusing Civil–Military Relations'.
19 Such an untrammelled definition of war-fighting, second guessing of political guidance, and the persistent search for professional military space insulated from higher-level political judgement and scrutiny was a key part of General Douglas MacArthur's later military career; it led to his dismissal on 11 April 1951. For the civil–military implications, see the classic statement by Finer, *The Man on Horseback*, 22–23.
20 Whether Smith exaggerates the end of industrial war, and thus the need for Western states to continue to prepare for them, is seen by some academic analysts as well as military practitioners as a moot point. See later discussion and also Dandeker, 'The End of War?'.
21 Rieff, *At the Point of a Gun*; Parmar, 'Foreign Policy Fusion'.
22 Holmes, *Dusty Warriors*, 135.
23 Holmes, *Dusty Warriors*, 74; 252; 329–330.
24 For example, while Smith developed his thinking on the idea of condition, other analysts focused on the ways in which, in international missions, intervening states (as in Bosnia during the 1990s) might hold the strategic initiative, and could influence what warring parties did; but how it was up to those parties within the territory in which states had intervened to actually reach a resolution of their conflicts. See Dandeker and Gow, 'The Future of Peace Support Operations'.
25 Ignatieff, *Virtual War*, 208.
26 Ignatieff, *Virtual War*, 162.
27 See Shaw, *The New Western Way of War.*
28 It is important to note that Shaw's book was published before the more serious Coalition casualties that arose after the fall of Baghdad and the premature Bush announcement of 'Mission accomplished'. It was completed in the summer of 2004.
29 See Boene *et al.*, *Facing Uncertainty.*
30 Dannatt, *Transformation in Contact.*
31 It also broaches a wider question: is the future one of wars amongst the people as Smith thinks or is there still a significant risk of interstate war which cannot be ignored? See Gray, *Another Bloody Century.* Dannatt, as CGS, was seeking, with some difficulty, to balance a need to be focused on intervention and stabilisation operations while also being aware of the possibility of needing to be capable of meeting the requirements of major interstate war. Whether the UK can afford to

do this in resource terms is a moot point. For a recent discussion, see Cornish and Dorman, 'Blair's Wars and Brown's Budgets'. After the next election, it is likely that these issues will form the backdrop of a Defence Review. In this author's opinion, given the economic context and crisis in state budgets, such a review might well be as far-reaching as the Sandys Review of 1957. A further complication is that future conflict may comprise 'hybrid wars', in which a state might encounter an opponent willing and able to mount state-based regular warfare, guerrilla and terror campaigns simultaneously in the same space(s) in which that state's interests can be threatened. This makes notions of choosing whether or not to prepare for interstate wars or wars amongst the people too tied to either/or kinds of thinking, which obscures the point that different kinds of war and conflict may co-exist rather than supplant one another as part of an overarching historical trend. The experience of Israel in the second Lebanon war (2006) is a case in point. Extrapolating from counter-insurgency in Afghanistan is a risky game to play. I am indebted to a conversation with Eyal-Ben Ari for this last point. See Hoffman, *Conflict in the 21st Century*; Michael, Kellen and Ben-Ari (eds), *The Transformation of the World of War and Peace Support Operations.*

32 Dannatt, *Transformation in Contact.* Commanders on the ground in Afghanistan have also made the point that a political settlement with the Taliban might be required to provide conditions for sustainable stability and the conditions in which intervening states might be able to withdraw their troops. This is an explicit shift away from the idea of victory. See the remarks of Brigadier Mark Carleton-Smith made in October 2008, reported in Coghlan and Evans, 'We Can't Defeat Taleban'.

33 Egnell, 'Between Reluctance and Necessity'.

34 Kiszely, 'Post-Modern Challenges for Modern Warriors'.

35 Shaw, *The New Western Way of War.* This issue is raised in Dandeker, 'The End of War?'. For a recent, interesting analysis of narrative coherence and the need for fleet-footedness in the West's 'war of ideas' with Islamic global jihad, see Betz, 'The Virtual Dimension of Contemporary Insurgency and Counter-Insurgency'.

36 See Goffman, *The Presentation of Self in Everyday Life.*

37 See Egnell, 'Between Reluctance and Necessity'.

38 The key point that military force is a necessary but not sufficient means of achieving peace and success in Afghanistan, often repeated by UK professional soldiers, for example by Brigadier Ed Butler, ex-Commander of British Forces in Helmand Province, Afghanistan.

39 Ricks, *Fiasco.*

40 Hastings, '"Soldier of Good Fortune"'.

41 This was the central message in Dandeker and Gow, 'The Future of Peace Support Operations'.

42 Ricks, *The Gamble.*

43 Stewart, 'What Worked in Iraq Won't Help Afghanistan'.

44 See Riedel, 'Combating Al Qaeda', 9–11.

45 Gow and Dandeker, 'Strategic Peacekeeping'.

46 One of the key problems in the pursuit of success is the complexity of measurement: how do we know we have succeeded? What metrics should be used and why? Recently, US Army General John Cradock (NATO SACEUR) informed a US Senate Panel that 'right now our assessments of progress [in Afghanistan] are anecdotal and they vary daily, weekly, with whoever makes the observation and where they are making them.... I could not agree more that we must have objective metrics.' He also indicated a need for a metric that 'tells us whether or not more or less of the country is secure', and also for more integrated databases on development issues. See Gray, 'NATO Can't Measure Afghan War Performance – General'.

47 See Evans, 'Commentary'.
48 Dandeker and Gow, 'The Future of Peace Support Operations'.
49 For a report that stresses the positive momentum achieved by this operation, see Evans, 'Commentary'. For a comment on the complexity of the notion of 'doing deals with the Taliban', see Bergen, 'Commentary'.
50 This might involve appealing to a complex mixture of reason and emotion – see Betz, 'The Virtual Dimension of Contemporary Insurgency and Counter-Insurgency', 515. In dwelling on the emotive and reasoned components of narrative compulsion, one is reminded of much earlier discussions in the philosophy and sociology of science concerning the rational and non-rational bases of paradigms and the conditions of paradigm change. See Kuhn, *The Structure of Scientific Revolutions*; Lakatos, 'The Methodology of Scientific Research Programmes'; Macintyre, 'Epistemological Crises, Dramatic Narrative, and the Philosophy Of Science'.
51 I am indebted to Eyal Ben-Ari for this observation about the relevance of the idea of victory to military culture. See also Dandeker and Gow, 'Military Culture and Strategic Peacekeeping'.
52 Shaw, *Civil Society and Media in Global Crises*; and *The New Western Way of War*; Bet-el, 'Media and Conflict'. In the past few months, the IDF spent considerable effort to limit media access to Gaza in the recent Operation Cast Lead. For a recent, perceptive discussion of the media and contemporary conflict, see Bet-el, 'Media and Conflict'.
53 Shaw, *The New Western Way of War*, 36–41, 47–70.
54 Dandeker, 'Surveillance and Military Transformation'.
55 Burk, 'Public Support for Peacekeeping in Lebanon and Somalia'; Smith, *The Utility of Force*.
56 Forster and Edmunds, *Out of Step*.
57 However long the ISAF mission lasts, the chronic legitimacy problems (and resource constraints) have led many to question whether any UK government will have the inclination to mount a similar intervention for a generation.
58 Krulak, 'The Strategic Corporal'; Dandeker, 'Surveillance and Military Transformation'.
59 The reputational damage of Abu Ghraib is the most obvious example. The UK's version of this phenomenon was the subject of an official enquiry – the 'Aitken report'. See MOD, *The Aitken Report*. On Abu Ghraib itself, see US Senate Armed Services Committee (2008). The military's need for legal advice in the field as well as media spokespersons to manage narratives will increase in importance.
60 In doing so, Smith was building on the work of Charles Moskos on the transformation of roles of the contemporary military profession. See Charles C. Moskos' discussion of the soldier–scholar and soldier–statesman in Moskos, 'Toward a Postmodern Military', 19. For a restatement, which draws out the difficulties arising from politically assertiveness associated with these roles, see Williams, 'The Military and Society Beyond the Postmodern Era'.
61 Peters, *Seeds of Terror*.
62 For a sign that the US is reconsidering policy in this area, see Farmer, 'Afghanistan Opium Crackdown Doing Little'.

## Bibliography

Bergen, Peter. 'Commentary: Deals with Taliban Could Be Dangerous'. www.cnn.com/2009/POLITICS/03/09/bergen.taliban/index.html.

Bet-el, Ilana. 'Media and Conflict: An Integral Part of the Modern Battlefield'. In *The Transformation of the World of War and Peace Support Operations*, eds Kobi

Michael, David Kellen and Eyal Ben-Ari. Westport, CT: Praeger Security International, 2009.

Betz, David. 'The Virtual Dimension of Contemporary Insurgency and Counter-Insurgency'. *Small Wars and Insurgencies* 19, 4 (2008): 510–540.

Boene, Bernard. 'How "Unique" Should the Military Be? A Review of Representative Literature and Outline of a Synthetic Formulation'. *European Journal of Sociology/Archiv Europeen Sociologie* 31 (1990): 3–59.

Boene, Bernard, Dandeker, Christopher, Kuhlmann, Jurgen and van der Meulen, Jan. *Facing Uncertainty, Report No. 2, The Swedish Military in International Perspective.* Stockholm: Swedish National Defence College, 2000.

Bond, Brian. *The Pursuit of Victory: From Napoleon to Saddam Hussein.* Oxford: Oxford University Press, 1998.

Burk, James. 'Public Support for Peacekeeping in Lebanon and Somalia: Assessing the Casualties Hypothesis'. *Political Science Quarterly* 114, 1 (1999): 53–78.

Burk, James. 'Expertise, Jurisdiction and Legitimacy of the Military Profession'. In *The Future of the Army Profession,* eds Don M. Snider and Lloyd J. Mathews. New York: McGraw-Hill, 2002.

Coghlan, Tom and Evans, Michael. 'We Can't Defeat Taleban, Says Brigadier Mark Carleton-Smith'. *The Times Online,* 6 October 2008. www.timesonline.co.uk/tol/news/world/asia/article4887927.ece.

Cohen, Stuart. 'The Futility of Operation Cast Lead'. BESA Center [Begin-Sadat Center for Strategic Studies] Perspectives Papers, No. 68, 16 February 2009. www.biu.ac.il/SOC/besa/perspectives68.html.

Cornish, Paul and Dorman, Andrew. 'Blair's Wars and Brown's Budgets: From Strategic Defence Review to Strategic Decay in Less Than a Decade'. *International Affairs* 85, 2 (2009): 247–261.

Dandeker, Christopher. 'Surveillance and Military Transformation: Organizational Trends in Twenty-First Century Armed Services'. In *The New Politics of Surveillance and Visibility,* eds K. Haggerty and R.V. Ericson. Toronto: University of Toronto Press, 2006.

Dandeker, Christopher. 'The End of War? The Strategic Context of International Missions in the Twenty First Century'. In *The Transformation of the World of Warfare and Peace Support Operations,* eds Kobi Michael, Eyal Ben-Ari and David Kellen. Westport, CT: Praeger Security International, 2009.

Dandeker, Christopher and Gow, James. 'The Future of Peace Support Operations: Strategic Peacekeeping and Success'. *Armed Forces and Society* 23, 3 (1997): 327–348.

Dandeker, Christopher and Gow, James. 'Military Culture and Strategic Peacekeeping'. *Small Wars and Insurgencies* 10, 2 (1999): 58–79.

Dannat, General Sir Richard. Chief of General Staff, British Army. *Transformation in Contact,* speech given at the Institute of Public Policy Research (IPPR), 19 January 2009. www.mod.uk/DefenceInternet/AboutDefence/People/Speeches/ChiefStaff/20090119AddressToTheInstituteForPublicPolicyResearchLondon.htm.

Egnell, Robert. 'Between Reluctance and Necessity: The Utility of Military Force in Humanitarian and Development Operations'. *Small Wars and Insurgencies* 19, 3 (2008): 397–422.

Evans, Michael. 'Commentary: Army Can Celebrate Real Success With Kajaki Dam'. *The Times,* 3 September 2008. www.timesonline.co.uk/tol/news/world/asia/article4661302.ece.

Farmer, Ben. 'Afghanistan Opium Crackdown Doing Little, Says Us Envoy. US Attempts To Eradicate Opium Poppies From Afghanistan Have Been Criticised As "Wasteful And Ineffective", By the New US Super Envoy to the Region'. *Daily Telegraph*, 22 March 2009. www.telegraph.co.uk/news/worldnews/asia/afghanistan/5032711/Afghanistan-opium-crackdown-doing-little-says-US-envoy.html.

Feaver, Peter D. 'The Civil–Military Problematique: Huntington, Janowitz, and the Question of Civilian Control'. *Armed Forces & Society* 23, 2 (1996): 149–178.

Feaver, Peter D. *Armed Servants: Agency, Oversight and Civil–Military Relations*. Cambridge, MA: Harvard University Press, 2003.

Finer, Samuel, E. *The Man on Horseback: The Role of the Military in Politics*. Peregrine: Penguin Books, 1962.

Forster, Anthony. 'Breaking the Covenant: Governance of the British Army in the Twenty-First Century'. *International Affairs* 82, 6 (2006): 1043–1057.

Forster, Anthony and Edmunds, Tim. *Out of Step: The Case for Change in British Armed Forces*. London: Demos, 2007.

Goffman, Erving. *The Presentation of Self in Everyday Life*. New York: DoubleDay Anchor, 1959.

Gow, James and Dandeker, Christopher. 'Strategic Peacekeeping: Military Culture and the Defining Moment'. In *Aspects of Peacekeeping*, eds Stuart Gordon and Francis H. Toase. London: Frank Cass, 2001.

Gray, Andrew. 'NATO Can't Measure Afghan War Performance – General', Reuters, 25 March 2009. www.alertnet.org/thenews/newsdesk/N24381642.htm.

Gray, David R. 'New Age Military Progressives: U.S. Army Officer Professionalism in the Information Age'. In *Army Transformation: A View from the U.S. Army War College*, ed. Williamson Murray. Carlisle, PA: The Strategic Studies Institute, 2001.

Gray, Colin S. *Another Bloody Century: Future Warfare*. London: Weidenfeld and Nicolson, 2005.

Hastings, Max. '"Soldier of good fortune", Review of Thomas E. Ricks, *The Gamble: General Petraeus and the Untold Story of the American Surge in Iraq 2006–08*'. *Sunday Times*, 23 February 2009.

Hoffman, G. Frank. *Conflict in the 21st Century: The Rise of Hybrid Wars*. Arlington, VA: Potomac Institute for Policy Studies, 2007.

Holmes, Richard. *Dusty Warriors: Modern Soldiers at War*. London: HarperCollins, 2006.

Huntington, Samuel. *The Soldier and the State: The Theory and Practice of Civil–Military Relations*. Harvard, MA: Belknap Press, 1957.

Ignatieff, Michael. *Virtual War: Kosovo and Beyond*. London: Chatto and Windus, 2000.

Janowitz, Morris. 'Military Elites and the Study of War'. *Journal of Conflict Resolution* 1, 1 (1957): 9–18.

Janowitz, Morris. *The Professional Soldier*. New York: Free Press, 1960.

Kiszely, John. 'Post-Modern Challenges for Modern Warriors'. *The Shrivenham Papers* 5, December 2007. www.isn.ethz.ch/isn/Digital-Library/Publica tions/Detail/?ots591=0C54E3B3-1E9C-BE1E-2C24-A6A8C7060233&lng=en&id=90360.

Krulak, Charles. 'The Strategic Corporal: Leadership in the Three Block War'. *Marines Magazine*, January 1999. www.au.af.mil/au/awc/awcgate/usmc/strategic_corporal.htm February 12, 2008.

Kuhn, Thomas S. *The Structure of Scientific Revolutions.* Chicago: Chicago University Press, 1962.

Lakatos, Imre. 'The Methodology of Scientific Research Programmes'. In *Philosophical Papers,* Volume 1, eds J. Worrall and G. Currie. Cambridge: Cambridge University Press, 1977.

Larson, Arthur D. 'Military Professionalism and Civilian Control'. *Journal of Political and Military Sociology* (Spring 1974): 57–72.

Macintyre, Alistair. 'Epistemological Crises, Dramatic Narrative, and the Philosophy of Science'. *Monist* 60 (1977).

Michael, Kobi, Kellen, David and Ari, Eyal Ben, eds. *The Transformation of the World of War and Peace Support Operations.* Westport and London: Praeger Security International, 2009.

MOD, UK. *The Aitken Report: An Investigation into Cases of Deliberate Abuse and Unlawful Killing in Iraq in 2003 and 2004,* 25 January 2008. www.mod.uk/NR/rdonlyres/7AC894D3-1430-4AD1-911F-8210C3342CC5/0/aitken_rep.pdf.

Moskos, Charles C. 'Toward a Postmodern Military: The United States'. In *The Postmodern Military: Armed Forces after the Cold War,* eds Charles Moskos, John Allen Williams and David R Segal. New York: Oxford University Press, 2000.

Parma, InderJeet. 'Foreign Policy Fusion: Liberal Interventionists, Conservative Nationalists and Neoconservatives – The New Alliance Dominating the US Foreign Policy Establishment'. *International Politics* 46, 2 (2009): 177–209.

Peters, Gretchen. *Seeds of Terror: How Heroin is Bankrolling the Taliban and Al Qaeda.* London and New York: St. Martin's Press, 2009.

Reid, Tim and Evans, Michael. 'Obama Says US is Losing War in Afghanistan and Hints at Taleban Talks'. *The Times,* 9 March 2009. www.timesonline.co.uk/tol/news/world/asia/article5869476.ece.

Ricks, Thomas E. *Fiasco: The American Military Adventure in Iraq.* New York: Penguin Press, 2006.

Ricks, Thomas E. *The Gamble: General Petraeus and the Untold Story of the American Surge in Iraq 2006–08.* London: Allen Lane, 2009.

Ricks, Thomas E. 'A Military Tactician's Political Strategy In Defending the Troop Increase, Commander Paved Way for a "Long War"'. *Washington Post,* 9 February 2009. www.washingtonpost.com/wp-dyn/content/article/2009/02/08/AR2009020802321. html.

Riedel, Bruce. 'Combating Al Qaeda: Strategies for the Future'. Transcript, *International Peace Institute,* 12 May 2009.

Rieff, David. *At the Point of a Gun: Democratic Dreams and Armed Intervention.* London: Simon & Schuster, 2005.

Rubin, Gerry, R. 'United Kingdom Military Law: Autonomy, Civilianisation, Juridification'. *The Modern Law Review* 65, 1 (2002): 36–57.

Segal, David R. and Schwartz, Janet S. 'Professional Autonomy of the Military in the United States and Soviet Union'. *Air University Review* (September–October 1981).

Shaw, Martin. *Civil Society and Media in Global Crises.* London: Pinter, 1996.

Shaw, Martin. *The New Western Way of War: Risk-Transfer War and its Crisis in Iraq.* Cambridge: Polity Press, 2005.

Smith, Hugh. 'What Costs Will Democracies Bear? A Review of Popular Theories of Casualty Aversion'. *Armed Forces & Society* 31 (July 2005): 487–512.

Smith, General Sir Rupert. *The Utility of Force: The Art of War in the Modern World.* London: Allen Lane, 2005.

Stewart, Rory. 'What Worked in Iraq Won't Help Afghanistan'. Chatham House, London, 17 March 2009, extract in *The Times*, 17 March 2009.

Ulrich, Marybeth Peterson. 'Infusing Civil–Military Relations in the Officer Corps'. In *The Future of the Army Profession*, ed. Lloyd J. Mathews. Boston: McGraw-Hill, 2002, 245–270.

US Senate Armed Services Committee. *Senate Armed Services Committee Inquiry into the Treatment of Detainees in U.S. Custody*, 2008. http://armed-services.senate.gov/Publications/EXEC%20SUMMARY-CONCLUSIONS_For%20Release_12%20December%202008.pdf.

Weber, Max. 'Parliament and Government in a Reconstructed Germany (A Contribution to the Political Critique of Officialdom and Party Politics)'. *Economy and Society*, 2 volumes. Translation by Guenther Roth and Claus Wittich. University of California Press, 1977, Vol. 2, 1381–1469.

Williams, John Allen. 'The Military and Society Beyond the Postmodern Era'. *Orbis* 52, 2 (2008): 199–216.

# 3 The international private security company

## A unique and useful actor?

*Christopher Spearin*[1]

Back in 1991, Martin van Creveld contended that non-state actors and non-interstate warfare would increasingly be prevalent in the post-Cold War world.[2] These developments would pose significant challenges regarding how state militaries were to be organized and successfully utilized. Writing in 2006, van Creveld reasserts that the earlier trends have continued, if not accelerated, in the post-9/11 world. State militaries are still unprepared to the degree that "[a]s of the opening years of the twenty-first century, the mightiest, richest, best-equipped, best-trained armed forces that have ever existed are in full decline and are, indeed, looking into an abyss".[3] Many factors speak to this seemingly dire situation. Victory, as traditionally understood, is an elusive concept in the sense that strategic success, especially through the quick application of kinetic force, is unlikely. Instead, operations, almost entirely conducted in the complex weak/failed state context, take long periods of time, require considerable patience, and involve tasks frequently not suited to the sharp application of military might.[4] Indeed, the things sought after are rebuilding efforts, stabilization, and effective policing such that an acceptable "condition" is achieved.[5] State militaries often buckle under the weight of these tasks, as they are often not sufficiently trained and prepared for them, nor do they often have the requisite number of personnel. Moreover, they frequently confront non-state adversaries such as terrorists, militias and rebel groups that disrespect human-rights conventions, show contempt for humanitarian obligations, or do not follow the Law of Armed Conflict. Indeed, their success, in part, rests on this disregard. "Playing" by accepted rules puts military personnel at a disadvantage and makes them vulnerable to casualties, a sensitive issue both militarily and politically in the developed world context, one that can easily sap the resolve of intervening states to seek an acceptable condition over the long haul.

During the turbulent years bracketing van Creveld's critical assessments, a new actor came to the fore alongside state militaries and other armed non-state actors: the international private security company (PSC). While PSCs predate the 1990s, it was the Cold War's demise that triggered the release of millions of personnel from uniformed service, some of whom formed the initial cohort of PSC personnel. While this significant shift is now almost 20 years old, opportunities have continued to be had such that personnel are still drawn to PSCs, and the industry remains vibrant. For instance, at the time of writing, in Afghanistan and Iraq respectively there are approximately 28,000 and 50,000 armed contractors.

A shift as substantial as this and on such sensitive issues, the management and control of violence, rightly attracted considerable analytic attention.[6] One line of study looked at the supply factors that have advanced the industry's development. In addition to the creation of a relevant and skilled manpower pool, the arms industry and civilian suppliers together offered private enterprise the weaponry, necessary kit and logistical support. Another line of study looked at the demand side, largely in geopolitical terms, created by the removal of what has been termed "overlay".[7] With the superpowers reducing the size of their militaries in order to capitalize on the peace dividend, and with the ideological impetus no longer existing to engage in many parts of the developing world, many conflicts were left to fester without outside intervention. Third, in terms of ideas, the privatization of security fit nicely into the larger studies about the appropriate role of the state vis-à-vis private or non-state actors.[8] From the developed-world perspective, this concerned the embrace of the neoliberal economic ethos, especially in the United States and the United Kingdom. From the developing-world perspective, this related to the imposition of economic and political reforms by international organizations such as the International Monetary Fund and the World Bank.[9]

Largely missing from these studies was a consideration of the types of conflicts in which PSCs were to be applied, and why firms might or might not be appropriate.[10] Therefore, does the increasing prevalence of this particular armed non-state actor in the contemporary security milieu indicate that PSCs can do things for which state militaries are not well suited either in terms of tasks required, political saliency or manpower available? Put differently, is the PSC an actor that is, or will continue to be, one well adapted to the nebulous conditions of current operations in which success is conceived of differently, timelines are longer and violence is still employed, but in a much different fashion?

In answering these questions, this chapter focuses on the dual, and sometimes overlapping, issues of PSC uniqueness and utility. The chapter argues, first, that PSCs, because of their desire for continued acceptance and economic viability, will not do activities that state military cannot, namely those that involve the explicit abrogation of rules and standards regarding how armed force is to be employed and to what end. Instead, PSCs openly seek to integrate themselves with state-led initiatives while, at the same time, states and their militaries increasingly seek to manage the day-to-day conduct of firms for the sake of achieving success over both the immediate and longer terms. Second, because of the security challenges present and the number of responsibilities involved, contemporary operations are highly labour-intensive and thus cause military leadership great angst. PSCs, however, are not necessarily the panacea to state military personnel woes, especially in the US context. This is because, from one perspective, firms themselves drain militaries of specialized and relevant expertise and, from another perspective, the political impact of indigenous private labour – as viewed here through the Afghanistan case – is likely to be negative. Like the efforts of intervening state militaries, those regarding the development of an acceptable condition through an augmented reliance on PSCs are not necessarily assured or straightforward.

## Deniability versus integration

With the evolution of the state system over the past several centuries, one of the appeals of mercenary troops has been the plausible deniability they allow for their employers.[11] They have allowed state clients, or at the very least state supporters, to achieve the objectives they desire, however nefarious, sometimes through means that are equally objectionable. For instance, the soldier of fortune mercenaries that plagued post-colonial Africa served as indirect proxies for masters, both corporate and governmental, in Europe and the United States. Robert Denard, one soldier of fortune of longstanding, said this of France, his home country: "I would receive the yellow light which meant there was no opposition."[12] In their fights to prevent the spread of communism in Africa and/or to regenerate colonial linkages, the methods soldiers of fortune employed earned them the label of *les Affreux* – the terrible ones.

In the present context, as Peter Singer contends, PSCs provide their state clients with a similar plausible deniability: "If an operation goes awry, the activities of a firm are easier for a government to deny and the blame simpler to shift."[13] Deniability becomes especially relevant in contemporary conflicts because non-state adversaries do not feel themselves bound to norms and conventions of warfare. Yet state militaries, due to cultural, legal and historical factors, are reluctant to respond in kind.[14] One salient argument, therefore, is that PSCs are ideal for state clients because they can be extremely ruthless by engaging in their extra-legal and relatively unmonitored application of violence: "[They] may have the greatest advantage because they can seamlessly operate below the surface using a wide range of coercive tools to oppose these enemies."[15] For Malcolm Patterson, this current stance towards private commercial violence is in keeping with the historical record: "[P]owerful governments of both the past and the present have found it expedient to keep some distance from the operations of companies where these firms exercise (or misuse) coercion or violence on their behalf."[16]

Increasingly, however, the issue of deniability is becoming less salient because, PSCs, given whom they work for and for what purposes, cannot be wished away. While there is an expectation, held by both private and public actors, that PSCs will do certain tasks, they are not ones that militaries *will not do* for legal and human rights reasons nor are they ones divorced from the official activities of states.

Firms have generally sought to distance themselves from the sorts of activities and methods distinctive of soldiers of fortune and eschew the pejorative characteristics of the mercenary label. The industry conducts a wide variety of activities, including security sector reform to humanitarian demining, activities which are seemingly beneficial and well-regarded, and that also generate revenue and diversify PSC service lines. The industry has a range of clients, from states to international organizations (including some United Nations bodies), non-governmental organizations, and corporations with whom it wishes to stay on good terms and not embarrass for the sake of acceptance and continued economic success. Some of the industry's major trade associations, like the International Peace Operations Association (IPOA) based in the United States and the British Association of Private Security Companies (BAPSC), have codes of conduct for their members to obey. In fact, BAPSC has a tiered membership based on independently audited adherence to professional standards, ranging from the operational to the financial.

Moreover, with respect to the actual utilization of violence, the industry as a whole presents itself as offering, in contractual terms, only the defensive or reactive application of violence.[17] This is a significant shift away from a few companies in the early 1990s that did work alongside indigenous forces to apply violence offensively. Though not part of any dictionary or legal definition regarding the word "mercenary", PSCs contend that offensive privately employed violence is the work of mercenaries. Hence for ArmorGroup, as but one example, "[w]e don't do military replacement. We won't take part in offensive action of any kind. We protect people on the ground."[18] In this light, ArmorGroup advertises its long-term pedigree appealing to a variety of client audiences: "For over 25 years ArmorGroup has been recognised as a leading provider of defensive, protective security services to national governments, multinational corporations and international peace and security agencies operating in hazardous environments."[19] For IPOA, as expressed by Doug Brooks, the association's president, the industry rests on a defensive footing: "[Earlier firms] were supporting offensive combat operations. I don't think that'll happen again, and certainly not that way."[20] Similarly, for Andy Bearpark, BAPSC's director general, firms signing the association's code of conduct "pledge to avoid any armed exchange in their operations, except in self-defence ... and to decline to accept contracts that might conflict with international human rights legislation or potentially involve criminal activity".[21] As such, the international private security industry as a whole disassociates itself from both the words "mercenary" and "private military company".

Leaving the word "military" to state militaries also suggests that they are increasingly focused on offensive actions, combat duties, and the proactive threat of violence, distinctions which many military personnel are comfortable with and in fact wish to maintain. Organizationally, this relates to what militaries are trained to do and, normatively, to what they think makes them different from other actors. In the Canadian case in Afghanistan, PSCs provide "perimeter security" for the Canadian Provincial Reconstruction Team's base in Kandahar City, and they create security cordons following incidents involving Canadian convoys. In 2007, a Canadian military spokesperson made clear the difference between the types of tasks:

> The Canadian Forces does not use any private security contractors to conduct offensive operations.... Using private security contractors for specific tasks permits Canadian Forces personnel to focus their efforts on those duties where they bring the greatest value to the mission.[22]

Similarly, British military analysts in their 2008 report to the Chief of General Staff viewed the utilization of private security by British forces in Iraq for defensive duties as an effective distribution of responsibilities: "[T]he value of freeing soldiers for wider operational tasks, while not relinquishing responsibility for overall unit/base security, was warmly welcomed."[23] With respect to the United States, the Government Accountability Office determined in 2006 that many of the protective activities PSCs conducted in Iraq stood outside of the US military's "stated mission" in the country.[24]

On the one hand, it is the case that recently developed doctrine and policy frequently reveal that militaries can and should have a significant number of roles in contemporary operations. Complex environments beget complex responses. For example, for the United States, important guiding documents such as the 2006 *Quadrennial Defense Review* and the 2006 *National Security Strategy* indicate that stability operations and reconstruction efforts are no longer to be considered secondary tasks. This assertion stands in sharp contrast to arguments made by the then US Secretary of State, Condoleezza Rice, back in 2000, that "[c]arrying out civil administration and police functions is simply going to degrade the American capability to do the things America has to do. We don't need to have the 82nd Airborne escorting kids to kindergarten."[25] Similar shifts in approach can be found in the doctrine and policy of the United Kingdom, Canada and Australia, amongst other countries.

On the other hand, considerable doubt exists in military and policy circles about the overall utility of this shift. For Christopher Dandeker, this concern is based on realistic assessment and stands apart from the military institution's usual conservatism.[26] While complex intrastate challenges will continue to exist, so too will interstate threats likely involving "traditional" uses of military force that cannot be underplayed. Hedging too much towards one type of threat environment and developing the requisite force structure and capabilities is often seen in zero-sum terms.

Also, there are limitations, physical and mental, that analysts contend limit a military from being all things to all people. It is difficult enough for many to imagine a deployed military body, as an organization, concurrently conducting the three relevant functions, or operating fluidly between them, as described in the "Three-Block War" concept: high intensity combat, peacekeeping and humanitarian relief delivery.[27] It is another step still to imagine the individual soldier and his/her unit constantly making the jump back and forth between activity types.[28] While arguments such as these may simply be meant to reify the military ethos, point towards the sort of wars militaries wish to fight (or at least the types of tasks they wish to conduct), and/or avoid asking the question of how far militaries should move down the path of intrastate/weak state operations, they are ones that the presence of a PSC industry can inherently provide support to through a division of tasks.

Obviously, though, this division of tasks is not complete – state militaries continue to do all the tasks currently conducted by PSCs. Moreover, many PSCs work in the same operational environments as state militaries, often in support of the larger official presence of an intervening state. Thus, not only are PSC activities not deniable, but how they conduct these activities can have repercussions for the state military presence. Just as PSCs may relieve state militaries of certain tasks, they also risk liabilities such that there can be negative operational and strategic implications as seen from two angles.

One angle is that contractor activity may lead to changes in military timetables, methods and objectives. In the wake of the brutal deaths of personnel from Blackwater USA (now called Blackwater Worldwide) in Fallujah, Iraq, in March 2004, US Marines launched in April 2004 what became known as the First Battle of Fallujah. This response revealed how contractors are seen as part of the overall official presence, not as actors that are deniable. Indeed, US military and civilian leaders explicitly dealt with the issue in the media and were bellicose in their comments. Brigadier General Mark Kimmitt, the Deputy Director for Coalition Operations, indicated on 1 April 2004 that the US response to the killings was to be both "deliberate" and "overwhelming". This military response, however, led to the deaths of hundreds of Iraqis, changed the perception of conflict of many, both within and outside of Iraq, and triggered violence elsewhere in the country. What is more, as asserted by Lieutenant General James T. Conway, the commander of the 1st Marine Expeditionary Force, the response put an end to the ongoing plans to manage the city through less disruptive and more diplomatic means:

> We felt like we had a method that we wanted to apply to Fallujah: that we ought to probably let the situation settle before we appeared to be

> attacking out of revenge.... Would our system have been better? Would we have been able to bring over the people of Fallujah with our methods? You'll never know that for sure, but at the time we certainly thought so.[29]

In this regard, the US response came about exactly the way the insurgents wanted it to be – an overwhelming response following insurgent incitement: "[T]he underlying purpose is really to provoke a violent reaction, or preferably an escalation of violence ... to show the people of Iraq what a ruthless lot the US-led invaders are."[30] In a similar way, the US military's aggressive response to the May 2007 kidnapping of British PSC personnel in Baghdad again played into these dynamics.[31] The US military may have fallen into this strategic trap set by the insurgents, but it was the recognition of a PSC as part of the team that proved to be the bait.

The second angle is that, while PSCs may indicate that they respect human rights standards and humanitarian obligations, as do state militaries, actual practice may not universally hold up to scrutiny. PSCs can alienate indigenous populations, which can rebound negatively upon the foreign military presence and the larger political endeavours they support. Surveys conducted in Afghanistan, for instance, determined that respondents could not distinguish the boundaries, organizationally and operationally, between PSC personnel and foreign military personnel.[32] While this may reveal a lack of respondent savvy, the issue is moot; PSCs and foreign militaries are both representative of the overseas presence and are, in many instances, integrated actors. The larger endeavour of winning hearts and minds in a place like Iraq, therefore, becomes incredibly challenging when some PSC personnel are alienating in their actions:

> The problem is in protecting the principal they had to be very aggressive, and each time they went out they had to offend ... [Iraqis], forcing them to the side of the road, being overpowering and intimidating, at times running vehicles off the road, making enemies each time they went out.[33]

If PSC personnel, even in the conduct of defensive duties, employ violence irresponsibly or create negative repercussions, then achieving an acceptable condition on the ground through synergistic, multi-actor effort is less likely.

Therefore, in order to minimize operational complications and prevent indigenous upset, PSCs are now becoming even more closely linked to state militaries, thus making their activities even less deniable. In the Canadian case, for instance, as of 2008 new contractual provisions indicate that PSCs hired by Canada in Afghanistan must abide by the International Committee of the Red Cross' code of conduct for combatants. The provisions also command that PSCs must subject themselves to weapons and ammunition inspections to ensure that the Law of Armed Conflict is not violated. For the United States, in recognizing that PSC efforts may be working at cross-purposes with the overall US presence, both military and diplomatic, the US Departments of Defense and State in October 2007 homogenized their approaches towards PSC usage (e.g. intelligence exchange, shared operational awareness, rules on the use of force, etc.). These developments occurred in the wake of the 16 September 2007 Nisoor Square incident when Blackwater personnel opened fire in what they thought was an ambush, killing 17 Iraqis. General David Petraeus, the Commander of Multi-National Force – Iraq, testified before the US Senate Armed Services Committee in May 2008 that as a result of these initiatives, interaction between US forces and PSCs had increased even more:

> [T]here has been significant progress also in the coordination and cooperation between private security contractors and those forces that, if you will, "own the train", are responsible for the areas. There are much closer efforts between the contractual units and our forces ... [a]nd the incidence of escalation of force from private security contractors has been reduced very dramatically.

Additionally, before this incident, the United States Congress amended the Uniform Code of Military Justice (UCMJ) in October 2006 to extend the authority of military law over certain civilians "serving with or accompanying" the military in a "contingency operation". This course of action allows for the UCMJ to be employed should the US Department of Justice not wish to prosecute an individual for an alleged transgression. Without a doubt, substantial criticism has been directed towards these policies that present certain legal roadblocks of their own.[34] Yet, while certainly an important matter, legal efficacy is not the focus here; recognizing that PSCs in contemporary operations, unlike mercenaries in the past, do not operate "under the radar" in terms of what they do and how they do it as seen from the client's perspective.[35]

Many PSCs, in fact, do not wish to "go to ground". Regulation by states, for them, is a mechanism that indicates even greater acceptance. It is also a restrictive, competitive vehicle which will hopefully, as many PSC personnel often put it, weed the "cowboys" out from the business. Indeed, the actions of one firm can impact detrimentally on others. For example, a backlash against the industry following the Nisoor Square incident caused contracts to be cancelled for ArmorGroup, saw its share price tumble by 40 per cent, and helped to instigate a takeover of the PSC by G4S in March 2008. The business associations, to the degree that they can act as industry-focused bodies with limited voluntary memberships, also try to enforce compliance. After the Nisoor Square incident, Blackwater terminated its IPOA membership in October 2007 following a compliance assessment regarding IPOA's Code of Conduct.[36] Recognizing, however, that more than self-policing is required, BAPSC has even suggested that the UK government should appoint an ombudsman for oversight of BAPSC members.[37] Similarly, IPOA has generally welcomed actions taken by US law-makers to better govern the PSC industry.

Overall, actions taken by clients and PSCs suggest that, while PSCs may have certain utility, they operate in niche areas. Moreover, PSCs are not wholly unique actors to be let loose in the contemporary security environment. While they are non-state actors, what they do, how and for whom, prevents them from operating in a totally different way from state militaries in complex operations.

## US interaction with PSCs in confronting contemporary security challenges

### *Positive aspects for the United States*

With respect to this analysis, it is difficult to over-emphasize the impact of the United States. First, the United States is the world's foremost spender on defence. US defence expenditures are greater than the combined expenditures of the next 13 countries.[38] Second, this defence spending, along with spending in diplomatic and development affairs, informs, in part, why the United States is the main consumer of PSC services, as reflected in the financial amounts of contracts, the number of firms employed, and the range of tasks PSCs perform on behalf of the US government. Indeed, so great is the US government's impact upon the PSC industry that Deborah Avant contends that "its choices have therefore had a large impact on the market's ecology".[39]

The current nature of the international diplomatic landscape and the long-term roles required in taskings variously described as nation-building, stability operations and low-intensity conflict also point towards the US embrace of the private option. Whereas the previous section identified the different (and sometimes overlapping and interlinked) fields of endeavour for state militaries and PSCs in confronting contemporary security challenges, here it is appropriate to note that the US military specifically is often seen as particularly unsuited (and perhaps unwilling) to perform all of the necessary activities.

From one standpoint, the so-called American Way of War, and the "Revolution in Military Affairs (RMA)" or "Transformation" which supports it, possesses a number of characteristics, an important one here being the favouring of technology. This is a throwback to Cold War times when qualitative technological superiority was seen by US military planners as a way to counter the quantitative lead held by Soviet Bloc forces. Yet, applied to the present context, an overemphasis on technology in confronting contemporary security challenges has revealed limitations in how force is applied and expectations about the result. In his analysis of

US military conduct, Frederick Kagan contends that an evolving mentality facilitated by technological advancement oversimplifies the efficacy of applying violence vis-à-vis the political and social complexity of the environments in which force is applied: "[US leaders] see the enemy as a target set and believe that when all or most of the targets have been hit, he will inevitably surrender and American goals will be achieved."[40] This approach, however, is asynchronous with other observations made at the start of this chapter: that formal surrender is rare in contemporary conflicts, that victory is a much more elusive concept, and that military force is but one of the instruments utilized to achieve political objectives, often applied in tandem with other instruments. As Rupert Smith also observes, a country like the United States is not oblivious to the changed context, but expectations about the utility of force, especially applied by the US military, remain fixed in an earlier era:

> [T]here is an acceptance in many circles that we now conduct operations rather than wars, but we still expect them to deliver a definitive military victory in its own right that will resolve a political problem, rather than one that contributes and supports the resolution of the matter by other means.[41]

Linked to this approach heavily based on technology is an expectation of speed which does not correspond well with the "long-haul" presence necessary for sufficient nation-building to occur, and for low-intensity conflict to be eventually suppressed.[42] This approach creates expectations, amongst military officials and government leaders alike, that success is quickly achievable, a point made clear in President Bush's premature "Mission Accomplished" speech aboard the USS *Abraham Lincoln* on 1 May 2003, less than two months after the invasion of Iraq. What is more, insurgents, rebels, terrorists, and those who oppose an intervening presence or rebuilding efforts, will attempt to slow things down and take an attrition approach that plays to neither US strengths nor expectations.

Similarly, the American Way of War's favouring of technology comes at the expense of manpower. This, in part, means that funds once spent on manpower are now directed towards the development and purchase of high technologies. It also assumes that an individual supported by the technological prowess of the US military can do more. This assumption, alongside pursuit of the post-Cold War peace dividend, informs the reduction of the US Army from 18 to ten divisions between 1989 and 2000. However, in contemporary operations, the blunt contention made by Ralph Peters is that "[m]ass is back".[43] This leads Jeffrey Record to suggest that the American Way of War "is of limited value in situations requiring the conquest, occupation, and administration of territory; these missions require 'boots on the ground' in sizable numbers".[44] Moreover, mass is back to the degree that, given the necessary troop-to-population ratios, most militaries, that of the United States included, will be severely taxed.[45] Increasing the numbers of quality personnel takes time, as General Peter Schoomaker noted when he was the Chief of Staff of the Army in 2005:

> [T]here will be people that want us to grow mesquite and pine and all kinds of stuff, you know, that might grow faster, but that's not the way to do it. We've got to go back and invest in oak trees.

What is more, there is some thinking that, in both a geopolitical and functional sense, there should be a rough division of responsibilities between the United States, which applies force in a quick and highly kinetic fashion, and other allied countries which provide the troops and expertise for other follow-on operations. Perhaps described best by John Hillen's claim that "[s]uperpowers don't do windows", this approach favours the United States as the global enforcer rather than as a global cop.[46] This division, in turn, makes many traditional US allies, such as many European countries, the so-called "cleaning lady to American intervention".[47] Additionally, qualitative divisions rising between the US military and European militaries, particularly those of NATO countries, are making interoperability in kinetic operations seemingly more difficult. Lessening the defence expenditure gap between the US and European countries is a tall order because caring for an aging European population will increasingly redirect resources away from military investments and more into social programming. The United States, for its part, will not want to risk its personnel due to the operational frailties of others, nor wish to compromise the prospects for operational success. Thus, while the dividing lines between what the United States does and other countries do will never be distinct, the gap will continue to exist, by default if not by design. Taking advantage of this gap, as Joel Sokolsky argues, is not necessarily a bad thing:

> [Success] will very much depend upon being able to carry out post-conflict stabilization and restoration activities as well as other forms of operations other than war. Military units are essential for this and can be quite effective even if they are not on the cutting edge of the RMA.[48]

Whether the United States wishes to take advantage of this gap, and whether traditional allies wish to fill these roles, are points that increasingly stand apart from the issue of actual capabilities. Charles Kupchan and Peter Trubowitz suggest that great powers have traditionally met their goals and covered resource requirements by devolving responsibilities to others.[49] To this end, with specific respect to the United States, maintaining cooperative relationships is not a moot point: "It does seek out allied contributions as part of its policy of maintaining its dominant global position and for the purposes of establishing the legitimacy of its use of power abroad and securing support at home."[50] However, it increasingly seeks out assistance that will be performed largely on US terms in order to minimize the quid pro quos and political limitations placed on troop-contributing countries. While functionality is an important factor, this approach also explains the developing penchant of the United States for ad hocism and coalitions of the willing, at least in military matters, rather than necessarily favouring longstanding and heavily institutionalized mechanisms such as NATO. Even members of the coalition of the willing are to follow US direction, as bluntly assessed by the now former Secretary of Defense, Donald Rumsfeld:

> [W]ars can benefit from coalitions of the willing, to be sure, but they should not be fought by committee. The mission must determine the coalition, the coalition must not determine the mission, or else the mission will be dumbed down to the lowest common denominator.[51]

Similarly, the 2006 *Quadrennial Defense Review* does not focus especially on the role of allies, and when it does, it demands cooperation with the United States largely on its terms and in areas/responsibilities the United States does not wish to apply itself.[52] Given the ongoing challenges the United States faces in maintaining the commitment of allies to operations in Afghanistan, the eagerness of many allies to oblige is questionable.

In contrast, PSCs help to flesh out the US presence abroad by performing duties that supplement the US effort in the absence of sufficient allied commitment and/or allow the US military to concentrate on certain tasks. Indeed, as Barry McCaffrey, a retired US Army general and now a member of DynCorp's board of directors, contends, PSCs are integral to ensuring success in US operations abroad: "[W]e've got an armed forces in uniform that is incapable of carrying out the current national-security strategy.... They [PSCs] see themselves as part of the war effort.... Without them, our war effort collapses."[53] This argument is echoed by the now former Chief of Staff for NATO's International Security Assistance Force in Afghanistan, General Bruno Kasdorf: "If you don't have enough military forces, very often that is a way out then, to count on private security companies."[54] The importance of PSCs is emphasized in the 2006 *Quadrennial Defense Review* which indicates the private presence is the fourth component in the Pentagon's Total Force concept flanking the active forces, the reserves and civilian government employees. What is more, turning to the private sector, rather than to allies, seemingly has an expediency that does not require the United States to spend political capital or to make compromises for the sake of allied participation: "When you can hire people to go to war, there's none of the grumbling and political friction."[55]

At first blush, then, PSCs seemingly offer clients, especially the United States, the resilience to carry on long-term tasks, applied quickly from the marketplace without the political complications and potential demands posed by allies. True, it is not sufficient to simply argue that as long as the money is forthcoming, PSCs will remain engaged and stay in theatre. For instance, in June 2005, the US Embassy in Iraq stepped in when Global Risk Security, which had the contract through the Iraqi Ministry of Transport to secure Baghdad's International Airport, went on strike due to payment arrears. This, in turn, briefly shut down the airport, a major hub for the US presence in Iraq. Members of Control Risk Group similarly threatened to strike in July 2006 when new contracts signed by the PSC entailed pay reductions of at least 20 per cent for its personnel.[56] Some analysts have even contemplated nefarious schemes regarding publicly traded PSCs whereby shareholder buyouts and hostile takeovers, perhaps instigated by adversaries (whether corporate, state or non-state actor), result in the removal of a PSC from a particular marketplace, region or conflict environment.[57]

These assessments, however, should be balanced alongside the resilience of PSCs in comparison to many states that have formed the so-called coalitions of the willing led by the United States. In Iraq as one example, PSCs like Blackwater, ArmorGroup and DynCorp International have together suffered more than 80 fatalities at the time of writing, yet these firms have remained in theatre and not developed risk-adverse strategies that might somehow compromise their contractual obligations. This level of loss is much higher, for instance, than for many of the states that provided troops to support the US endeavour in Iraq and that have since pulled out of theatre.

Additionally, because the United States has such a presence in the industry in terms of being a prominent customer and having many firms headquartered in the United States, it is likely to be more resilient to other states interfering in the marketplace by attempting to control "their" firms. This has been the case in the past, for instance, in the field of transport and logistics. In September 2003, political interference stemming from differences between France and the United Kingdom regarding the Iraqi invasion, prevented the French firm Corsair from completing its contract to transport British soldiers to Basra.[58] Similarly, another fear was that Canada might be denied access to heavy-lift aircraft owned by Russian firms in the midst of the Kosovo crisis because of Moscow's support of Belgrade. However, if anything, it is the United States that has used power generally, and its clout in the industry specifically, to employ PSCs as vehicles for policy. One contention is that the Clinton administration pressured the Angolan government in 1995 to drop its contract with South Africa-based Executive Outcomes and replace the PSC with Military Professional Resources Incorporated headquartered in Alexandria, VA.[59] Similarly, reports indicate that the US government convinced the Steele Foundation, a PSC located in San Francisco, to not bolster its protective detail for its client Haitian President Jean Bertrand-Aristide during the last days of his government in 2004.[60]

There are, however, two issues, located at different ends of the financial spectrum, which suggest that deriving sufficient numbers of personnel from the private sector for usage over the long haul may not be a sustainable or appropriate solution for the United States.

### *Implications for Special Operations Forces*

Though they form only 2 per cent of manpower of the US military, inclusive of active and reserve units, Special Operations Forces (SOF) personnel are at the forefront of contemporary US military operations.[61] In addition to fulfilling domestic counter-terrorism duties, collectively the different organizations that make up the United States Special Operations Command (USSOCOM) perform several important duties, such as foreign security sector training, humanitarian assistance, liaison with indigenous actors, intelligence gathering, irregular warfare and covert operations. While large concentrations of regular forces geared for conventional warfare are increasingly viewed as instruments less appropriate for the more nuanced and discrete management and, when appropriate, application of violence in contemporary operations, SOF are seen as more likely to fit the bill. To this end, the 2006 *Quadrennial Defense Review* calls for US SOF to increase by 15 per cent. This emphasis on SOF is in keeping with President Bush's 2004 decision making USSOCOM the lead command in the "War on Terror". SOF, therefore, have become nothing less than a "tool of choice" for US officials.[62] Such prominence, unsurprisingly, has led to strains between USSOCOM and the leadership of much larger coventional forces of the US military. This is because USSOCOM recruits much of its personnel from the conventional forces, particularily those of considerable capability, and because of the different understandings as to how violence is to be employed and for what purposes.

These strains are exacerbated by the fact that SOF personnel are increasingly attracted to working in the private sphere. Remuneration is often considerable for former SOF operatives. In July 2005, the Government Accountability Office assessed that monthly PSC salaries for retired SOF personnel were in a very high range – from US$12,000 to as much as much as US$33,000 per month. Robert Gates, the US Defense Secretary, has made the link between money and departure clear: "I worry that sometimes the salaries that they are able to pay in fact lure some of our soldiers out of the service to go to work for them."[63]

The lure of this high level of pay is doubly attractive when USSOCOM's operational tempo is taken into account. Rotation periods have sometimes been decreased, while time abroad remains the same. Total numbers deployed have also increased substantially from 31 per cent in fiscal year 2000 to 80 per cent in both fiscal years 2004 and 2005.[64] In the wake of this high operational tempo comes a desire for a less-demanding schedule, a higher degree of choice and a greater amount of home time that working in the private sphere can provide. As a result, one sees personnel from all tiers of US SOF going to work in PSCs. On the one hand, in some cases, the move to the private sphere seemingly allows SOF personnel to utilize their specific skill sets. For instance, Army Special Forces groups have familiarization with the language and cultures of certain regions – the 3rd Group and the 5th Group focus upon the Middle East and Central Asia. This familiarization would be a seeming asset for a PSC and the individual contractor alike when applied to Afghanistan or Iraq. On the other hand, individuals with much different skill sets, such as US Air Force combat controllers, are also being attraced to PSC work, even though their particular responsibilities in state uniform are not employed by the private sphere.

In this vein, the PSC industry has a general requirement for former SOF personnel, whatever their stripe – a fact that speaks to the dynamics of the current PSC marketplace. One aspect is that in lieu of consistent standards as to what skills sets and attributes are required for PSC work (compared, for instance, to increasing attention paid to how PSC personnel should conduct themselves), having SOF capabilities on staff suggests a high level of quality. Indeed, only between 10–30 per cent of SOF recruits successfully make it into SOF units. Given that many PSCs are founded and/or managed by former SOF personnel, this furthers the qualitiative attraction still. Another aspect is that having former SOF personnel serves as a good marketing tool for PSCs given the way SOF capabilities are portrayed in the media, by public officials and in popular culture. SOF personnel implicitly offer both a company and its clients a competitive edge. Finally, in their interactions with the rank and file of the US military, PSCs contend that SOF expertise plays into the larger issue of credibility. One US Army colonel explicitly sought out this credibility in his decision to engage a PSC: "I'm hiring the elder statesmen of the combat arms community.... They won't have to go through any sniff test when they arrive in the field as consultants. They'll be instantly looked up to."[65]

This private attraction to SOF along with high operational tempo, however, comes at a quantitative and qualitative cost. As an example of manpower challenges, many of the small, close-knit teams of US Army Special Forces deployed in Iraq had often as little as six personnel, down from the standard of 12 personnel.[66] As a whole, since the turn of the century, the US SOF community has been underfilled between 5 per cent and 86 per cent, depending on the specialty.[67] Maintaining quality becomes even more challenging when dealing with manpower gaps comes at the same time as force expansion. Already, the balance between training and operations has been altered. In fiscal year 2000, 61 per cent of SOF personnel were in training mode compared to only 16 per cent in fiscal year 2003 and 17 per cent in fiscal years 2004 and 2005.[68] In this light, in 2002, US Army Special Forces started to recruit directly "from the street" in order to hurry manpower injects and avoid increasing further tensions between the SOF and conventional military communities. Yet this move has been described as a "not a one-for-one trade" as experienced senior personnel moved to PSC employment and younger direct entry SOF personnel, without previous military experience, took their places. As a result, "[t]here have been some catastrophic failures where a lack of experience has led to bad things".[69] Additionally, increasing manpower levels as indicated by the 2006 *Quadrennial Defense Review* will reduce further the experience level in the field as senior operators engage in training duties.[70] Overall, the "SOF Truth" that "quality is better than quantity" is stretched in one direction as SOF struggle to keep critical mass, and in another direction as one looks at what actually constitutes sufficient critical mass from a quality standpoint.

In the face of high operational tempo demands and the lure of private employment, efforts to keep SOF personnel, especially senior operators, in uniform have met only limited success. Stop-loss policies, implemented between 2002 and 2004, could only be temporary measures because of their controversial nature. An incentive package, introduced in 2004, linked particularly to extended service, had only limited results; approximately half of those eligible signed on. Also, a more substantive package was not offered due to the fear of exacerbating tensions between the SOF community and the (more costly) conventional forces. While Secretary Gates has mused about introducing non-compete clauses to prevent PSCs from poaching SOF talent, the feasibility of such a clause has come into question, by US officials and PSCs alike. This might also serve as a dissatisfier because SOF, ironically, would be put in a too-special class. This line of argument is utilized, for instance, by Blackwater's Eric Prince:

> I think it'd be upsetting to a lot of soldiers if they didn't have the ability to go use the skills that they've accumulated in the military to go work in the private sector, because you could make the same case about aviation mechanics, jet engine mechanics, guys that work on a reactor on a submarine.[71]

In sum, SOF as a tool of choice is a double-edged sword. Should SOF retention measures in the United States eventually be successful, PSCs will be denied some of the manpower that would help to maintain their presence in contemporary operations over the longer haul. In this regard, SOF in other countries such as the United Kingdom, Australia and Canada are the subject of similar retention measures to stem the bleeding of SOF personnel to PSCs that work, in many cases, for US clients. This is not an insurmountable issue, as we shall see below, but a challenge nonetheless for the PSC industry. Should these measures continue to garner only limited benefit, the US military, and especially its SOF, will continue to be strained and possibly limited in what it can achieve in complex operations.

***Use of indigenous personnel***

While PSCs have an affinity towards the employment of former miltiary personnel with SOF experience, they also have an eye towards the bottom line in what is a very competitive industry. As a result, profits can be heightened or contract bids reduced by relying upon cheaper manpower pools. This can take the form of hiring so-called Third Country Nationals (TCNs). For instance, a wide range of TCNs (such as South Africans, Chileans, Ugandans and Nepalese) support the PSC presence in Iraq. Increasingly, this also takes the form of hiring local employees – the so-called "colonial model".[72] From one standpoint, this is nothing new. PSCs operating in Africa in the 1990s often employed indigenous personnel who were managed by foreigners. What is different now, however, is the scale. Approximately half of the PSC personnel in Iraq, as many as 25,000 people, are Iraqis. In Afghanistan, the focus for the rest of this section, the level is even higher. Of the approximately 28,000 PSC personnel working for 140 firms, about 20,000 of them are Afghans. Collectively, in comparison with foreign state military contingents, this number stands second to the numerical presence of the US military in Afghanistan. Upwards of 10,000 Afghan PSC personnel are based in Kabul alone, the home for many international organizations, non-governmental organizations, corporations, let alone foreign embassies, working to stabilize the country. Similarly, in Kandahar City, a "commando district" has sprung up in order to service the marketplace provided by NATO contigents conducting operations in the south of Afghanistan.[73]

While some firms have avoided using local personnel because of, as we shall see, potential negative implications, others have substantially engaged indigenous manpower.[74] For instance, US Protection and Investigation employs 3,600 locals, Saladin employs more than 2,000 and ArmorGroup employs 850.[75] As one PSC owner put it, economics play an important role in determining his hiring patterns in Afghanistan: "For the price of three ex-pats, I can hire 15 guys."[76]

This reliance, in fact, is in keeping with the "American Way of War", which emphasizes the use of indigenous forces as proxies. On the one hand, in addition to representing a significant quantitative contribution, indigenous forces have qualitative characteristics too. They possess knowledge of the adversaries, of the local populations and of the country's terrain. Also, because they are indigenous, these forces will most likely have a commitment to seeing their interests achieved.[77] Recent examples include reliance upon the Kosovo Liberation Army in the former Yugoslavia and the anti-Taliban Northern Alliance in Afghanistan. On the other hand, the introduction of proxies can complicate matters for the United States because their approaches and goals may not always be synonymous. The ability of the United States to rein in indigenous initiatives, therefore, as Jeffrey Record observes, may be limited:

> Proxy political agendas are rarely identical to that of the intervening external power, and since proxy ground forces fight in their own countries and absorb most of the casualties, they can insist on pursuing their own local objectives even if those objectives run afoul of their external patron's desires.[78]

One of the further complications is that this particular private reliance on indigenous personnel comes at a time when disarmament, demobilization and reintegration (DDR) activities are underway. Though DDR efforts commenced in 2003, considerable work still needed to be done. While militia membership may have been as high as 300,000, the United Nations reported in 2007 that Afghanistan still had 2,000 militias with 120,000 people in their ranks.[79] Thus, given the level of fighting in Afghanistan over the past several decades and the considerable activities of militias, it is unsurprising that PSCs in Afghanistan are impacted by militias. One estimate by the NGO Swiss Peace found that approximately 80 per cent of PSC personnel had militia backgrounds.[80] The PSC industry has, however, become a mechanism through which some militia leaders have bypassed DDR processes. Whereas some firms explicitly link their hiring methods to DDR processes, others go directly to local commanders. For firms, hiring on militias, in whole or in part, is a quick way to obtain a relatively cohesive group of personnel. For militia leaders, this avenue is a way to earn money, keep their influence, and legitimise their forces in post-Taliban Afghanistan. As a result, a message is sent to the populace that not much may have changed and that old power structures remain in place.[81] For one Afghan observer, "[t]hey are the people who fought for money earlier.... Now with their guns they enrolled in private companies to make money. They still think they rule."[82] In short, the PSC industry serves as a disincentive for militia members to hand in their weaponry and for warlords to concentrate their efforts elsewhere.

An additional complication, and the quid pro quo to DDR efforts, is that state security sector actors are concurrently being reformed and bolstered. This is necessary to ensure that indigenous armed forces will be able to take over the tasks conducted by outside forces. Indeed, in Afghanistan, the United States and other coalition participants have indicated that their forces will withdraw from the country as the Afghan National Army (ANA) and the Afghan National Police (ANP) acquire the necessary expertise, resiliency and capacity, let alone a common sense of identity. This challenging and longer-term project is made all the more difficult by the draw of PSCs which stand collectively as a rival force, if only in manpower terms. Here again the PSC industry stands as a disincentive. PSC pay is often double that for ANA and ANP members, and even with a cut perhaps taken by militia commanders, PSC pay is still higher than for public security sector employment.[83] Coupled with the problems that have plagued the timely delivery of pay and the dangers the ANA and ANP routinely confront, the lure of PSC employment becomes even stronger. NATO officials now worry that PSCs are drawing personnel away from the ANA and ANP, which undermines the Afghan government and sets back the horizon for the withdrawal of foreign forces.[84] For the Afghan government, a considerable PSC presence is required as long as further foreign troops are not forthcoming while the ANA and ANP develop: "In a compromise with the large international community, and its legitimate and high demand for security protection, the [government of Afghanistan] has allowed for limited PSC operations and activities."[85] However, in terms of image, the government's reliance upon outside forces is problematic enough in terms of creating and sustaining legitimacy. It becomes all the more challenging for government legitimacy and security sector development when the government relies, if only implicitly, upon PSCs stocked with Afghan personnel and sometimes with ties to warlords. For one Afghan, "the government's power is shown as being weak, making the ambit of the government's security forces smaller".[86] Similarly, as Swiss Peace found, with "the existence of many PSCs, Afghan security forces [ANA and ANP] did not feel themselves accountable for the security of civil society".[87]

In total, the reliance upon indigenous PSC employees, while economically and competitively valuable, comes at the expense of hampering the creation and nurturing of an acceptable condition, as the Afghanistan case suggests. From one angle, the state security sector, the development of which is the focus for external forces and the domestic government, may be denied the necessary manpower required to bring about the "when-appropriate" departure of foreign troops. It may also be limited in terms of serving as a legitimacy-enhancing vehicle for the government. From another angle, should security sector development eventually be successful, this suggests that the PSC presence will decline accordingly to a more modest level. However, whereas PSCs stocked with either personnel from the developed world or with TCNs would simply depart, a considerable indigenous component would remain in-country. As such, if considerable DDR efforts or a well-managed private-to-public manpower transfer was not conducted, the former PSC personnel could very well disrupt stable conditions given their skills, capabilities and linkages to powerful actors.

## Concluding remarks

As Rupert Smith puts it, the need for state militaries to adapt to contemporary challenges and adversaries, those different from interstate warfare, is determined by many factors: "[T]he decisions of the opponent, the choice of objectives, the way or method force is applied, and the forces and resources available."[88] As many of the chapters in this volume reveal, however, making the necessary changes is easier said than done. Different expectations exist as to what are the appropriate activities of military forces, ones that separate them from other governmental and non-governmental actors. Tradition, history, organizational ethos and military culture often combine to deny those desiring change much traction. It takes considerable time to change the mindsets of those trained and educated in a certain manner on how to achieve certain goals through particular ways and means.

Greater involvement of PSCs in contemporary operations does not necessarily provide a quick fix to the challenges noted above. Certainly, on the one hand, PSCs would seem to be ideal. They ostensibly can achieve objectives in certain ways, over longer time periods, and at manpower levels that state forces do not wish to contemplate. The demand for plausible deniability, the ends justifying the means approach of mercenaries, and adaptability and commercial resiliency over the long haul, especially when viewed through the American lens, all appear to point towards the PSC solution. On the other hand, based on this study, it would appear that PSCs do not represent a unique solution; they have their own inherent limitations that can impact upon their own efforts, those of state militaries, and the ability of military leaders to bring about (or resist) change. Many assessments have indicated that the PSC industry is not going away anytime soon, that commercial opportunities continue to be had as states generally, and the United States in particular, grapple with the changing international landscape and different types of threat. This is likely to be true, but *caveat emptor*, the PSC is not the panacea.

## Notes

1 The author would like to thank Kaitlin Bardswich and Donald Spearin for their helpful assistance in the preparation of this chapter.
2 Van Creveld, *The Transformation of War.*
3 Van Creveld, *The Changing Face of War*, ix.
4 Mandel, *The Meaning of Military Victory*; Smith, *The Utility of Force.*
5 Dandeker, "The End of War?", 27; Smith, *The Utility of Force*, 182.
6 See, for example, Singer, "Corporate Warriors"; Avant, *The Market for Force*; Kinsey, *Corporate Soldiers and International Security.*
7 Buzan, *People, States and Fear*, 198.
8 See, for example, Rosenau, *Turbulence in World Politics*; Strange, *Retreat of the State.*
9 Reno, "War, Markets, and the Reconfiguration of West Africa's Weak States".
10 An exception would be Mandel, "Fighting Fire with Fire".
11 Mandel, *Armies Without States*, 31.
12 Cited in Arnold, *Mercenaries*, p. 61.
13 Singer, "Corporate Warriors", 18.
14 De Wijk, "The Limits of Military Power", 80.
15 Mandel, "Robots, Mercenaries, and Soldiers", 23. See also 12, 18–19.
16 Patterson, "Private Military Actors in United Nations Peacekeeping and Humanitarian Operations", 169.
17 Emphasis upon *defensive contractual terms* is important because of the general blurring between the offensive and defensive in highly insecure areas. Blurring can occur due to the level of firepower PSCs employ, both in an absolute sense and relative to the level of force employed by attackers. It can occur because of the length of engagements. It can occur because PSCs often operate as quick reaction

forces that apply coercion at a sufficient level to extract/protect those assets and individuals in need. Similarly, as PSCs also operate from fixed positions, violence is often applied to push back attackers before support arrives from state militaries, or to facilitate an escape should timely support not be forthcoming. With respect to explicitly going on the offensive, one main exception would be the utilization of PSCs to destroy opium and cacao crops as part of anti-narcotics measures. This can be conceived as an inherently aggressive and proactive activity.

18 Cited in Keilthy, "Private Security Firms in War Zones Worry NGOs".
19 ArmorGroup, "ArmorGroup International plc Wins UK Government Contract in Afghanistan".
20 Cited in Weiner, "Peace Corp.".
21 Bearpark, "Why Private Security Companies Are Essential in Post-Conflict Iraq".
22 Cited in Mayeda, "MPs Seek Cost of Private Contractors", A5.
23 Cited in Brady, "Britain Could Hire 'Mercenaries' to Guard Bases in Combat Zones".
24 US, Government Accountability Office, *Rebuilding Iraq*, 1; Patterson, "Private Military Actors in United Nations Peacekeeping and Humanitarian Operations", 263.
25 Cited in Gordon, "The 2000 Campaign".
26 Dandeker, "The End of War?", 31.
27 See, for example, van Creveld, *The Changing Face of War*, 270.
28 Mandel, "Robots, Mercenaries, and Soldiers", 4.
29 Cited in Chandrasekaran, "Key General Criticizes April Attack in Fallujah", A17.
30 Smith, *The Utility of Force*, 277.
31 Peev, "Warning of Security Lapses in Iraq as Hunt for Abducted Britons Goes On".
32 Swiss Peace, *Private Security Companies and Local Populations*, 28.
33 Cited in Luban, "Blackwater Pays Price for Iraqi Firefight".
34 Lanigan, "US Military Court-Martialing Civilian Contractor Ali While DOJ Slumbers"; Ries, "Contingency Contractor Personnel".
35 Blackwater, for its part, announced in July 2008 that it was shifting away from providing close protection to clients and was going instead to concentrate on logistics, training and aviation services. For Blackwater's Chief Executive Officer, Erik Prince, "[t]he experience we've had would certainly be a disincentive to any other companies that want to step in and put their entire business at risk"; cited in Apuzzo, "Blackwater Plans Shift From Security Business".
36 International Peace Operations Association, "IPOA Statement Regarding the Membership of Blackwater USA".
37 Patterson, "Private Military Actors in United Nations Peacekeeping and Humanitarian Operations", 242.
38 Peña, "A Smaller Military to Fight the War on Terror", 292.
39 Avant, *The Market for Force*, 220.
40 Kagan, "War and Aftermath", 4.
41 Smith, *The Utility of Force*, 371.
42 Echevarria, "Toward An American Way of War", 16.
43 Peters, "A Grave New World".
44 Record, "Collapsed Countries, Casualty Dread, and the New American Way of War", 19.
45 Quinlivan, "Force Requirements in Stability Operations".
46 Hillen, 1997; Sokolsky, "Between a Rock and a Soft Place"; Peña, "A Smaller Military to Fight the War on Terror".
47 Cited in Bronson, "When Soldiers Become Cops", 126.
48 Sokolsky, "Between a Rock and a Soft Place", 316.

49 Kupchan and Trubowitz, "Grand Strategy for a Divided America", 82.
50 Sokolsky, "Between a Rock and a Soft Place", 313–314.
51 Rumsfeld, "Transforming the Military", 31.
52 Cordesman, *Salvaging American Defense*, 408.
53 Cited in Robberson, "Contractor's War Role Debated Here, Abroad".
54 Cited in Straziuso and Abrashi, "Afghans Cracking Down on Security Firms".
55 Cited in Steinberg, "Rohatyn, Shultz, Cheney 'Privatization' Scheme to Wreck U.S. National Security".
56 Fidler, "Baghdad Bubble Over as Work Dries Up".
57 Patterson, "Private Military Actors in United Nations Peacekeeping and Humanitarian Operations", 186; Singer, "Corporate Warriors", 168.
58 Samuel and Smith, "French Block Airlift of British Troops to Basra".
59 Silverstein, "Privatizing War".
60 Tamayo, "U.S. Allegedly Blocked Extra Bodyguards".
61 Peña, "A Smaller Military to Fight the War on Terror", 302.
62 Litt, "Special Ops Forces are 'Tool of Choice' ", 20.
63 Cited in Whitelaw, "Big Demand for a Very Specialized Set of Skills".
64 US, Government Accountability Office, *Special Operations Forces*, 31.
65 Cited in Kaplan, "Outsourcing Conflict".
66 Naylor, "The Spec Ops Stretch".
67 US, Government Accountability Office, *Special Operations Forces*, 4.
68 US, Government Accountability Office, *Special Operations Forces*, 31.
69 Cited in Naylor, "The Spec Ops Stretch".
70 Naylor, "The Spec Ops Stretch".
71 Cited in Whitelaw, "Big Demand for a Very Specialized Set of Skills".
72 Swiss Peace, *Private Security Companies and Local Populations*, 17.
73 Blanchfield, "The Civilian Presence in Modern Warfare".
74 Swiss Peace, *Private Security Companies and Local Populations*, 41.
75 Swiss Peace, *Private Security Companies and Local Populations*, 18.
76 Cited in Butler, "How a Nice Quebec Firm Found Itself in a War Zone".
77 Byman, "Remaking Alliances for the War on Terrorism", 781.
78 Record, "Collapsed Countries, Casualty Dread, and the New American Way of War", 18.
79 Mallaby, 2002; Nawa, 2007.
80 Swiss Peace, *Private Security Companies and Local Populations*, 19.
81 Swiss Peace, *Private Security Companies and Local Populations*, 17, 43.
82 Cited in Mojumdar, "Nobody Guarding Afghanistan's Guards".
83 Swiss Peace, *Private Security Companies and Local Populations*, 21. See also Institute for International Studies, "Complex Irregular Warfare", 415.
84 Swiss Peace, *Private Security Companies and Local Populations*, 39.
85 Cited in Abrashi and Straziuso, "Afghans Crack Down on Private Security".
86 Cited in Mojumdar, "Nobody Guarding Afghanistan's Guards".
87 Swiss Peace, *Private Security Companies and Local Populations*, 38.
88 Smith, *The Utility of Force*, x.

## Bibliography

Abrashi, F. and Straziuso, J., "Afghans Crack Down on Private Security", Associated Press, 11 October 2007. Online, available at: www.msnbc.msn.com/id/21246718/ (accessed 30 July 2008).

Apuzzo, M., "Blackwater Plans Shift From Security Business", *International Herald Tribune*, 21 July 2008. Online, available at: www.iht.com/articles/2008/07/21/africa/blackwater.php (accessed 31 July 2008).

ArmorGroup, "ArmorGroup International plc Wins UK Government Contract in Afghanistan", press release, 29 November 2006. Online, available at: www.armorgroup. com/files/news/4381/FCO_win___final.doc (accessed 30 July 2008).

Arnold, G., *Mercenaries: The Scourge of the Third World*, New York: St. Martin's Press, 1999.

Avant, D., *The Market for Force: The Consequences of Privatizing Security*, Cambridge: Cambridge University Press, 2005.

Bearpark, A., "Why Private Security Companies Are Essential in Post-Conflict Iraq", *Business Intelligence Middle East*, 2 March 2008. Online, available at: www.bi-me.com/main.php?id=17863&t=1 (accessed 30 July 2008).

Blanchfield, M., "The Civilian Presence in Modern Warfare", *The Ottawa Citizen*, 19 November 2007. Online, available at: www.canada.com/components/print.aspx?id=7f5dd6de-fe30–42f1–8cfd-de5e1a1d 4173 (accessed 30 July 2008).

Brady, B., "Britain Could Hire 'Mercenaries' to Guard Bases in Combat Zones", *Independent*, 6 July 2008. Online, available at: www.independent.co.uk/news/uk/home-news/britain-could-hire-mercenaries-to-guard-bases-in-combat-zones-860882.html (accessed 30 July 2008).

Bronson, R., "When Soldiers Become Cops", *Foreign Affairs*, 81 (2002): 122–132.

Butler, D., "How a Nice Quebec Firm Found Itself in a War Zone", *The Ottawa Citizen*, 4 June 2007. Online, available at: www.canada.com/ottawacitizen/news/story.html?id=24511a9c-8dc6–4aae-9eb4–1b5801332ae4&k=17618 (accessed 30 July 2008).

Buzan, B., *People, States and Fear: An Agenda for International Security Studies in the Post-Cold War Era*, 2nd edition, Boulder: Lynne Rienner, 1991.

Byman, D., "Remaking Alliances for the War on Terrorism", *The Journal of Strategic Studies*, 29 (2006): 767–811.

Chandrasekaran, R., "Key General Criticizes April Attack in Fallujah", *Washington Post*, 13 September 2004, A17.

Cordesman, A.H., *Salvaging American Defense: The Challenge of Strategic Overstretch*, Westport: Praeger, 2007.

Dandeker, C., "The End of War? The Strategic Context of International Missions in the Twenty First century", in M. Chritiansson (ed.) *Eight Essays in Contemporary War Studies*, Stockholm: Militarhogskolan Karlberg, Acta Politica II, 2007.

de Wijk, R., "The Limits of Military Power", *Washington Quarterly*, 25 (2002): 75–92.

Echevarria, A., "Toward An American Way of War", Strategic Studies Institute Monograph, US Army War College, 2004.

Fidler, S., "Baghdad Bubble Over as Work Dries Up", *Financial Times*, 31 July 2006. Online, available at: www.uslaboragainstwar.org/article.php?id=11308 (accessed 30 July 2008).

Gordon, M., "The 2000 Campaign: The Military; Bush Would Stop U.S. Peacekeeping in Balkan Fights", *New York Times*, 21 October 2000. Online, available at: http://query.nytimes.com/gst/fullpage.html?res=9C07E4DE1E3EF932A15753C1A9669C8B63&sec=&spon=&pagewanted=2 (accessed 30 July 2008).

Hillen, J., "Superpowers Don't Do Windows", *Orbis*, 41 (1997): 241–258.

*Independent*, The, "Top 10 Firms Profiting from Iraq", *Independent*, 13 March 2006. Online, available at: http://news.independent.co.uk/business/analysis_and_features/article350930.ece (accessed 31 July 2008).

Institute for International Studies, "Complex Irregular Warfare: The Privatisation of Force", *Military Balance 2006*, London: Institute for International Studies, 2006.

International Peace Operations Association, "IPOA Statement Regarding the Membership of Blackwater USA", press release, 12 October 2007. Online, available at: http://ipoaonline.org/php/index.php?option=com_content&task=view&id=156&Itemid=80>%20&task=view&id=156&Itemid=80 (accessed 31 July 2008).

Kagan, F.W., "War and Aftermath", *Policy Review*, August–September (2003): 3–27.

Kaplan, R.D., "Outsourcing Conflict", *The Atlantic Monthly*, September (2007). Online, available at: www.theatlantic.com/doc/200709u/kaplan-blackwater (accessed 30 July 2008).

Keilthy, P. "Private Security Firms in War Zones Worry NGOs", Reuters, 11 August 2004. Online, available at: www.alertnet.org/thefacts/reliefresources/109223838271.htm?_lite_=1 (accessed 30 July 2008).

Kinsey, C., *Corporate Soldiers and International Security: The Rise of Private Military Companies*, New York: Routledge, 2006.

Kupchan, C.A. and Trubowitz, P.L., "Grand Strategy for a Divided America", *Foreign Affairs*, 86 (2007): 71–83.

Lanigan, K., "US Military Court-Martialing Civilian Contractor Ali While DOJ Slumbers", *Jurist*, 19 May 2008. Online, available at: http://jurist.law.pitt.edu/hotline/2008/05/us-military-to-court-martial-ali-while.php (accessed 30 July 2008).

Litt, D., "Special Ops Forces are 'Tool of Choice'", *National Defense*, 87 (2003): 20–22.

Luban, D., "Blackwater Pays Price for Iraqi Firefight", *Asia Times*, 19 September 2007. Online, available at: www.atimes.com/atimes/Middle_East/II19Ak04.html (accessed 30 July 2008).

Mallaby, S., "... And Their Armies", *Washington Post*, 1 April 2002, A15.

Mandel, R., *Armies Without States: The Privatization of Security*, Boulder: Lynne Rienner Publishers, 2002.

Mandel, R., "Fighting Fire with Fire: Privatizing Counterterrorism", in R.D. Howard and R.L. Sawyer (eds) *Defeating Terrorism: Shaping the New Security Environment*, Guilford: McGraw-Hill/Dushkin, 2004.

Mandel, R., *The Meaning of Military Victory*, Boulder: Lynne Rienner Publishers, 2006.

Mandel, R., "Robots, Mercenaries, and Soldiers: The Comparative Utility of Future Fighting Forces", paper presented at the International Studies Association Annual Meeting, San Diego, California, March 2006.

Mayeda, A., "MPs Seek Cost of Private Contractors", *The Ottawa Citizen*, 26 November 2007, A5.

Mojumdar, A., "Nobody Guarding Afghanistan's Guards", *Asia Times*, 21 November 2007. Online, available at: www.atimes.com/atimes/South_Asia/IK21Df01.html (accessed 30 July 2008).

Nawa, F., "The Gunmen of Kabul", 2007. Online, available at: www.corpwatch.org/article.php?id=14863 (accessed 2 January 2010).

Naylor, S.D., "The Spec Ops Stretch", *Armed Forces Journal*, October 2006. Online, available at: www.afji.com/2006/10/2174369 (accessed 31 July 2008).

Patterson, M.H., "Private Military Actors in United Nations Peacekeeping and Humanitarian Operations", unpublished thesis, Cambridge University, 2007.

Peev, G., "Warning of Security Lapses in Iraq as Hunt for Abducted Britons Goes On", *The Scotsman*, 1 June 2007. Online, available at: www.redorbit.com/news/international/951391/warning_of_security_lapses_in_iraq_as_hunt_for_abducted/index.html (accessed 30 July 2008).

Peña, C.V., "A Smaller Military to Fight the War on Terror", *Orbis*, 50 (2006): 289–306.

Peters, R., "A Grave New World – 10 Lessons From the War in Iraq", *San Diego Union-Tribune*, 17 April 2005. Online, available at: www.signonsandiego.com/uniontrib/20050417/news_mz1e17peters.html (accessed 31 July 2008).

Phinney, D., "Mercenaries to Peacemakers", *Corpwatch*, 29 November 2005. Online, available at: www.theroughcut.net/Articles/article6_recent.html (accessed 31 July 2008).

Quinlivan, J., "Force Requirements in Stability Operations", *Parameters*, 25 (1995): 59–69.

Record, J., "Collapsed Countries, Casualty Dread, and the New American Way of War", *Parameters*, 32 (2002): 4–23.

Reno, W., "War, Markets, and the Reconfiguration of West Africa's Weak States", *Comparative Politics*, 29 (1997): 493–510.

Ries, M., "Contingency Contractor Personnel", *Army Lawyer*, January (2008): 96–100.

Robberson, T., "Contractor's War Role Debated Here, Abroad", *San Diego Union-Tribune*, 29 December 2006. Online, available at: www.uniontrib.com/union-trib/20061229/news_1b29dyncorp.html (accessed 30 July 2008).

Rosenau, J.N., *Turbulence in World Politics: A Theory of Change and Continuity*, Princeton: Princeton University Press, 1990.

Rumsfeld, D., "Transforming the Military", *Foreign Affairs*, 81 (2002): 20–32.

Samuel, H. and Smith, M., "French Block Airlift of British Troops to Basra", *Telegraph*, 16 September 2003. Online, available at: www.telegraph.co.uk/news/worldnews/europe/france/1441608/French-block-airlift-of-British-troops-to-Basra.html (accessed 30 July 2008).

Silverstein, K., "Privatizing War", *The Nation*, 28 July 1997. Online, available at: www.mtholyoke.edu/acad/intrel/silver.htm (accessed 30 July 2008).

Singer, P.W., "Corporate Warriors: The Rise of the Privatized Military Industry and its Ramifications for International Security", *International Security*, 26 (2002): 186–220.

Singer, P.W., *Corporate Warriors: The Rise of the Privatized Military Industry*, Ithaca: Cornell University Press, 2003.

Smith, R., *The Utility of Force: The Art of War in the Modern World*, London: Penguin Books, 2005.

Sokolsky, J.J., "Between a Rock and a Soft Place: The Geopolitics of Canada–US Security Relations", in H. Segal (ed.) *Geopolitical Integrity*, Montreal: Institute for Research on Public Policy, 2005.

Steinberg, J., "Rohatyn, Shultz, Cheney 'Privatization' Scheme To Wreck U.S. National Security", *Executive Intelligence Review*, 31 March 2006. Online, available at: www.larouchepub.com/other/2006/3313rohatyn_privatiz.html (accessed 30 July 2008).

Strange, S., *Retreat of the State: The Diffusion of Power in the World Economy*, New York: Cambridge University Press, 1996.

Straziuso, J. and Abrashi, F., "Afghans Cracking Down on Security Firms", *USA Today*, 11 October 2007. Online, available at: www.usatoday.com/news/world/2007-10-11-3134857502_x.htm (accessed 30 July 2008).

Swiss Peace, *Private Security Companies and Local Populations: An Exploratory Study of Afghanistan and Angola*, November 2007.

Tamayo, J., "U.S. Allegedly Blocked Extra Bodyguards", *Miami Herald*, 1 March 2004. Online, available at: www.commondreams.org/headlines04/0301–06.htm (accessed 30 July 2008).

United States, Department of Defense, "Coalition Provision Authority Briefing", 1 April 2004. Online, available at: www.defenselink.mil/transcripts/transcript.aspx?transcriptid=2400 (accessed 30 July 2008).

United States, Government Accountability Office, *Rebuilding Iraq: Actions Needed to Improve Use of Private Security Providers*, GAO-05–737, July 2005.

United States, Government Accountability Office, *Rebuilding Iraq: Actions Still Needed to Improve the Use of Private Security Providers*, GAO-06–865T, June 2006.

United States, Government Accountability Office, *Special Operations Forces: Several Human Capital Challenges Must Be Addressed to Meet Expanded Role*, GAO-06–812, July 2006.

United States, Senate, "Hearing of the Senate Armed Services Committee on the Nomination of General David Petraeus to be Commander of U.S. Central Command and the Nomination of General Raymond Odierno to be Commander of Mulitnational Force Iraq", 22 May 2008. Online, available at: www.centcom.mil/images/multimedia/2008–05–22%2C%20senate%20armed%20services%20committee%20full%20transcri pt.pdf (accessed 30 July 2008).

van Creveld, M., *The Transformation of War*, New York: Free Press, 2001.

van Creveld, M., *The Changing Face of War*, New York: Ballantine Books, 2006.

Weiner, R.U., "Peace Corp.", *Boston Globe*, 23 April 2006. Online, available at: www.boston.com/news/globe/ideas/articles/2006/04/23/peace_corp (accessed 30 July 2008).

Whitelaw, K., "Big Demand for a Very Specialized Set of Skills", *US News & World Report*, 5 October 2007. Online, available at: www.usnews.com/articles/news/iraq/2007/10/05/big-demand-and-big-pay-for-security-contractors.html (accessed 30 July 2008).

# 4 Great expectations

## The use of armed force to combat terrorism

*Isabelle Duyvesteyn*

In the current war on terror the military instrument enjoys great prominence. For the casual observer, the use of this instrument might even be seen as the dominant means to deal with terrorism. Wars are fought in Afghanistan and Iraq to capture and eliminate terrorists. The United States was quick to militarize its approach after 9/11. It is astounding '[t]o think that the leading world power would have found itself in the extraordinary position of needing to mobilize four aircraft carriers and 400 planes to oppose a nonstate threat'.[1] But this is indeed what occurred. The United States was not alone in its choice for military instruments; many of its allies followed suit. The United Nations chose to see the confrontation with terrorists in similar terms. It legitimated unilateral military action against the perpetrators and their sponsors. In the not-too-distant past, however, the same United Nations had condemned military action after terrorist activity: when the United States used force against Libya in 1986; against Iraq in 1993 after plans were uncovered revealing it aimed to assassinate President George H.W. Bush and the Emir of Kuwait; and in 1998 against Afghanistan and Sudan after the US Embassy bombs in Tanzania and Kenya.[2]

The objective of this contribution is twofold. It aims to answer the question of whether the military instrument is indeed the most used and dominant instrument to deal with terrorism. Other instruments that can be used to deal with terrorist activity are the police and security services, the judiciary and socio-economic and political measures, and any combination of these. Rather than the military, the police and judiciary approaches are indicated to have been important in the past. However, the historical development of these practices cannot be gleaned from presently available studies. The second objective of this contribution is to focus on the question of whether the use of the military instrument contributes to achieving the desired effect of stopping terrorism. Do terrorists end their activities when the military is used against them? We find that there are both strong indications and arguments that the use of the military actually works counterproductively and encourages terrorism. Other instruments can be more clearly linked to reductions in terrorism. More investigation, however, is warranted because we know surprisingly – perhaps shockingly – little about the effectiveness of past counterterrorism campaigns.

## The use of the military instrument to fight terrorism

The application of military force after a significant terrorist attack is understandable in light of the acute threat to security. There is, according to Sir Michael Howard a clear and 'thunderous political imperative: "something must be done"'.[3] It is difficult to resist this call for forceful action: 'the illusion of the offensive is critical in such situations. A liberal, democratic polity must be seen to respond to protect the life and property of its citizens.'[4] A terrorist attack challenges the state in the exercise of one of its core functions. According to Benyamin Netanyahu, former Israeli prime minister and well-known hawk, there should not even be a hesitation. Democracies do not have a choice when it comes to fighting terrorism.[5] The reason, according to Daniel Byman, is that: 'The painful answer might be that "doing something" is needed to reassure people after a massive attack ... because a perception that the government was passive could contribute to a massive overreaction. Reacting may be necessary to prevent overreacting.'[6]

Current prescriptions for dealing with terrorism stress the importance of doing something, including using force. The recipe, formulated by two researchers of the RAND Corporation, and emblematic of the available prescriptive literature, looks as follows: 'first crushing the current terrorist threat and then bringing about changes to make terrorism's reemergence less likely.'[7] Furthermore, '[h]istory also suggests that the crushing phase should be accomplished first (in part to avoid encouraging terrorism), but the groundwork for the more positive features must be begun early if the strategy is to bear fruit later'.[8] This prescription is used by many states in order to curb terrorism; not only the United States but also many European states and other allies practice these prescriptions.

It is not clear, however, on what empirical evidence the claims for successful counterterrorism are based. The literature on counterterrorism, at present, is replete with prescriptive and normative arguments, with a rather limited or non-existent empirical base. As others have investigated, of the more than 20,000 titles on counterterrorism, there are only six evidence-based studies.[9] The majority of the literature is essayistic or polemic, and heavily influenced by the very specific example of the case of Israel.[10] This case, even its most ardent supporters must admit, is not an unqualified counterterrorism success. How does the present approach of quickly crushing the terrorists and gradually changing the supportive environment compare to past approaches? What measures have been popular in the past?

## Debate about past approaches

There is disagreement, but at the same time little discussion in the existing literature, about what the most popular past approaches have been. There are experts who see the military approach as insignificant in counterterrorist policies since the 1960s: 'In the past, counter-terrorism has been fought with minimal military involvement.'[11] A report from the Congressional Research Service claims that 'there have been relatively few instances where foreign militaries have been used to deal with terrorism, [therefore] this brief paper does not deal with the military option'.[12] This perspective is confirmed for the case of the United States. There has only been one study presenting empirical evidence to substantiate claims about past counterterrorism campaigns. Michele Malvesti, investigating American responses to terrorism between 1983 and 1998, has found that judicial approaches were dominant.[13] In total, the number of military retaliations paled in comparison to law-enforcement activities. Of the 2,400 terrorist incidents in this period, as compiled by the American State Department's 'Patterns of Global Terrorism' reports, only three (0.1 percent) in the period of investigation triggered a military response.

The idea that the military response is negligible in comparison to the judicial approach is challenged by several other studies. According to Ronald Crelinsten and Alex Schmid, in the 1970s the criminal justice model was dominant and shifted in the 1980s toward a more military approach.[14] The 1980s and the Reagan administration's stance toward Libya are seen as forming the watershed. The complete opposite, again only for the case of the United States, has been argued by Laura Donahue: 'Before the 1986 Iran-Contra affair, the United States emphasized the military response to terrorism. Afterward, however, the NSC [National Security Council] became more removed from counterterrorist policy, and the government concentrated on a judicial response.'[15] This position finds confirmation in a study by Gregory Miller.[16] Again, others stress that no such shift occurred in either direction and the dominant approach has consistently been judicial.[17] Martha Crenshaw argues that it is not so much time periods that distinguish the two approaches of military force and judiciary measures, but geography: 'Military responses to terrorism are more common in the Third World, where terrorism is usually linked to rural insurgency and popular grievances.'[18]

With the exception of the study by Malvesti, this debate has not clearly been grounded in empirical investigation. It can also be criticized on other grounds. First, ideas about the period before the 1960s have been completely absent. Thinking of noteworthy cases, we might see indications of military measures being adopted in earlier periods – for example, the decolonization period and France's activities in Algeria in the 1950s. Second, strategies or approaches emphasizing other measures than the military or judicial seem non-existent.[19] The concessionary model and the do-nothing approach have no place in the discussions, while in practice they did repeatedly occur. As Davis and Jenkins note, the 'score card' for responses to terrorism since the 1980s ranged from no response at all (Al Kohbar Towers, USS *Cole*), to concessions (American hostages to Lebanon), to withdrawal (Beirut).[20] Adrian Guelke is a lone voice in the desert arguing that, before the advent of specialized forces to deal forcefully with terrorism, concessions and negotiations were the standard options:

> The initial impulse of Western governments faced by kidnapping of diplomats or the taking hostage of civilians was to negotiate with the perpetrators and to make concessions to secure the release of the victims. There was criticism of this response, but it was muted in the absence of alternative courses of action. It took the creation of specialist forces ... to provide Western governments with the option of attempting to rescue hostages by force.[21]

Experts have made several other claims about the adoption of a particular approach to counterterrorism.

First, the adoption of measures does not follow a logical selection process. Rather, the 'ad hoc adoption of counterterrorist policies' has been very common.[22] It is clear that ideas and conceptions held about terrorism influence the choice for a particular counterterrorism strategy.[23] When terrorism is seen as war, counterterrorism favors repressive measures; when terrorism is seen as criminal, it favors judiciary responses; and when terrorism is a sign of irrationality, it calls for psychological approaches. The perspective a state holds on terrorism has important implications for counterterrorism practice:

> If terrorism is treated as a crime, the response is limited to the methods of standard law enforcement: infiltration, intelligence, arrest and detention. While nations may consider placing terrorists outside the law by classifying them as armed combatants, they run the risk of granting terrorists the legitimacy and protection afforded by the Geneva Convention.[24]

There are several authors who explicitly study terrorism from the perspective of warfare.[25]

Second, counterterrorism has a strong tendency for reactive rather than proactive measures.[26] This fits in with the 'something must be done' calls after terrorist attacks. The police and the military might be logical choices in the absence of clear policy.

Currently available databases are thoroughly unhelpful in solving the puzzle of past trends and patterns in counterterrorism. They hardly focus on counterterrorism or the ending of terrorism. Out of the five reputable databases in the field, only two contain some information about countermeasures or responses. The database on International Terrorism Attributes of Terrorist Events (ITERATE), the Study of Terrorism and Responses to Terrorism (START) and the State Department's 'Patterns of Global Terrorism' do not contain any information on responses to terrorist acts.[27] The Memorial Institute for the Prevention of Terrorism (MIPT) database, established after the Oklahoma bomb attack in 1995, contains information exclusively regarding judiciary processes in the United States.[28] The most promising database, at least at first sight, is the Terrorism in Western Europe database (TWEED), compiled by Jan Oskar Engene.[29] It focuses on terrorist activity within Western European states since 1950.

The TWEED database, however, has several shortcomings. Not only does it focus on domestic terrorism in Western Europe and does not include the category of transnational or internationally operating terrorist organizations, which is arguably more representative of the phenomenon; but also, and more problematically, the information pertaining to countermeasures is only available for a small fraction of the total of the 11,245 coded events in the database. For only 17 percent of these cases, countermeasures have been included. The picture this very limited number of cases presents is that the use of the police forces to combat terrorist activity has been important in Western Europe. Half of the recorded cases were primarily dealt with by the police. We can, furthermore, see that the second most popular approach is the judiciary procedure, which covered almost 40 percent of the cases. Armed confrontations account for about 10 percent of the responses and the category of 'secret service' forms a tiny fraction of state activities.

If we treat these data as just indications – based on their limits, little else is possible – several aspects are striking. First, in the existing literature sketched above, the police option has been consistently downplayed. Previous scholarship has focused too narrowly on two dominant counterterrorism paradigms, the war model and the criminal justice model.[30] It turns out that the first and foremost response can rather be found in the activities of the police. Paul Wilkinson has argued in the past that the police is an appropriate instrument to deal with what he calls 'spasm terrorism', but in most other cases the police force is not forceful enough to deal with terrorist activity.[31] The activities of the police can overlap but are not necessarily and completely commensurate with judicial procedures. The police can, for example, carry out surveillance and patrolling operations, can strengthen potential targets (infrastructure protection and target hardening), institute or increase security checks, such as at airports and stations, and the police can use force. All of which do not need to lead to criminal procedures. It can be suspected that the category of police might also contain activities that have a link with the secret and intelligence services, which have been severely underrepresented in the database (one coded event). The activities of these services are necessarily difficult to detect and therefore do not gain a place in the database.

Based on the data available in the MIPT database, we can gather a little more about the types of criminal proceedings that follow in cases of terrorist activity. This data relies on Federal Bureau of Investigation (FBI) information and is therefore only concerned with terrorist activity in the United States. The largest category of outcomes is the guilty plea to one or more charges brought before court.[32] This is followed by a second category of conviction of terrorists. Together, these categories present the majority of the cases. This trial and conviction outcome is followed by dismissal of the case on government motion and acquittal. The last category is the suspects being on the run. It would be interesting to find out to what extent this American record is reflected in other states dealing with terrorism. Data are, unfortunately, not available at present.

The most notable aspects of these quantified findings are, first, that the consistent stress on judicial process and the rule of law in the larger debate about counterterrorism is not reflected to the same extent in these findings. The judicial process by extension often includes the work of the police, but a particular focus on the police forces itself is uncommon. Second, the use of military instruments in terms of armed action and exchange is not negligible as has been claimed by several participants in the debate. Perhaps surprisingly, in the TWEED database, armed force was employed in about 20 percent of the coded cases. Even though we must account for the possibility that armed exchanges have occurred between counterterrorism forces that fall under the jurisdiction of the police instead of the armed forces, in a substantial proportion of the cases force has been involved. This indication also goes against the often-heard claim that Europeans, the focus of the TWEED database, are generally considered to be soft on terrorism.[33] Third, the track record of the United States in dealing with terrorism affecting the American mainland is highly favorable to the judiciary process.

Overall, we must conclude that our knowledge on counterterrorism models and their application in the past shows serious shortcomings. In the data we have available, the force option is not the predominant instrument it sometimes seems today, and it is not negligible as some have claimed in the debate. In light of our limited knowledge, this logically leads to a plea for more, and more thorough, research. We do not really know with much certainty which approach has been most common in the past and where. Can we be more certain about the effects of the counterterrorism measures that have been instituted? The judiciary approach in the United States has had success in the past. Does the use of the military instrument bode well for gaining the upper hand over terrorism?

## The effectiveness of the use of the military instrument in counterterrorism

All modern, developed states have military forces available to deal with security problems. The available literature presents at least six ways in which the military instrument has been used to counter terrorism. First, military force can be used to assist civilian authorities in times of crisis; it can be employed to restore law and order. Second, military force can be used to preempt terrorist attacks; this has been a popular term in the vocabulary of the Bush administration's war on terror. Third, the military can also function as a deterrent; with enough credibility of striking back, potential terrorists might be deterred from carrying out attacks. Fourth, force can be used to retaliate; when a state has been attacked, military retaliation can be an option. Fifth, assassination is another method of employment of military force to deal with terrorist organizations; killing the leadership of terrorist organizations has been argued to lessen the effective operation of these groups. Sixth, wars can be fought to reduce or eliminate terrorist activity. All these uses of military force and the existing ideas about their effectiveness will be discussed in turn.

### *Assistance and protection*

Military force can be used to support the civilian authorities confronted with terrorist violence. Restoring law and order, riot control, securing critical infrastructure and ending hostage situations are several of the tasks that can and have been carried out by military forces. The effectiveness of armed force in hostage-rescue missions seems almost unquestioned: 'in responding to international terrorism, the most effective role for military/para-military forces is the hostage rescue assault. It is morally defensible; it is clearly defined; can be conducted within a legal, constitutional framework; and, finally is likely to succeed.'[34]

However, the use of troops on the streets can play into the hands of the terrorists:

> the wide deployment of armed and uniformed troops implies an extremely serious terrorist threat which could easily be used by the terrorists to their own propaganda advantage. This is even more true if armoured vehicles are involved in the operation.[35]

Terrorists aim to provoke the government into overreacting. By responding excessively, the government runs the risk of delegitimizing itself. Overreaction can take place when the state employs large-scale force. It complies with the terrorists' desire and their painted image of the state as illegitimate and repressive.

Not only is the deployment of troops potentially problematic, but also the use of force itself can complicate matters. Grant Wardlaw has argued:

> As a rule the enemy is trained to apply the maximum force that is necessary to take an objective and eliminate an enemy. The army need not usually be worried about causing damage or loss of life, gaining or

> maintaining public support or avoiding confrontation. It seems obvious that, in a society which is not accustomed to the sight of heavily armed detachments, on public-order duties supported by armoured vehicles and with little usual contact with the public, the army is unsuited, both by training and doctrine, to an internal security role.[36]

The armed forces can also be used to guard and protect potential targets of terrorist activity, such as important installations and individuals. In a comparative quantitative study, Lum, Kennedy and Sherley conclude that: 'the fortification of embassies, and efforts to protect diplomats have been effective in reducing terrorist attacks on these targets.'[37] However, we should not discount the possibility that attacks have switched in focus as a result.[38]

In light of the favorable appraisals of the use of force to end hostage crises, it might be the case that the conditions under which force is used are important. As Peter Chalk argues:

> a process of militarization will only be tolerated within the state if two conditions are met; first, it must be apparent that anti-terrorist commando teams have been created for a readily identifiable and necessary purpose. Second, it must be clear that any force these units subsequently employs will be used in a controlled and precise manner and only after all other possibilities have been exhausted.[39]

Military force can be useful in dealing with particular expressions of terrorist violence, but the employment of force needs to be carefully controlled, for it can easily play into the hands of the terrorists and become counterproductive.

### *Preemption*

Preemption has been defined as 'striking in advance of hostile action to prevent its occurrence and to avoid suffering injury'.[40] In contrast to prevention, a long-term approach, preemption is a short-term pro-active response. 'Preemption is the quintessential proactive policy in which terrorists and their assets are attacked to curb subsequent terrorist campaigns.'[41] Preemption has been stressed by the Bush administration in its national security strategy published in 2002. Jeffrey Ross and Robert Ted Gurr identify preemption as conducive for ending terrorism. The other condition they stress is deterrence.[42] Both rely on the military instrument. Preemption itself can possess a deterrent effect.[43]

The problem with demonstrating that preemption works is that it relies on counterfactuals. We will never know how successful preemption is because it will be impossible to prove something that did not happen.

There are those who do claim, on the basis of contentious evidence, that in particular Western security services have been successful at prevention.[44] Gal Luft argues that assassination, discussed in more detail below (pp. 75–76), works well as a mechanism of preemption:

> Hundreds, if not thousands, of Israelis go about their lives without knowing that they are unhurt because their murderers met their fate before they got the chance to carry out their diabolical missions. This silent multitude is the testament to the [assassination] policy's success.[45]

There is a fundamental problem with measuring and proving that preemption can work as an effective instrument to counter terrorism.

A second problem with preemption is the ethical dimension. States using preemption 'run the risk of relinquishing the moral high ground that derives from being a victim of an attack and therefore entitled to some form of redress'.[46] A final problem is international justification. It will be extremely difficult not only to prove but also justify preemption when it involves breaking international laws and conventions, not in the least the sovereignty of other states. We can therefore not conclude that preemption is an unproblematic and effective counterterrorism instrument.

### *Deterrence*

The issue of deterrence of terrorists by adopting a strong military posture is a heavily debated aspect in the counterterrorism literature. Deterrence relies on the presence of military force that will be credibly used in case of a threat of, or actual, attack.[47] Several scholars have argued that deterrence against terrorists cannot work. The main argument is that terrorists do not have a center of gravity that can be attacked. 'An adversary that prefers escalation regardless of the consequences cannot be deterred.'[48] Others have argued that the concept of deterrence should not be applied wholesale to terrorism but to parts of the terrorist system. 'The higher the value terrorists place on what is not gained through terror relative to what is, the more compelling is a threat against the former, and the more likely deterrence will succeed.'[49] Clients, supporters and associates can be deterred from performing their function in terrorist activity thereby undermining the force of the phenomenon. Examples are state sponsors of terrorism and 'societal elements' sustaining terrorism.[50] Deterrence should be aimed at what the terrorists value, namely their financiers, sponsors and local interests.

There are several problematic aspects to the idea of deterring terrorists. First, if deterrence is successful, it can promote a substitution effect: 'Efforts to deter terrorist events often displace the attack to other venues, modes of attack (e.g. from skyjacking to a kidnapping), countries, or regions, where targets are relatively softer.'[51] There is strong empirical evidence for this substitution effect of terrorism.[52]

A second problematic aspect is the fact that deterring part of the terrorists' system, in particular state sponsors of terrorism, has not had a record holding promise. While the United States, for example, initially adopted a strong military posture against Libyan sponsorship of terrorism, '[t]here ... [is] little disagreement that this state sponsorship of terrorism increased during the 1980s'.[53] The activities against Libya did not deter others imitating its path, for example Iran sponsoring Hezbollah. Preemption, retaliation and the other military measures discussed here are all expected to deter future terrorist activity. Data to illustrate that this is actually the case is scarce. There are few reasons, neither from the perspective of substitution, nor from the limited effect on state sponsors, to see deterrence as an entirely successful approach to counterterrorism.

### *Retaliation*

Retaliation is an important driving force in counterterrorism. As Charles Eppright has argued, retaliation has been particularly popular against 'politically vulnerable state sponsors of terrorism'.[54] We know a little about the conditions that make the use of retaliatory measures likely. The study by Malvesti, referred to above, identifies six factors that explain the likelihood of retaliation after a terrorist attack.[55] First, the perpetrator of the attack can be immediately identified. Second, perpetrator repetition; the responsible actor has repeated an attack or is likely to do so. Third, the target of the attack is an American citizen and working in an official capacity for the American government. Fourth, the fait accompli nature of the incident; damage has been done and a response is called for. Fifth, the attack can be linked to strong anti-American sentiments. Sixth, the actor needs to be susceptible to countermeasures. A target needs to present itself that is suitable for attack to influence the actor's behavior. Strikingly, the number of casualties has not been found to be a factor in the decision to retaliate.

An interesting debate has developed around the American retaliations against Libya from 1986 onwards. The United States used armed force against Libya in April 1986, after the bombing of a Berlin discotheque frequented by American military personnel. The attack killed three and injured 229. Some scholars have concluded that the American retaliation led to a decrease in terrorist activity.[56] However, other research has found that the American raid led to an increase of terrorist attacks, at least in the immediate aftermath.[57] It has been found that the subsequent military retaliation against Libya 'resulted in a statistically significant increase in the number of terrorist attacks in the short run'.[58] Attacks increased against American, British and Israeli targets.[59] These attacks allegedly culminated in Libyan agents downing an airliner over Lockerbie, Scotland, on 21 December 1988, killing 270 people.[60]

A slightly contradictory finding has been presented by Bryan Brophy-Baermann and John Conybear for the case of Israeli retaliation after Palestinian attacks. They have found that retaliation does not possess a significant escalatory effect. Neither does the use of military force for retaliation act as a long-term deterrent in this particular case.[61] The authors conclude that there is only a very temporary effect of retaliation under the condition that the forceful action exceeds the expectations of the terrorists.

Based on the two cases of Libya and Israel, we do not really know whether retaliation leads to escalation and/or can act as a deterrent. It does seem clear, at least as is claimed by several experts, that the window of opportunity for effective retaliation is rather short. Neil Livingston suggests that it is no more than 72 hours after the terrorist incident. Retaliation should be 'swift and discriminating'.[62] Others have observed that retaliation can lead to the strengthening of group cohesion, thereby making the counterterrorism task more difficult.[63] What should perhaps be investigated is what the conditions are under which retaliation escalates, as the study by Malvesti has attempted for the likelihood of retaliation in the case of the United States.

### *Decapitation and assassination*

Taking out the leadership of terrorist organizations is also a military measure suggested as effective by several experts.[64] According to Donald Hanle, the terrorist leadership presents the center of gravity in the counterterrorism struggle. He sees two successful examples of taking out the leadership that lead to the stifling of terrorism: 'the ambush of the Black Panther leaders in Chicago in December 1969 and the attack on the Symbionese Liberation Army headquarters in Los Angeles in May 1974. In both cases the "revolution" was terminated by a single lethal act.'[65] Others point to the arrest of the leadership of Sendero Luminoso and the Kurdish Workers Party (PKK).[66]

The killing of the opposing leadership has been a policy implemented by Israel in dealing with Palestinian terrorism. The leadership of Hamas and Hezbollah has been a repeated target of assassination of the Israeli government. There seems to be large-scale agreement among the mainly (pro-)Israel authors writing on this subject that, despite drawbacks, assassination should remain squarely in the counterterrorism tool box.[67] As Gal Luft argues, there is a 'profound cumulative effect of targeted killing on terrorist organizations. Constant elimination of their leaders leaves terrorist organizations in a state of confusion and disarray.'[68] This cumulative effect is favorable for suppressing terrorist activity, these authors argue.

While there is a strong strand of positive appraisal of the practice, all authors recognize that after targeted killings there is always an upsurge in terrorist or other violent activity.[69] There is a cycle of violence and counter-violence. A second problem is that the policy creates martyrs: 'Assassination may create martyrs and thus stimulate imitation with a greater effect on total commitment than the deterrence resulting from the assassination.'[70] Third, it increases terrorist legitimacy by showing what the terrorists want to demonstrate, namely that the state breaches its own laws, disrespects the right to life and is repressive. Fourth, it complicates negotiation processes by eliminating leaders and compromising trust. Perhaps the largest weakness is the ethical dimension of collateral damage in assassination policies: 'If terrorism is condemned because it kills the innocent, how can one justify counterterrorism tactics that kill them too?'[71] States using assassination as an instrument enter a moral slippery slope that plays into the hands of their opponents.

### *War*

Fighting war to curb terrorism, as has allegedly been taking place in the war on terror, has also occurred in the past. Peter Sederberg has argued that the Israeli invasion of Lebanon in 1982 was for a large part inspired to put an end to cross-border terrorism, targeted against Israel.[72] However, in this argument, the dividing lines between (counter)terrorism, (counter) insurgency and civil war become dangerously thin. More frequently than a chosen confrontation, as in the case of the Israeli invasion, inadvertent escalation into civil or even international war occurs: 'In a few cases, for example, Lebanon and Sri Lanka, civil wars were either sparked or sustained by acts of terrorist violence.'[73] In terms of insurgencies, 'many of them began with low-level violence like kidnapping, assassination, raids, or provocative explosions. However, these events commonly escalated to often massively destructive civil wars primarily because of the foolish – and often breathtakingly incompetent – overreaction of governments.'[74] There is a sliding scale between terrorist activity and insurgency. In cases of international terrorism, there is even a risk of international war – for example, the assassination in Sarajevo in 1914.[75]

So far it has become clear that there are serious problems attached to the use of the military instrument to fight terrorism. However, despite its arguable counterproductive and escalatory effect, and the resulting civilian victims, Colin Gray argues that force remains necessary.[76] He concludes, together with several others, that despite problems, no effective counterterrorism strategy is complete without military force.[77] In particular, if states are not willing to use military force against terrorism, they run the risk of losing the deterrent power of the armed forces altogether.[78] However, above we have already questioned the effectiveness of armed force as a deterrent in cases of terrorist activity (see pp. 73–74).

The most extreme position in this plea for military force in the fight against terrorists is the argument that current uses of force are too limited.

First, Donald Hanle and Gayle Rivers have argued in the past that elite Special Forces units are the most appropriate but are also severely underutilized against terrorist activity.[79] Second, Charles Eppright also sees the use of the military as most feasible, were it not for political constraints: 'Recent history and current events [in 1997] reveal how the U.S. national leadership's natural devotion to political matters limits the full extent of the effectiveness of a military response to terrorism.'[80] What is needed is more force, and force unencumbered by political constraints and dictates of international humanitarian law: 'the kinds of actions that might theoretically be effective against al-Qaeda are unthinkable, given American values – even American values in total war.'[81] It is not less but rather more force that is necessary to 'out-terrorize the terrorists'.[82]

What can be concluded, at this stage, regarding the effectiveness of military counterterrorism measures? Military measures against terrorist activities have been indicated to possess a high risk of escalation. This is an impression that is fortified by several general observations from other scholars. As Andrew Silke has argued:

> Ultimately harsh, aggressive policies in response to terrorism fail so often in their stated aims, because they so badly misunderstand and ignore the basic psychology of the enemy and of observers. Strength and power alone are not enough to defeat terrorism.[83]

If a state uses force, it could, by this very act, encourage terrorists and contribute to the further radicalization of the terrorist movement.[84] As Robert Art and Louise Richardson conclude: 'the military was too blunt an instrument to be successfully deployed against terrorism.'[85] Military force 'fails to achieve political objectives'.[86]

These observations, however, are too-often anecdotally substantiated, theoretically flimsy and/or incompletely argued-through or quantified. It is unfortunate that, to date, they cannot be substantiated by large-scale database analysis or comparative case studies, with the exception of the Richardson and Art volume. We simply know too little about whether force works against terrorists, apart from these indications that there are problems and drawbacks.

It seems most productive that research efforts should concentrate on the conditions under which military force might be useful and effective. Some experts have suggested that the effectiveness of counterterrorism is dependent on the type of terrorist activity.[87] As Peter Sederberg has cogently formulated: 'If we consider terrorism to be a particular form of coercive tactic selected by a variety of groups pursuing different purposes, then context becomes more important in determining appropriate response.'[88] Military force, Gregory Miller concludes – based on a quick scan of a series of case studies – has been effective against Sendero Luminoso in Peru and the Egyptian Islamic Jihad. However, in his list of suggestions, he does not include military force as a tool in effective counterterrorism. Miller proposes that the geographical and temporal conditions, the existing support base, its funding and the age of the group might all be related to the effectiveness of the measures. Furthermore, Peter Sederberg suggests that the role terrorism plays in the general strategy adopted by the radical movement should be taken into account, as, similarly, should the role of external sponsorship.[89] It can also be proposed that the development phase or the generation of terrorists involved can be a factor related to the effectiveness of counterterrorist measures.

Apart from the conditions, we should also consider the order in which the instruments are employed, It has been argued above that there exists a very strong counterterrorism imperative in the shape of 'something must be done'. 'Something' all-too-often turns out to be the use of force. When you approach your opponent in a forceful manner, the effectiveness of subsequent soft measures can be questioned. It could be argued that the order of the hard followed by the soft approach reduces the effectiveness of the latter. When positions have polarized, and attitudes hardened by forceful action, to what extent can negotiations and concessions, for example, bridge the gap of understanding between the actors?

## How does terrorism end?

When looking at discussions in the literature about how terrorism ends, few uncontested examples seem to exist in which military force has led to the ending of terrorist campaigns. Several authors have studied the final phases of terrorism, but again the most rigorous scientific methods have not been applied everywhere.[90] Indications from past cases show that terrorist organizations can be successful or they can transform themselves into legitimate political parties. They can also be defeated, or they can become criminal organizations.

### *Success*

Terrorists can end their activities because their aims have been (partially) achieved. Recurrent examples of successful campaigns are Irgun in British Palestine, the National Liberation Front in Algeria, the Quebec Liberation Front in Canada and the African National Congress in South Africa. What these cases have in common is the specific nature of the contests: decolonization or nationalist struggles.[91] It might be the case that the non-absolute terms of many of these wars allow for subsequent compromise and solutions compatible with terrorist aims.

A second feature of these successful cases is that many are linked to insurgency struggles. Terrorism forms a tactic, or one of the instruments employed, in the confrontation with the state opponent. This might indicate that more than just terrorist tactics are needed to achieve the desired end. Terrorism can be successful in the context of a larger insurgency struggle. Terrorist tactics, rather than a terrorist strategy, can make a difference.

Another route to the suspension of terrorism is a successful transformation into a political party. Martha Crenshaw suggests that political engagement might be needed to achieve success for groups using terrorism.[92] The challenge is, as formulated by Louise Richardson, 'to delegitimize the resort to terrorism as a means of effecting political change and to reduce the opportunities and incentives for doing so. It is to channel the effort to redress grievances into conventional politics.'[93] There might be a few conditions that could precipitate this transformation. Not only does to what extent the grievances are shared by large parts of the population matter, the concrete nature of the demands also plays a role.[94] This political challenge in counterterrorism has been successful in several cases, among others for the Palestinian Liberation Organization and the Irish Republican Army. The drawback of this route to ending terrorism is that in many cases the road to negotiation and concession causes hardliners to step up campaigns of violence.[95]

### *Defeat*

Many terrorist organizations experience defeat, but not all are bound to fail, as some have argued.[96] Defeat can both be caused by external pressure applied through the counterterrorist instruments or through internal dynamics. Internally, groups can fall apart when individual members simply quit or the group as a whole becomes less successful in attracting new membership. Popular support can dissipate for several reasons: fear of government reprisal, alternative ways to air grievances or terrorist attacks provoke revulsion.[97] Group think, group pressure, clashing individuals, internal suspicions, decreasing ideological motivation can all be reasons for groups to collapse internally.[98] Organizational dynamics demand that:

> Terrorist organizations … must constantly work to maintain themselves. If recruiting dries up, or if funding becomes unavailable, the organization may be unable to sustain itself. On the other hand, self-preservation may in fact force organizations to continue terrorist activities even if the leadership otherwise wishes to give them up. It may be that the only way for the organization to continue to attract new recruits and financial support is to continue to gain publicity for its terrorist actions.[99]

Military defeat is possible according to some experts. Audrey Kurth Cronin argues that Russian populist revolutionary group Narodnaya Volya in the late nineteenth century, the Peruvian Shining Path[100] and the Kurdish

Workers Party in the late twentieth century have all been reduced in importance as a result of military action.[101] Gregory Miller has argued that Egyptian Islamic Jihad has been dealt with successfully in a military manner.[102] However, several qualifications seem to be necessary. The agenda of the Narodnaya Volya or 'People's Will Organization' was taken over by those leading the Russian Revolution. Similarly, we cannot exactly argue, at this point in time, that the Shining Path, the PKK or Egyptian Islamic Jihad have suspended their activities.[103]

Another often-quoted example of military defeat of terrorism is the case of Uruguay.[104] The campaign to eradicate Tupamaros terrorism led to the suspension of democratic rule, and rule by a military junta between 1972 and 1985. Perhaps this example explains the preoccupation of scholars with studying the alleged tenuous relationship between counterterrorism and democracy.[105] The counter argument, that there is no such clash, has not received a lot of attention.[106] This is not to discount the use of force, but the examples are debatable or possess serious drawbacks.

Defeat by way of the criminal justice system is said to have been highly effective in the case of the German Red Army Faction and the Italian Red Brigades in the 1970s and 1980s respectively. Laws and judiciary procedures were adapted to deal harshly with terrorist crimes and, according to Wilkinson and others, they were defeated.[107] This idea of defeat does not explain the continued existence of the Red Army Faction into the 1990s and the activities of the Red Brigades in the twenty-first century. While the authors use the word 'defeat', they might actually mean 'containment'.

Some terrorist organizations transform themselves into criminal undertakings. The link with the criminal underworld is and has been strong in the history of modern terrorism. Funding through bank robberies and drug smuggling is common. Some terrorist organizations focus in their struggle on enrichment rather than ideology. The Revolutionary Armed Forces (FARC) in Columbia and the IRA in Northern Ireland have largely become criminal organizations. The FARC apparently is quite successful in its new undertaking, grossing an income of over US$1 billion annually.[108] We cannot properly argue that this is a form of defeat; it is, however, a transformation, which has occurred several times in the past.

While different studies have looked at these possible routes for transformations, the conditions under which they occur remain to date under-investigated. More research is necessary to map the processes that lead to terrorist organizations changing shape and color in the course of their existence.

## Conclusion

Why is the use of force to fight terrorism such a popular and prominent notion? Experts have concluded time and again that fighting insurgencies is highly political.[109] Why is it that, for dealing with terrorism, arguably an even more surgical use of force than the insurgents', we have never drawn a similar conclusion? As we are learning and relearning the lessons of counterinsurgency, it seems the lessons of fighting terrorism have not even been properly formulated.

From the perspective of the state, the effectiveness of the military instrument can be called into question. Based on currently available insights from non-structured case study analysis, but also on theoretical discussions and statistical indications, this leads us to believe that there are serious problems with the military approach.

First, the use of the military instrument complies with the terrorist aims. The use of the military instrument complies with the logic of terrorism: provoke the opponent into overreacting. As this contribution has illustrated, this mechanism is at work in many past cases of terrorist activity. When the state is provoked into using force in defense of its interests, it plays into the hands of the terrorists. This would indicate that terrorism can be effective in its operational aim of provocation.

Second, the use of force can lead to a counterterrorist spiral. Terrorist attacks create 'a false sense of insecurity' or an 'insecurity dilemma'.[110] The threat of becoming a victim of terrorism is generally hugely overestimated. This fear of possible attack leads to all kinds of measures that can be seen as excessive. This counterterrorist spiral works by evoking 'fear of possible attack, the introduction of measures, the allocation of greater resources, increased publicity, and increased fear'.[111] Furthermore, there are:

> [f]our further elements … [which influence] this spiral: a multiplier effect (wherein the measures introduced piled up without reconsideration or repeal), incomplete information that accompanied matters related to terrorism, the moral ascription associated with the term 'terrorist' and the reluctance of individuals in authoritative positions to risk under-preparation for terrorist attacks. Each of these encouraged the adoption of more counterterrorist measures.[112]

The measures start leading a life of their own and are difficult, if not impossible, to halt.

Third, the use of the military instrument leads to substitution. One avenue that was open for terrorists was closed off; other avenues are opened up to pursue the aims. This complicates the counterterrorism process: 'the very effectiveness of local antiterrorism efforts may even turn a local movement into a global one. When primary local groups are put out of reach, militants may shift their focus to secondary global goals.'[113] Not only local and global changes can be witnessed; substitution occurs between regions, types of attacks and sometimes even between groups. The seemingly successful actions of counterterrorism today create problems for the future:

> The danger now lies in the fact that the looser the operational connections become and the less Islamic terrorism is instigated by a single figure, the harder it will be to uncover exploitable links and the more likely that the instigators of future terrorist attacks will escape the notice of US intelligence.[114]

Fourth, the use of the military instrument leads to ethical dilemmas: what is the difference between terrorism and counterterrorism? As Martin van Creveld argued almost 20 years ago, 'when you fight terrorism, you become a terrorist'.[115] Out of necessity, 'one must use those same weapons and tactics belonging to the terrorists' inventory – psychology, stealth, speed, surprise, and cunning – against the terrorists themselves'.[116] Laura Donahue therefore provokingly argues:

> when the … essential qualities of an act of terrorism[117] are shared by an act of counter-terrorism, the line between the two fades, with similar consequences borne by both entities in relation to their domestic and international legitimacy. If, then, such actions are indistinguishable in their execution and effect, the stark issue with which we are faced is one of moral equivalence; is there a difference between a liberal, democratic state engaging in these actions and a non-state actor doing so?[118]

Current forceful counterterrorism solutions create problems for the future. These problems attached to counterterrorism, including the use of the military instrument, has elicited the following comment from one critical observer: 'We should fear counterterrorism more than we fear terrorism.'[119] As long as we do not have more evidence to solidly base counterterrorism policies on, we should indeed be wary. For researchers, the challenge today is to find out what works – and, importantly, what works under which conditions.

## Notes

This chapter has been previously published as Isabelle Duyvesteyn, 'Great Expectations: The Use of Armed Force to Combat Terrorism', *Small Wars and Insurgencies*, 19, 3 (2008): 328–351. We would like to thank Taylor and Francis for permission to reprint it here.

1 Delpech, 'The Imbalance of Terror', 37.
2 De Jonge Oudraat, 'Combating Terrorism', 168.
3 Howard, 'What's in a Name?', 8–20; Wilkinson, *Terrorism Versus Democracy*, 90.
4 Donahue, 'In the Name of National Security', 47.
5 Netanyahu, *Fighting Terrorism*, 6.
6 Byman, 'A Corrective That Goes Too Far?', 512.
7 Davis and Jenkins, *Deterrence and Influence in Counterterrorism*, xi.
8 Davis and Jenkins, *Deterrence and Influence in Counterterrorism*, 28.

9 Lum, Kennedy and Sherley, 'The Effectiveness of Counter-Terrorism Strategies', 3.
10 Byman, 'Do Targeted Killings Work?', 106; Finlan, 'Trapped in Dead Ground', 12; Livingston, 'Proactive Responses to Terrorism', 225.
11 Delpech, 'The Imbalance of Terror', 35; Taillon, *The Evolution of Special Forces in Counter-Terrorism*, xiii.
12 Serafino, 'Combating Terrorism', 2, note 2.
13 Malvesti, 'Explaining the United States' Decision to Strike Back at Terrorists', 404–425.
14 Crelinsten and Schmid, *Western Responses to Terrorism*, 333, 315.
15 Donahue, 'In the Name of National Security', 37.
16 Miller, 'Confronting Terrorisms', 342.
17 Harmon, *Terrorism Today*, 131.
18 Crenshaw, 'How Terrorism Declines', 84.
19 Wilkinson, 'Politics, Diplomacy and Peace Processes', 69.
20 Davis and Jenkins, *Deterrence and Influence in Counterterrorism*, 26–27.
21 Guelke, *The Age of Terrorism and the International Political System*, 185.
22 Donahue, 'In the Name of National Security', 17.
23 Crelinsten and Schmid, *Western Responses to Terrorism*, 315.
24 Gross, 'Fighting by Other Means in the Mideast', 352.
25 Silke, 'Terrorism and the Blind Men's Elephant', 12–28; Carr, *The Lessons of Terror*; Neumann and Smith, *The Strategy of Terrorism*; Duyvesteyn, 'Paradoxes of the Strategy of Terrorism'.
26 Arce and Sandler, 'Counterterrorism', 183–200.
27 ITERATE: http://ssdc.ucsd.edu/ssdc/ite00001.html; START: www.start. umd.edu/; US Department of State, *Patterns of Global Terrorism 1996*. Available at: www.state.gov/s/ct/rls/pgtrpt/
28 MIPT database. Available at: www.mipt.org/.
29 Engene, 'Five Decades of Terrorism in Europe', 109–122. The TWEED database is available at: www.uib.no/People/sspje/tweed.htm, University of Bergen, Norway.
30 Steven and Gunaratna, *Counterterrorism*; Crelinsten and Schmid, *Western Responses to Terrorism*.
31 Wilkinson, *Terrorism and the Liberal State*.
32 MIPT database, available at: www.mipt.org/.
33 Rees and Aldrich, 'Contending Cultures of Counterterrorism', 905–923.
34 Charters, *The Deadly Sin of Terrorism*, 220, italics in original; Taillon, *The Evolution of Special Forces in Counter-Terrorism*.
35 Wardlaw, *Political Terrorism*, 94; Wilkinson 'Politics, Diplomacy and Peace Processes'.
36 Wardlaw, *Political Terrorism*, 90.
37 Lum, Kennedy and Sherley, 'Are Counter-Terrorism Strategies Effective?', 504–505.
38 Enders and Sandler, 'Is Transnational Terrorism Becoming More Threatening?', 327.
39 Chalk, 'The Liberal Democratic Response to Terrorism', 26.
40 Livingston, 'Proactive Responses to Terrorism', 222.
41 Arce and Sandler, 'Counterterrorism', 184.
42 Ross and Gurr, 'Why Terrorism Subsides', 405–426.
43 Sederberg, 'Responses to Dissident Terrorism', 270.
44 Chalk, 'The Liberal Democratic Response to Terrorism', 28.
45 Luft, 'The Logic of Israel's Targeted Killing'.
46 Livingston, quoted in Sederberg, 'Responses to Dissident Terrorism', 270–271.

47 It has even been argued that it is not only the presence of military force but also the training that can act as deterrent:

> 'soft targets' are less likely to be found in states whose special forces are assisted by, or include, members of internationally reputed, elite counter-terrorist forces from abroad. It is clear that the level of terrorism has lessened in those countries whose troops have received training from the SAS, for example.
>
> (Taillon, *The Evolution of Special Forces in Counter-Terrorism*, 136–137)

48 Whiteneck, 'Deterring Terrorists', 187; Freedman, *Superterrorism.*
49 Trager and Zagorcheva, 'Deterring Terrorism', 96.
50 Whiteneck, 'Deterring Terrorists', 198.
51 Arce and Sandler, 'Counterterrorism', 184.
52 Enders and Sandler, 'What Do We Know About the Substitution Effect in Transnational Terrorism?', 119–137.
53 Siqueira and Sandler, 'Terrorists Versus the Government', 894.
54 Eppright, 'Counterterrorism and Conventional Military Force', 339.
55 Malvesti, 'Explaining the United States' Decision to Strike Back at Terrorists'.
56 Prunckun and Mohr, 'Military Deterrence', 267–280; Hanle, *Terrorism*, 218.
57 Enders, Sandler and Cauley, 'UN Conventions, Technology, and Retaliation in the Fight Against Terrorism', 83–105; Silke, 'Fire of Iolaus', 241–255; Lum, Kennedy and Sherley, 'Are Counter-Terrorism Strategies Effective?', 505.
58 Lum, Kennedy and Sherley, 'Are Counter-Terrorism Strategies Effective?', 506, italics in original.
59 Silke, 'Fire of Iolaus'.
60 Mueller, 'Six Rather Unusual Propositions about Terrorism', 491.
61 Brophy-Baermann and Conybear, 'Retaliating Against Terrorism', 196–210.
62 Livingston, 'Proactive Responses to Terrorism', 220.
63 Kegley, *International Terrorism*, 190.
64 Harmon, *Terrorism Today*; Art and Richardson, *Democracy and Counterterrorism*, 564; Cronin, 'How al-Qaida Ends', 17–48; Hanle, *Terrorism.*
65 Hanle, *Terrorism*, 216.
66 Harmon, *Terrorism Today*; Cronin, 'How al-Qaida Ends'.
67 Byman, 'Do Targeted Killings Work?'; Gross, 'Fighting by Other Means in the Mideast'; Luft, 'The Logic of Israel's Targeted Killing'.
68 Luft, 'The Logic of Israel's Targeted Killing'.
69 Byman, 'Do Targeted Killings Work?'; Gross, 'Fighting by Other Means in the Mideast'; Luft, 'The Logic of Israel's Targeted Killing'.
70 Heymann, 'Dealing with Terrorism', 29.
71 Byman, 'Do Targeted Killings Work?', 101.
72 Sederberg, 'Responses to Dissident Terrorism', 272.
73 Eubank and Weinberg, 'Terrorism and Changes in Political Party Systems', 125.
74 Mueller, 'Response', 526; Wilkinson, *Terrorism Versus Democracy*, 190.
75 Wilkinson, 'Pathways Out of Terrorism for Democratic Societies', 458.
76 Gray, 'Combating Terrorism', 17–23.
77 Daalder and Lindsay, 'Nasty, Brutish, and Long', 408; Cronin, 'How al-Qaida Ends'.
78 Gray, 'Combating Terrorism'.
79 Hanle, *Terrorism*; Rivers, *The War Against the Terrorists.*
80 Eppright, 'Counterterrorism and Conventional Military Force', 334. Walter Laqueur argues that counterterrorism has become less effective in the twentieth century because of limitations on the police force compared to in the nineteenth century. Laqueur, *The New Terrorism*, 45.

81 Davis and Jenkins, *Deterrence and Influence in Counterterrorism*, 59.
82 Luttwak, 'Dead End'.
83 Silke, 'Fire of Iolaus', 253; Hewitt, *Effectiveness of Anti-Terrorism Policies.*
84 Crenshaw, *How Terrorism Ends.*
85 Art and Richardson, *Democracy and Counterterrorism*, 571; Eppright, 'Counterterrorism and Conventional Military Force', 341.
86 Trager and Zagorcheva, 'Deterring Terrorism', 121.
87 Miller, 'Confronting Terrorisms'; Laqueur, 'Reflections on the Eradication of Terrorism', 207–212.
88 Sederberg, 'Responses to Dissident Terrorism', 265.
89 Sederberg, 'Responses to Dissident Terrorism', 264.
90 Crenshaw, 'How Terrorism Declines' and *How Terrorism Ends*; Cronin, 'How al-Qaida Ends' and *Ending Terrorism*; Wilkinson, 'Pathways Out of Terrorism for Democratic Societies', 'How Terrorism Ends', 'Politics, Diplomacy and Peace Processes', *Terrorism Versus Democracy*; Harmon, *Terrorism Today.*
91 Duyvesteyn, 'The Role of History and Continuity in Terrorism Research'; Sederberg, 'Responses to Dissident Terrorism', 277; Wilkinson, 'Pathways Out of Terrorism for Democratic Societies', 460.
92 Crenshaw, 'How Terrorism Ends'.
93 Richardson, *The Roots of Terrorism*, 9.
94 Sederberg, 'Responses to Dissident Terrorism', 277.
95 Sederberg, 'Responses to Dissident Terrorism', 278; Wilkinson, 'Pathways Out of Terrorism for Democratic Societies', 461; Hewitt, *The Effectiveness of Anti-Terrorism Policies.*
96 Carr, *The Lessons of Terror*; Abrahms, 'Why Terrorism Does Not Work'.
97 Cronin, 'How al-Qaida Ends', 28–29.
98 Cronin, 'How al-Qaida Ends', 18; Charters, *The Deadly Sin of Terrorism*, 214; Guelke, *The Age of Terrorism*, 180.
99 Crenshaw, 'How Terrorism Ends', 3.
100 Miller, 'Confronting Terrorisms'.
101 Cronin, 'How al-Qaida Ends', 30.
102 Miller, 'Confronting Terrorisms'.
103 See, for recent activities, the MIPT database. Available at: www.mipt.org/.
104 Wilkinson, 'Pathways Out of Terrorism for Democratic Societies'.
105 Wilkinson, *Terrorism Versus Democracy*; Schmid, 'Terrorism and Democracy', 14–25.
106 Quan Li, 'Does Democracy Promote Or Reduce Transnational Terrorism Incidents?', 278–297; Charters, *The Deadly Sin of Terrorism.*
107 Wilkinson, 'Pathways Out of Terrorism for Democratic Societies' and 'Politics, Diplomacy and Peace Processes'; Harmon, *Terrorism Today.*
108 Cronin, 'How al-Qaida Ends', 31.
109 Kilcullen, 'Countering Global Insurgency', 597–617.
110 Mueller, 'Six Rather Unusual Propositions About Terrorism', 487.
111 Donahue, 'In the Name of National Security', 18.
112 Donahue, 'In the Name of National Security', 18.
113 Trager and Zagorcheva, 'Deterring Terrorism', 121.
114 Pillar, 'Counterterrorism After Al Qaeda', 104.
115 Van Creveld, 'The New Terrorism', 19.
116 Taillon, *The Evolution of Special Forces in Counter-Terrorism*, xiv.
117 Essential qualities are: violence, fear, aimed at a broader audience, purposive act, political nature, aimed at non-combatants and instrumental act.
118 Donahue, 'In the Name of National Security', 34, italics in original.
119 Wolfendale, 'Terrorism, Security and the Threat of Counterterrorism', 753.

## Bibliography


Abrahms, Max. 'Why Terrorism Does Not Work'. *International Security* 31, 2 (2006): 42–78.

Arce, Daniel G. and Todd Sandler. 'Counterterrorism: A Game-Theoretic Analysis'. *Journal of Conflict Resolution* 49, 2 (2005): 183–200.

Art, Robert J. and Louise Richardson, eds. *Democracy and Counterterrorism: Lessons From the Past.* Washington, DC: United States Institute of Peace, 2007.

Brophy-Baermann, Bryan and John A.C. Conybear. 'Retaliating Against Terrorism: Rational Expectations and the Optimality of Rules Versus Discretion'. *American Journal of Political Science* 38, 1 (1994): 196–210.

Byman, Daniel. 'A Corrective That Goes Too Far?'. *Terrorism and Political Violence* 17, 4 (2005): 511–516.

Byman, Daniel. 'Do Targeted Killings Work?'. *Foreign Affairs* 85, 2 (2006): 95–111.

Carr, Caleb. *The Lessons of Terror.* London: Little Brown, 2002.

Chalk, Peter. 'The Liberal Democratic Response to Terrorism'. *Terrorism and Political Violence* 7, 4 (1995): 10–47.

Charters, David A., ed. *The Deadly Sin of Terrorism: Its Effect on Democracy and Civil Liberty in Six Countries.* Westport, CT: Greenwood Press, 1994.

Crelinsten, Ronald D. and Alex Schmid, eds. *Western Responses to Terrorism.* London: Frank Cass, 1993.

Crenshaw, Martha. 'How Terrorism Declines'. *Terrorism and Political Violence* 3, 1 (1991): 69–87.

Crenshaw, Martha. *How Terrorism Ends.* Special Report 48, Washington, DC: United States Institute of Peace, 1999.

Cronin, Audrey Kurth. 'How al-Qaida Ends: The Decline and Demise of Terrorist Groups'. *International Security* 31, 1 (2006): 17–48.

Cronin, Audrey Kurth. *Ending Terrorism: Lessons for Defeating Al–Qaeda.* Adelph: Paper 394. London: Routledge, 2008.

Daalder, I.H. and J.M. Lindsay. 'Nasty, Brutish, and Long: America's War on Terrorism'. *Current History* 100, 650 (2001): 403–408.

Davis, Paul K. and Brian Michael Jenkins. *Deterrence and Influence in Counterterrorism: A Component in the War on Al Qaeda.* Santa Monica, CA: RAND, 2002.

De Jonge Oudraat, Chantal. 'Combating Terrorism'. *Washington Quarterly* 26, 4 (2003): 163–177.

Delpech, Thére'se. 'The Imbalance of Terror'. *Washington Quarterly* 25, 1 (2002): 31–40.

Donahue, Laura K. 'In the Name of National Security: US Counterterrorist Measures 1960–2000'. *Terrorism and Political Violence* 13, 3 (2001): 15–60.

Duyvesteyn, Isabelle. 'Paradoxes of the Strategy of Terrorism'. In *Understanding Victory and Defeat in Contemporary War,* eds J. Angstrom and I. Duyvesteyn. London: Routledge, 2007.

Duyvesteyn, Isabelle. 'The Role of History and Continuity in Terrorism Research'. In *Mapping Terrorism Research,* ed. M. Ranstorp. London: Routledge, 2007.

Enders, Walter and Todd Sandler. 'Is Transnational Terrorism Becoming More Threatening? A Time-Series Investigation'. *Journal of Conflict Resolution* 44, 3 (2000): 307–333.

Enders, Walter and Todd Sandler. 'What Do We Know About the Substitution Effect in Transnational Terrorism?'. In *Research on Terrorism: Trends, Achievements and Failures,* ed. A. Silke. London: Frank Cass, 2004.

Enders, Walter, Todd Sandler and Jon Cauley. 'UN Conventions, Technology, and Retaliation in the Fight Against Terrorism'. *Terrorism and Political Violence* 2, 1 (1990): 83–105.

Engene, Jan Oskar. 'Five Decades of Terrorism in Europe: The TWEED Dataset'. *Journal of Peace Research* 44, 1 (2007): 109–122. Available at: www.uib.no/People/sspje/tweed.htm.

Eppright, Charles T. 'Counterterrorism and Conventional Military Force: The Relationship Between Political Effect and Utility'. *Studies in Conflict and Terrorism* 20, 4 (1997): 333–345.

Eubank, William and Leonard Weinberg. 'Terrorism and Changes in Political Party Systems'. *Terrorism and Political Violence* 4, 2 (1992): 125–140.

Finlan, Alastair. 'Trapped in Dead Ground: US Counter-Insurgency Strategy in Iraq'. *Small Wars and Insurgencies* 16, 1 (March 2005): 1–21.

Freedman, Lawrence, ed. *Superterrorism.* London: Blackwell, 2002.

Graeme, S.C.S. and R. Gunaratna. *Counterterrorism: A Reference Handbook.* Santa Barbara, CA: ABC Clio, 2004.

Gray, Colin. 'Combating Terrorism'. *Parameters* 23, 3 (Fall 1993): 17–23.

Gross, Michael. 'Fighting by Other Means in the Mideast: A Critical Analysis of Israel's Assassination Policy'. *Political Studies* 51, 2 (2003): 350–368.

Guelke, Adrian. *The Age of Terrorism and the International Political System.* London: I.B. Tauris, 1995.

Hanle, Donald J. *Terrorism: The Newest Face of Warfare.* Washington, DC: Pergamon Brassey's, 1989.

Harmon, Christopher. *Terrorism Today.* London: Routledge, 2008.

Hewitt, Christopher. *The Effectiveness of Anti-Terrorism Policies.* Lanham, MD: University Press of America, 1984.

Heymann, Philip B. 'Dealing with Terrorism: An Overview'. *International Security* 26, 3 (2001): 24–38.

Howard, Michael. 'What's in a Name?'. *Foreign Affairs* 81, 1 (2002): 8–20.

Kegley, Charles, ed. *International Terrorism: Characteristics, Causes, Controls.* New York: St Martin's Press, 1990.

Kilcullen, David J. 'Countering Global Insurgency'. *Journal of Strategic Studies* 28, 4 (August 2005): 597–617.

Laqueur, Walter. 'Reflections on the Eradication of Terrorism'. In *International Terrorism: Characteristics, Causes, Controls,* ed. C. Kegley. New York: St Martin's Press, 1990.

Laqueur, Walter. *The New Terrorism: Fanatics and the Arms of Mass Destruction.* London: Phoenix, 2001.

Livingston, Neil. 'Proactive Responses to Terrorism: Reprisals, Preemption and Retribution'. In *International Terrorism: Characteristics, Causes, Controls,* ed. C. Kegley. New York: St Martin's Press, 1990.

Luft, Gal. 'The Logic of Israel's Targeted Killing'. *Middle East Quarterly* 10, 1 (2003): 3–7.

Lum, Cynthia, Leslie Kennedy and Alison S. Sherley. 'Are Counter-Terrorism Strategies Effective? The Results of the Campbell Systematic Review on Counter-Terrorism Evaluation Research'. *Journal of Experimental Criminology* 2, 4 (2006a): 489–516. Available at: www.campbellcollaboration.org/doc-pdf/Lum_Terrorism_Review.pdf.

Lum, Cynthia, Leslie Kennedy and Alison S. Sherley. 'The Effectiveness of Counter-Terrorism Strategies: A Campbell Systematic Review' (2006b). Available at: www.campbellcollaboration.org/doc-pdf/Lum_Terrorism_Review.pdf

Luttwak, Edward. 'Dead End: Counterinsurgency Warfare as Military Malpractice'. *Harpers Magazine,* February 2007.

Malvesti, Michele. 'Explaining the United States' Decision to Strike Back at Terrorists'. In *Terrorism and Counterterrorism: Understanding the New Security Environment, Readings and Interpretations,* eds R.D. Howard and R.L. Sawyer. New York: McGraw-Hill, 2003.

Miller, Gregory D. 'Confronting Terrorisms: Group Motivation and Successful State Policies'. *Terrorism and Political Violence* 19, 3 (2007): 331–350.

Mueller, John. 'Response'. *Terrorism and Political Violence* 17, 4 (2005): 523–528.

Mueller, John. 'Six Rather Unusual Propositions about Terrorism'. *Terrorism and Political Violence* 17, 4 (2005): 487–505.

Netanyahu, Benjamin. *Fighting Terrorism: How Democracies Can Defeat Domestic and International Terrorism.* New York: Farrar, Strauss, Giroux, 1995.

Neumann, Peter R. and M.L.R. Smith. *The Strategy of Terrorism: Why it Works, Why it Fails.* London: Routledge, 2007.

Pillar, Paul. 'Counterterrorism After Al Qaeda'. *Washington Quarterly* 27, 3 (2004): 101–113.

Prunckun, H.W. Jr and P.B. Mohr. 'Military Deterrence: An Evaluation of Operation El Dorado Canyon'. *Studies in Conflict and Terrorism* 20, 3 (1997): 267–280.

Quan Li. 'Does Democracy Promote or Reduce Transnational Terrorism Incidents?'. *Journal of Conflict Resolution* 49, 2 (2005): 278–297.

Rees, Wynn and Richard J. Aldrich. 'Contending Cultures of Counterterrorism: Transatlantic Divergence or Convergence?' *International Affairs* 81, 5 (2005): 905–923.

Richardson, Louise, ed. *The Roots of Terrorism.* London: Routledge, 2006.

Rivers, Gayle. *The War Against the Terrorists: How To Win It.* New York: Charter, 1986.

Ross, Jeffrey Ian and Ted Robert Gurr. 'Why Terrorism Subsides: A Comparative Study of Canada and the United States'. *Comparative Politics* 21, 4 (1989): 405–426.

Schmid, Alex P. 'Terrorism and Democracy'. *Terrorism and Political Violence* 4, 4 (1992): 14–25.

Sederberg, Peter. 'Responses to Dissident Terrorism: From Myth to Maturity'. In *International Terrorism: Characteristics, Causes, Controls,* ed. C. Kegley. New York: St Martin's Press, 1990.

Serafino, Nina. 'Combating Terrorism: Are There Lessons to Be Learned From Foreign Experiences?'. Congressional Research Service Report. Washington, DC: Library of Congress, 2002.

Silke, Andrew. 'Terrorism and the Blind Men's Elephant'. *Terrorism and Political Violence* 3, 3 (1996): 12–28.

Silke, Andrew. 'Fire of Iolaus: The Role of State Countermeasures in Causing Terrorism and What Needs to be Done'. In *Root Causes of Terrorism,* ed. T. Bjorgo. London: Routledge, 2005.

Siqueira, Kevin and Todd Sandler. 'Terrorists Versus the Government: Strategic Interaction, Support and Sponsorship'. *Journal of Conflict Resolution* 50, 6 (2006): 878–898.

Steven, Graeme C.S. and Rohan Gunaratna. *Counterterrorism: A Reference Handbook.* Santa Barbara, CA: ABC Clio, 2004.

Taillon, J. Paul de B. *The Evolution of Special Forces in Counter-Terrorism.* Westport, CT: Praeger, 2001.

Trager, Robert F. and Dessislava P. Zagorcheva. 'Deterring Terrorism: It Can Be Done'. *International Security* 30, 3 (2006): 87–123.

US Department of State. *Patterns of Global Terrorism 1996.* Washington, DC: Department of State, 1997. Available at: www.state.gov/s/ct/rls/pgtrpt/.

van Creveld, Martin. 'The New Terrorism'. *The Economist,* 15 August 1989.

Wardlaw, Grant. *Political Terrorism: Theory, Tactics and Counter-Measures.* Cambridge: Cambridge University Press, 1982.

Whiteneck, Daniel. 'Deterring Terrorists: Thoughts on a Framework'. *Washington Quarterly* 283 (2005): 187–200.

Wilkinson, Paul. *Terrorism and the Liberal State.* New York: New York University Press, 1986.

Wilkinson, Paul. 'Pathways Out of Terrorism for Democratic Societies'. In *Contemporary Research on Terrorism,* eds. P. Wilkinson and A.M. Stewart. Aberdeen: Aberdeen University Press, 1987, 453–465.

Wilkinson, Paul. 'How Terrorism Ends', Special Report 48, Washington, DC: United States Institute of Peace, 1999.

Wilkinson, Paul. 'Politics, Diplomacy and Peace Processes: Pathways Out of Terrorism'. in *The Future of Terrorism,* eds. M. Taylor and J. Horgan. London: Frank Cass, 2000, 66–82.

Wilkinson, Paul. *Terrorism Versus Democracy: The Liberal State Response.* London: Frank Cass, 2006.

Wolfendale, Jessica. 'Terrorism, Security and the Threat of Counterterrorism'. *Studies in Conflict and Terrorism* 29 (2006): 753–770.

# 5 Why does peacekeeping succeed or fail?

## Peacekeeping in the Democratic Republic of Congo and Sierra Leone

*Sarah Kreps*

Why do some UN peacekeeping missions succeed and others fail? Why did the mission in Sierra Leone produce peace and the mission in the Democratic Republic of Congo (DRC) did not? These two UN missions have had unambiguously divergent outcomes. In 1999, Sierra Leone was in the midst of a brutal civil war that had killed or displaced about half its population. But by 2002, three years after the UN deployment, violence had ended in Sierra Leone. By 2005, the UN Mission in Sierra Leone (UNAMSIL) had exited the country, leaving behind self-enforcing peace, and becoming a peace-enforcement success story.[1] In contrast, seven years into the UN Mission in the Democratic Republic of Congo (MONUC), the UN's Under Secretary General for Humanitarian Affairs still equated the Congo with "the killing fields of our times".[2] The prospect of MONUC exiting and leaving a self-enforcing peace is remote.[3]

This chapter seeks to explain the factors that contribute to these starkly different outcomes, with an eye toward understanding the conditions under which force is likely to be effective in the peace-enforcement setting. As this analysis shows, MONUC and UNAMSIL had many situational factors in common – neither had had experience with democracy, both had low levels of economic growth, and each had concluded the civil war with a ceasefire and a peace agreement. Neither had resulted in a decisive victor and neither was characterized by a "mutually hurting stalemate" that would reduce incentives to keep fighting. While just suggestive, key differences in the timing and composition of the UNAMSIL and MONUC forces indicate that commonly cited situational factors may not account for all the variation in outcomes.

This analysis of two African peacekeeping cases points to peacekeeping capacity as a possible explanation for the divergent outcomes. In the UNAMSIL case, the intervention is better cast as two observations – within-case variation – because of the change in outcome over time: from hostile between October 1999–May 2000 to stabilizing and stabilized thereafter. Considering the case longitudinally helps to control for situational factors such as the country's history, society and economy, and suggests that any factors that changed over time may have contributed to the change in outcome. Process tracing suggests that an operative difference was peacekeeping capacity. In the early phase of the intervention, UNAMSIL consisted of poorly-equipped and wary UN forces, 500 of whom were captured in the spring of 2000, and none of which had been able to stem the killing, mutilation and displacement of Sierra Leonean civilians. In May 2000, the British rapidly deployed several hundred elite paratroopers that quickly established calm in Freetown and helped the Sierra Leonean army establish security throughout the rest of the country.

In contrast, MONUC's capacity may have increased since 1999, but an examination of its capacity relative to that of UNAMSIL suggests a significant deficiency in terms of the territory that the peacekeeping force is expected to cover. The vastness and ruggedness relegates patrol missions to six headquarters and leaves the vast interstices ungoverned and exploited by rebel forces. The DRC mission has degenerated into a regional conflagration with innumerable factions in conflict, several of which have not been incorporated into the reconciliation process.

The DRC case suggests an important qualification to the capacity argument: even a well-equipped, effectively mandated, abundant peacekeeping force is unlikely to succeed if there is no peace to keep.[4] Peace in the DRC, for example, has been more difficult to achieve because of a fundamentally flawed peace process that has left out factions that then acted as spoilers. This observation also implies that not all peace agreements are created equal. Both the intervention in Sierra Leone and the DRC "concluded" with negotiated settlements, but the former was inclusive of major factions whereas the latter was not. The nature of the peace agreement may therefore go some way to explaining why a large number of negotiated settlements (43 percent) break down within five years.[5]

Taken together, these two points suggest the following related caution on the utility of force. Just as too few and poorly equipped peacekeepers can limit the effectiveness of a peace-enforcement mission, so can an overly ambitious or ambiguous political objective. That peacekeeping capacity and the political environment are important to peacekeeping outcomes are not necessarily novel arguments. That said, they have yet to be confirmed as causes of success and failure amidst the litany of conceivable factors that have been identified in the peacekeeping literature. Which factors are most dominant and the pathways by which those factors affect outcomes are still open, debated empirical questions. Since the international community continues to undertake peacekeeping missions, better understanding the conditions amenable to success and failure is not just warranted but imperative.

The chapter proceeds in several parts. It first engages the peacekeeping literature and cites the situational factors that are thought to make peacekeeping missions more or less difficult. It then compares those factors for the two peacekeeping missions under consideration in this chapter – the DRC and Sierra Leone – and shows that most factors had similar values for both states, suggesting that those similarities cannot explain the divergent outcomes. It then offers background on the conflicts and traces the divergent factors – capacity and the political process – that appear to explain the divergent outcomes between the two cases. The chapter closes by discussing implications for the effectiveness of peace-enforcement missions.

## Existing accounts of peacekeeping success and failure

The question of whether peacekeeping missions help keep the peace after civil wars finds few dissenters in the current wave of peacekeeping scholarship.[6] On the contrary, a growing volume of support has coalesced around the argument that third-party intervention is an effective tool for maintaining peace after civil war.[7] Where the conclusions tend to diverge is in the situational factors that complicate peacekeeping operations and that therefore have an impact on whether particular operations are more or less able to keep the peace. George Downs and Stephen Stedman note that the peacekeeping literature contains "literally dozens of contextual variables" that affect the difficulty of the implementation environment.[8] These "preexisting conditions" create "baseline prospects for peace" even before peacekeepers enter the picture.[9]

Although the peacekeeping literature generally agrees that, on average, third-party intervention is helpful in creating durable peace, there is a distinct inconsistency in the situational factors that scholars find are more or less suited to peace.[10] Michael Doyle and Nicholas Sambanis, for example, find that wars that were identity-based, had high levels of death and displacement (a proxy for hostility), had a large number of factions fighting, and took place in a country with high resource dependence all contribute to lingering hostilities; these factors contributed to the structure or even the source of new wars and decrease the likelihood of peacekeeping success.[11]

Page Fortna, however, finds that the number of parties or factions has very little to do with the difficulty of forging sustainable peace, but agrees that peace is harder to keep the more deadly the civil war. High costs in terms of death tolls on either side engender animosity that lingers well past the formal termination of a civil war. On the other hand, decisive wars, long but not particularly deadly wars and high levels of economic development all produce conditions more conducive to enduring peace.[12] Barbara Walter's work on civil-war settlement concludes that duration of war, stalemate and magnitude of battle deaths affect the likelihood of successful outcomes.[13] Yet another account, that of Hartzell *et al.*, finds support for the argument that states that were previously democracies, wars that endured at low intensity but for a long period of time and peace agreements with provisions for territorial autonomy of threatened groups are all situational factors that affect duration of peace.[14] Still other accounts credit the presence of lootable resources – oil, diamonds, and minerals – as an enduring source of conflict that makes peace harder to keep.[15]

Offering perhaps the last major situational variable, Steve Stedman identifies the presence of spoilers as an impediment to peace. Spoilers are actors who view their interests as incompatible with the peace process and who consequently orient their behavior around preventing the implementation of peace accords. They may do this by not being a part of the peace process and sowing conflict from the outside. Spoilers may also be inside the peace process but view their interests as so threatened by the process that they act more insidiously to undermine the implementation of peace. Their presence creates a more difficult implementation environment, as peacekeepers not only must contend with actors whose ideal outcome is conflict rather than peace, but because it is not always possible to discern between limited and total spoilers, whose incentives and management are qualitatively different.[16]

In many cases, the situational factors advanced by scholars are contradictory, but they do offer a starting point for thinking about why these two cases of African peacekeeping might have different results. One reason to choose these two cases is that the cases are similar in many respects – while they are not perfectly matched, the two cases have many situational factors in common – but the outcomes are widely divergent. This divergence sets up a "method of difference" approach for identifying the factors that may be relevant to peacekeeping success and failure. Any situational factor that these two cases have in common is unlikely to explain the divergent outcomes. These factors can be eliminated as dominant causes. Process tracing then offers a way to show or identify the known or unknown explanatory variables that account for divergent outcomes.[17]

An additional reason to choose these cases is that, while many cases of peacekeeping have been the subject of voluminous research – Somalia, Bosnia, East Timor, Darfur, to name a few – the DRC has thus far been neglected in these scholarly studies despite the fact that it has been the largest UN operation in recent years and has been the deadliest war since the Second World War.[18] Understanding the source of its failure is clearly of great importance, and using Sierra Leone – one of the more successful recent peacekeeping missions – as a foil is therefore a useful approach.

As the specific cases of the DRC and UNAMSIL show, these situational factors are mostly problematic as explanations; with the exception of the peace agreements – in which Lomé incorporated major factions and Lusaka did not – they are fairly constant across a case of success and failure and therefore cannot explain the divergent outcomes. The question, then, is what does explain the variation? It is that question that this analysis turns to by examining the two cases in detail.

*Table 5.1* Comparison of situational factors for DRC/MONUC and Sierra Leone/UNAMSIL

| *Situational factors* | *DRC/MONUC* | *Sierra Leone/UNAMSIL* |
|---|---|---|
| Identity-based war | No | No |
| Number of factions | High[19] | Moderate[20] |
| Intensity/cost of war (casualties as % pop) | 2.2[21] | 2[22] |
| Level of economic development (GDP)[23] | $646 | $449[24] |
| Experience with democracy | No[25] | No[26] |
| War duration | 5 years (1998–2003)[27] | 8 years (1991–2000)[28] |
| Stalemate | No | No |
| Lootable resources | Yes (diamonds, gold, cobalt) | Yes (diamonds)[29] |
| War outcomes (decisive) | No | No[30] |
| Treaty/peace agreement | Lusaka; not all parties incorporated (possible spoilers)[31] | Lomé; all major parties incorporated[32] |
| Outcome | Failure | Success |

## The DRC mission

### *Background and escalation*

As with many civil conflicts in Africa, the conflict in the Democratic Republic of Congo has its roots in decolonization. When the DRC became independent from Belgium in 1960, conflict ensued as militias began attacking the European population, which elicited the intervention of Belgium; and the southern province of Katanga seceded, which drew the intervention of the UN; and the central province of South Kasai seceded, drawing the intervention of the Soviet Union and consequently the United States. The UN mission in the Congo was one of the most robust UN missions in the Cold War, with about 20,000 UN participants from a period of 1960 until 1964.[33]

Its mandate, however, was never clear, and the guidance on the degree to which peacekeepers could use force to mediate the conflict plagued its effectiveness.[34] Only once the UN was authorized to take "vigorous action" to remove foreign military actors after several periods of escalation was the UN able to prevent further secession, setting the stage for President Mobutu to assume power. His own tenure was characterized by mutinies and fracture, but he remained in power for three decades, until Cold War funding ran dry, his regime crumbled, and Laurent Kabila – backed by Rwandan Hutus who had fled to Congo after the Rwandan genocide in 1994 – assumed power, forced Mobutu into exile, and provoked the First Congo War between 1996 to 1997.[35]

The First Congo War constituted a rebellion against Mobutu, with Rwanda and Uganda joining the conflagration on the side of Kabila, the challenger to Mobutu. His Alliance of Democratic Forces for the Liberation of Zaire (AFDL) itself was thought to have killed 60,000 civilians, but ultimately assumed power in Kinshasa.[36] Kabila embarked on a campaign to centralize the government and undermine minority groups, while at the same time relying on the Rwandan government for support and trying to demand that the Hutu Interahamwe that had fled Rwanda after 1994 into the Great Lakes region of the DRC leave the territory in the East.[37]

Kabila's repression of rights antagonized opponents and triggered the Second Congo War in 1998, in which the rebel groups, backed by Rwanda and Uganda, took over eastern DRC, and Zimbabwe, Namibia and Angola contributed troops to bolster Kabila. Fighting attenuated by 1999 and allowed for the Lusaka peace accord and ceasefire in 1999, which the UN supported with a peacekeeping force that confronted almost immediately a breakout in conflict that intensified further with Kabila's assassination in 2001.

The assassination, however, allowed the peace talks to continue, as Laurent Kabila had stalled the process and his son, Joseph Kabila, continued the Inter-Congolese Dialogue that brought together the various factions with an interest in the DRC.[38] By early December 2002, several of the warring parties signed another peace agreement, the Pretoria Agreement[39] between Rwanda and the DRC that arranged for the withdrawal of Rwandan forces from the DRC in exchange for disarming the Interahamwe Hutus. This agreement, despite being referred to as the "All-Inclusive Agreement", did nothing to address the Ugandan involvement in the East. However, on 30 December 2002, the Ugandans and the DRC signed the Gbadolite Agreement that issued a ceasefire between three Ugandan-backed rebel groups.[40] The agreement called for rebel withdrawal to positions 40 kilometers apart and a demilitarization of Mambasa and Komanda (areas near Bunia in the East) enforced by MONUC. MONUC teams, however, observed and documented severe human rights violations, including executions, rapes and mutilation. Conditions clearly did not improve.[41]

Despite the constitutional progress and ostensible move toward "democracy" – the Pretoria Agreement created a transitional government that took over in 2003 – violence continued, particularly in the eastern province of North Kivu between Rwandan Hutus, Tutsi dissidents and government forces. Violence has also been sporadic in the capital Kinshasa,[42] and there has been government–militia violence in Katanga. In short, although the civil war technically ended in 2003, violence has continued, security conditions have deteriorated and the numerous factions remain hostile.[43] Thus, the UN peacekeeping mission has been renewed several times, most recently in May 2007, and extended until the end of 2008.

Although MONUC is now the largest UN peacekeeping mission, that distinction has only come over time. The need for a sizable peacekeeping force was evident from the outset, however. Even at the signing of the Lusaka Peace Accord in July 1999, security looked tenuous, since several rebel factions had elected not to sign the agreement. At the signing of the accord, Tunisia's foreign minister noted: "This is an African deal which begs international support. We cannot implement this on our own."[44] Yet international support was weeks away and only grew gradually over time. In the aftermath of the Lusaka Peace Agreement, the UN Security Council passed Resolution 1258 that established a 90-person observer mission, followed by Resolution 1279 that extended the mandate and augmented the number to 500 UN military observers to observe the Lusaka Ceasefire Agreement.[45] Resolution 1291 in February 2000 continued the escalation by authorizing 5,537 military personnel, which increased to 8,700 with Resolution 1445 in December 2002 and 10,800 with Resolution 1493 of July 2003; an additional 5,900 was added with Resolution 1565 in October 2004, and finally up to the peak of 16,661 troops, 1,011 police and 735 military observers (a total of 18,407) in October 2007.[46]

Despite the authorized numbers, actual deployment was considerably slower, delayed by sporadic conflict between the government and rebel groups. For example, the February 2000 authorization had permitted deployment of 5,537 military personnel, but by December 2000, only 224 had deployed. The first Blue Helmets did not arrive until March 2001, as a Uruguayan guard unit arrived in Kalemie. By July 2001, 2,366 soldiers had arrived from South Africa, Uruguay, Morocco, Senegal and Tunisia, and were charged with protecting MONUC installations. A year later, Blue Helmets numbered 3,804 and finally deployed in the restive Eastern part of the country. After the July 2002 Pretoria Agreement, military observers began monitoring withdrawal of Rwandan soldiers from Ituri, where they also identified growing ethnic violence.[47]

In the spring of 2003, security conditions degenerated further and the UN accordingly escalated its response. In response to the Luanda Agreement that called for the withdrawal of Ugandan forces from Bunia, the 7,000 Ugandan soldiers who had been occupying the area withdrew, causing chaos that the 700 Uruguayan Blue Helmets were unequipped to handle. They had been operating with a Chapter VI mandate and were quickly overwhelmed. On 30 May 2003, the Security Council called for

> the rapid deployment to Bunia of a highly trained and well-equipped multinational force, under the lead of a Member state, to provide security at the airport as well as to other vital installations in the town and to protect civilian population.

The emergency force, called Operation Artémis, was a French-led interim multinational force (IEMF) that replaced the 700 Uruguayan soldiers and military observers with 2,500 European and non-EU peacekeepers "on a strictly temporary basis".[48] The UN approved reinforcements on 28 July 2003, sending an additional brigade to Bunia and its environs, and the IEMF was authorized to use all necessary means. By 1 September 2003, however, the IEMF had transferred authority to MONUC, which then numbered 10,415 Uruguayan, Bangladesh, Nepalese, Pakistani and Indian soldiers.[49]

That stronger MONUC force helped stabilize conditions in Ituri, but also offered a target for militias. The killing of MONUC peacekeepers throughout 2004 triggered an additional strengthening of the mandate, allowing that they "proactively" contribute to pacification and improvement of security, provide support for conflict resolution and gather information on spoilers. By 2005, MONUC reached 16,000 peacekeepers, which continued to be the target of militia groups, and the Secretary General requested and was granted reinforcements.

Despite earning the label "Africa's first world war" when the magnitude of conflict-related mortality became clear in 2003,[50] the international response was glacial, reactive and insufficient. Perhaps, in its defense, the UN mission has not stopped the bloodshed but has arguably prevented the situation from degenerating into utter lawlessness. The UN did, for example, initially keep the parties in agreement, and has mediated the Inter-Congolese Dialogue on power-sharing in the DRC, a key step in advancing government reconciliation.[51] But the mission is nonetheless beset by considerable challenges, fundamentally having to do with its size relative to the size of the country. The DRC is about the size of Western Europe; it is five-times the combined size of Liberia, Sierra Leone and Côte d'Ivoire. Yet the peacekeeping size has been paltry in comparison. Though this is the most robust peacekeeping force the UN has deployed, it is roughly the same number of peacekeepers that participated in UNAMSIL, which was intended to police and provide security in an area about 1/33 the size of the DRC and a fraction of the population.[52]

That the force has been understaffed through most of the mission's history has manifested itself consistently but in different ways throughout the deployment. Early in the conflict, three rebel groups divided the country into thirds, with each of these groups and territories backed by neighboring regional states. The dearth of peacekeepers left wide areas of territory ungoverned and vulnerable to exploitation by the rebel groups of each territory, particularly since the patrols remained limited to the main UN headquarters and six sector headquarters, and tended not to venture into the interstices where the rebels take refuge. One victim of atrocities reported to the BBC that the MONUC forces sometimes chase the rebels but that the rebels go further into the forest and can use space to easily elude the MONUC forces.[53] As of 2007, after the 2006 election that optimists had hoped might mitigate the DRC's poor governance, "large swaths of the country" remain "authority-free zones where civilians are at the mercy of heavily armed groups who have made warfare a livelihood".[54]

Scarcity of troops also means that even if a large enough number can produce relative stability in one area, the fighting has migrated elsewhere. MONUC compounds are situated in the urban areas at one of six UN headquarters – Kinshasa, Kananga, Kisangani, Goma, Bunia and Buhavu – and have helped pacify the immediate surroundings through patrols, but instability has then tended to move to areas in between the compounds. Having created some stability around Bunia, for example, the militia activity moved to the surroundings where the international peace-enforcement presence was absent. Similarly, once Kinshasa was reasonably settled, fighting increased in North Kivu, where MONUC's presence was less prominent; moreover, the forces that were there were underequipped to deal with the Hutu and Congolese militias and therefore did not take part in direct combat.[55]

In addition to issues of troop levels and the vastness of the territory is the mandate itself, which is a dismal combination of ambiguous and weak. As one NGO report indicated, "ask ten employees of MONUC what is the mandate of their 4,300-strong military contingent and you are likely to get ten different answers".[56] Separately, Amnesty International reported that MONUC has been a "hostage to its weak mandate" and therefore the record in "promoting the security of the civilian population has been little short of disgraceful".[57] The mission is a challenged by the Chapter VII part of its mandate, which technically grants protection for civilians but in the vaguest sense:

> MONUC may take the necessary action, in the areas of deployment of its infantry battalions and as it deems it within its capabilities, to protect UN and co-located Joint Military Council personnel, facilities, installations and equipment, ensure the security and freedom of movement of its personnel, and protect civilians under imminent threat of violence.[58]

The imprecision in their mandate contributed to scant willingness and ability to keep the peace by giving the MONUC troops plausible deniability; they were able to assert quite correctly that they were fulfilling the mandate according to their interpretation since the resolution's language set unclear guidance and offered enormous loopholes. In one episode, a Congolese woman sought MONUC assistance from members of the RDC, but the 650 Moroccan and 550 Uruguayan troops remained at the base, offering no military response to an attack on civilians, claiming that they did not have the mandate to protect the citizens.[59]

Similarly, in the Bunia fiasco in 2003, the Uruguayan contingent took umbrage in what they believed was a Chapter VI mandate that only allowed them to act in self-defense. The UN Lessons Learned report defends Resolution 1291 of 2000, indicating that it authorized protection of civilians and even "preemptive self-defense against anticipated attack when supported by compelling evidence that hostile units or persons were committed to an immediate attack. These were the prevailing circumstances when Bunia was overrun by violence."[60] Though those may have been the UN's retrospective view of prevailing circumstances, the Uruguayan contingent held a different view and argued that its mandate was not nearly as expansive and therefore did not grant them authority to undertake offensive action against rebel groups. The fact that the Uruguayan unit only numbered 700 gave it further disincentive to confront the rebel group.

Insufficient troop numbers and the ambiguous mandate have been compounded by the fact that the DRC lacks highly trained and well-equipped peacekeepers. Notwithstanding the EU-led operation in 2003, Western countries have been absent from the list of troop-contributing countries. Uruguay, South Africa, Bangladesh, India and Pakistan have large contingents; Pakistan has actually established and run peace training centers funded through its own resources. The absence of developed-nation forces has signaled a lack of international interest to the local community.[61] The absence of Western forces and Western attention has real and perceived effects. The presence of its troops can produce a concrete, positive effect, as Operation Artemis showed; in a three-month period, the EU-led force was able to restore order in a region that had been engulfed by violence.

Those troops also have a signaling effect in that they show commitment from the international community. Referring to the struggles the mission has had in producing peace in the DRC, one senior MONUC officer in Kinshasa seemed to summarize the issue by saying: "Where's the P-5? Where are the Western troops?" Instead, MONUC's credibility has suffered because the composition of forces signals international apathy. Rebels can continue to destabilize the country knowing that retribution will be limited.

The effects of limited capacity are compounded even further by the DRC's infrastructure and geography. Not only is the country the size of Western Europe, but decades of civil war destroyed the DRC's infrastructure. Even military vehicles have difficulty moving within the DRC because the rugged geography makes the territory impassable.[62] As a result of the size, the absence of infrastructure and the impassable terrain, the UN essentially divided the country into sectors, with major urban areas – Kinshasa, Kananga, Kisangani, Goma, Bunia and Buhavu – acting as the hub of each sector.[63] The only way for MONUC forces to travel from one of the six headquarters to another is by airlift, which itself is difficult given the poor state of airfields in the DRC; patrols too far from the headquarters are prohibitively difficult because of the road conditions. Governance suffers from the combination of a large country with insufficient internal or external capacity; the DRC has no air force, no border security and poorly paid police, all of which have left much of the territory uncontrolled.[64]

Such size and prohibitive travel have an adverse effect on longer-term governability. Whereas the capital is in Kinshasa, many of the security problems have taken place in the eastern part of the country, in part because of the involvement of Ugandan and Rwandan soldiers. When MONUC troops sought to fill the security void after Ugandan soldiers withdrew from the Ituri region, MONUC – and particularly government forces – had trouble reaching the east. After several weeks of fighting, MONUC sought to fill the void and became the ad hoc administrator until the government could deploy an administrative and military capacity to the eastern part of the country. The size, geography and terrain make it more difficult to administer humanitarian services, more difficult to patrol and more difficult to govern.[65]

### *Factions, the peace process and spoilers*

With regard to one of the situational factors that might contribute to the DRC's instability, it is clear that not incorporating all factions into the peace process has created a number of spoilers. What makes this more of a challenge is the sheer number and shifting alignment of factions. Some rebel groups are in conflict with each other but also with the government; other factions, including both Tutsi- and Hutu-aligned forces, are in the

employ of neighboring countries such as Rwanda, Burundi and Uganda.[66] To the extent that these groups are funded by neighbors, the DRC conflict has at times resembled a regional interstate war more than an intrastate conflict. At one point during the conflict, Uganda sent in troops to support the Rwandan spoilers interfering in the East, Angola sent troops to side against Rwanda and Uganda, and Namibia and Zimbabwe deployed troops to fight against Angola and DRC President Kabila, turning the DRC into the field in which this broader regional conflagration was fought, with each state sending in troops and supporting ethnic militias within the DRC.[67]

The number of factions presents a considerable problem for negotiating and managing a comprehensive peace. The initial Lusaka Agreement established

> inter-Congolese political negotiations involving "les forces vives" that shall lead to a new political dispensation and national reconciliation in the DRC. The inter-Congolese political negotiations shall be under the aegis of a neutral facilitator to be agreed upon by the Congolese parties.

As MONUC has sought to supervise the "peace", the peace and reconciliation process continues under the aegis of the UN and South Korea. In other words, the Lusaka Agreement was incomplete and the Inter-Congolese Dialogue is ongoing. The problem, however, is that subsequent steps in the dialogue have either postponed difficult questions of power-sharing or left out key parties that have eventually played the role of spoilers. The Pretoria Agreement, for example, did not have participation from the Union des Patriotes Congolais (UPC), which had power in Bunia in the Eastern DRC. Rather, the Eastern DRC's Ituri region was largely bracketed and became the source of heavy fighting just one month later in August 2002 as the UPC played the role of spoiler to the Pretoria Agreement and made delivery of humanitarian aid and political reconciliation even more difficult.[68]

Similarly, in January 2008 the Congolese government concluded a series of peace talks and signed an agreement with 22 armed groups in which the parties would adhere to an immediate ceasefire, disengage forces and abide by international human-rights law. A fundamental flaw in the Goma Agreement, however, is that one of the major factions, the Democratic Forces for the Liberation of Rwanda (FDLR), was not a party to the agreement. The FDLR is Rwandan-backed and has contributed to the instability in the Kivus, a region that borders Rwanda in the eastern part of the country. The principle of the peace process and its provisions was commendable, but it is unsurprising that conflict continues in that region. The FDLR continues its attacks and the other rebel groups that were involved in the peace process, not wanting to lose ground, have contributed to ongoing hostilities.[69]

### *Lootable resources as an intervening variable*

Allowing the militias to fill the power vacuum left by insufficient forces has had a multiplier effect because of the role natural resources have played in fueling the rebel groups. A UN panel in 2008 found that rebel groups were buying arms with dividends from mines and taxation of minerals, a significant problem because of the high degree to which Congo's mines, particularly those in the East, are controlled by violent rebel groups. Indeed, a complementary UN Security Council report found that "control over territory has proven to be the key factor enabling armed groups in the DRC to profit from the exploitation of natural resources".[70] Mining is left unsupervised because neither the government nor MONUC can spare enough individuals to oversee the mines.[71]

Lootable resources in the form of diamonds, gold, copper and cobalt appear to be most important as an intervening rather than independent variable. They have ended up in the hands of rebels and fueled these rebel movements because the UN has not deployed an adequate peacekeeping force to intercept their exploitation. Were the troop ratios appropriate, the will and interest strong, and the forces equipped, the rebels would have been unable to access and exploit the resources. The absence of robust international presence means there are few obstacles to the ability of these groups to exploit resources for their benefit and to the detriment of stability.

### *International aid*

An obvious factor to fall under the heading of international capacity, and one that has been cited by Doyle and Sambanis, is that of foreign economic assistance. A look at the DRC suggests that capacity in the form of economic aid may be an important but insufficient condition, and matters more once the security conditions have been established. The DRC is one of the largest African recipients of aid – with $1.8 billion in annual foreign aid, it ranks below only Nigeria and the Sudan.[72] Salutary effects of foreign aid have trouble taking root in the country because stability is a necessary foundation for the aid programs. As the UN Children's Fund reports, peace is the "prerequisite for investment in vital basic services that have been limited by conflict . . . insecurity in the worst-affected areas in the east of the country has placed some of the victims beyond the reach of aid agencies" that are unwilling or unable to go into these conflict-ridden areas.[73] The problem is representative of the problems aid agencies across the DRC have encountered. Upstream in the causal arrow is the question not of development and assistance capacity, which are available in volumes, but in peace-enforcement capacity, which is scarce. Without the latter, the conditions for institution-building, development and reconciliation have been unable to take root.

International aid applied incorrectly or in the wrong sequence could actually be counterproductive for two reasons. First, while the UK has assisted in paying for the recovery of the rail system, the EU in repairing airfields, numerous other state, regional and international agencies assisting the stabilization and reconstruction of the war-torn country, one UN officer cited the possibility of "donor fatigue" if the situation did not improve more quickly than it had. Second, and this issue may be related to the first, too much assistance could actually lead to a moral hazard in which external capacity could provide a disincentive for organic production of capacity.[74]

## Sierra Leone: UNAMSIL

By way of historical background, the civil war in Sierra Leone has its origins in a broader regional conflict. Responding to deteriorating social and political conditions in the 1980s, a handful of students, supported by Muammar Gaddafi and Army Corporal Foday Sankoh, started training for guerrilla warfare in Benghazi, Libya, later in Liberia's civil war, and eventually into Sierra Leone in March 1991 for the start of the civil war. A guerrilla war followed in which the Revolutionary United Front (RUF), employing Liberian-style tactics, sought support from peasants and students throughout the country. These RUF insurgents, abetted by the Armed Forces Revolutionary Council, the so-called West Side Boys, and Liberia encountered loyal army officers, civil defense and, later, the Nigerian-led Economic Community of West African States Monitoring Group (ECOMOG) and the UN Mission in Sierra Leone. This conflict added between 50,000–75,000 fatalities and two million displaced people over an eight-year period.[75]

In response to the Sierra Leone civil war, the UN established an observer mission in June 1998 and conflict continued, with a decision to increase the UN presence to 210 military observers in August 1999. UNAMSIL was finally established on 22 October 1999 and provided for 6,000 troops, followed by an expansion to 11,100 military personnel in February 2000, and an escalation to 13,000 in May 2000.[76] UNAMSIL's objective was to help implement the Lomé Peace Accords that ostensibly ended a civil war that various forms of elections, peace agreements (Abidjan in 1996) and Western African peacekeepers had failed to terminate.

UNAMSIL was a Chapter VII mission charged with assisting in a multitude of tasks, from facilitating national elections, disarming rebel fighters, stemming illicit trading of diamonds, helping rebuild infrastructure, and reconciling local and national governments. Its mandate included protection of civilians, implementation of the peace agreement and security for UN peacekeepers. From its inception in October 1999 until May 2000, however, UNAMSIL failed to enforce the Lomé peace agreement, as it was confronted with belligerent RUF rebels that kidnapped UN peacekeepers.[77]

Part of the problem in the early phases is that UNAMSIL consisted of lightly armed, poorly trained Indian, Kenyan, Nigerian and Zambian troops that lacked logistics, communications, a well-understood mandate and adequate equipment for defense.[78] In light of these deficiencies, it is not surprising that they were easily resisted by the Revolutionary United Front (RUF), led by Corporal Foday Sankoh. The force almost collapsed as two events occurred. First, the UN tried to move into the diamond-mining area that was a rebel stronghold, only to meet with fierce resistance and unraveling of the peace process. Second, the RUF kidnapped 500 UN peacekeepers and 11 British peacekeepers, and the UN peacekeeping force was unable to rescue its personnel, leading to an enormous coup for the rebel group, discredit to the UN and additional brutal attacks on the Sierra Leonean population.[79] The tangible outcome of the kidnapping was that it bolstered the RUF and encouraged them to renounce the ceasefire, which UNAMSIL was relatively powerless to stop. [80] The fiasco led to predictions that UNAMSIL's failures would trigger the "end of UN peacekeeping".[81]

Responding to trouble in its former colony and the capture of about a dozen members of the United Kingdom's 2nd Battalion Parachute Regiment, the British launched Operation Palliser in early May 2000. The UK mission's objectives were to rescue its soldiers and provide logistical support that would finally facilitate compliance with the peace agreement. Carrying 13 Harriers, the British aircraft carrier *Illustrious* deployed off Freetown and met with the helicopter carrier *Ocean*, then the Navy's newest assault ship carrying 600 Royal Marines that joined the British Parachute Regiment which had already landed ashore. With 700 paratroopers, the mission included the largest British taskforce since the Falklands and produced what UN Secretary General Kofi Annan called a "critical impact".[82] He also noted: "One of the problems with peacekeeping has been the speed of deployment. With each delay the problems get worse."[83]

With the speed of British deployment, delay and further deteriorating conditions were quickly addressed and reversed. Where the rebels had been advancing on the capital prior to the British deployment, Operation Palliser was able to turn back and actually reverse that advance. Lasting just over five weeks, the British provided enough muscle and logistics support until the UN could deploy a more robust peacekeeping force by the middle of June.[84] As a UN Lessons Report acknowledged, the UK deployment gave UNAMSIL "breathing space to regroup, recover credibility, and reassert its presence", which it subsequently did by increasing its numbers, a move that further restored stability.[85]

Though the UN troop numbers, which were relatively sizable by the middle of 2000, were helpful, a key difference between MONUC and the mission in Sierra Leone was that the UK, with an advanced military, had the ability to quickly and unilaterally deploy an elite force to stabilize the situation.[86] Not only was Sierra Leone a former British colony, which produced an obvious affinity between the two states, but about a dozen British soldiers had been captured and were being held by RUF forces, which also contributed to strong incentives for securing the country and returning its soldiers.[87]

The UK force brought stability where UN contingents – too lightly equipped, disjointed, or unwilling to fight – had been unable to do. Moreover, the UN was not equipped, either politically or militarily, to produce a rapid deployment of soldiers to stem the instability after the RUF took UN soldiers hostage.[88] In contrast, British forces were rapidly deployable, armed, and more coherent than the motley UN group that had been beset by internal fighting between the Nigerian forces who were thought to be colluding with the rebels in trafficking diamonds and the disapproving Indian commander.[89]

According to a former UN military planner, the turnaround in Sierra Leone's security conditions after the British paratroopers landed in May 2000 was no coincidence. Sending in the British force sent a credible signal that key actors in the international community were serious about addressing the conflict in Sierra Leone and enforcing the Lomé peace agreement. As this military planner argued, the British are among the best in the world at stabilization and peacekeeping, which had real benefits, in that they are highly effective, and impacts on perceptions, in that sending these elite forces signaled a robust, defensible commitment to stabilization.[90]

Not only did the British deployment create short-term stability, it also created a sound foundation on which longer-term stability was later staged. Some 200 British troops stayed behind to train the Sierra Leonean army, which was then prepared for general elections in 2002 and 2004,[91] key affirmations of political stability, and to stand up Sierra Leone's security sector, support the Truth and Reconciliation Commission, and help the government reassert its control over diamond-mining, which has since become a boon rather than a liability for Sierra Leone.[92] They helped demobilize and disarm 45,000 RUF rebels, armed gangs, and militia groups who surrendered and were subsequently reintegrated into the Sierra Leonean army. Whereas the diamonds were used to buy weapons during the conflict, they became an engine of growth under stable security conditions, producing $160 million in annual income by 2004, up from $10 million in 2000. The economy has grown about 7 percent per year since 2002, and economic revitalization has helped attract returning refugees and young Sierra Leonean entrepreneurs, further propelling the economy. By 2005, the UN's peacekeeping mission was declared successful, having completed most of the mandate, helped the return of refugees and displaced persons, and restored government credibility.[93] Throughout this process, the UK was instrumental in the restructuring of the army, which was in many ways the vehicle for stability.

What is striking in comparing the peacekeeping force in Sierra Leone to that of MONUC in the DRC is its robustness. With one-tenth the population of the DRC and three percent of its size, the two had the same peacekeeping force of approximately 18,000 at their respective peaks. In terms of the forces that brought stability to Sierra Leone, however, those numbers were relatively scarce. Several hundred elite British troops were sufficient to restore and build peace in a state that had been ravaged by civil war for several years.

The relative scarcity of these British forces suggests that large numbers of troops may be helpful in producing peaceful outcomes, but may not be a necessary condition for success. Credible commitment came in the form of efficient, well-trained British forces that operated quickly and decisively, intervening independently and circumventing the UN's chain of command to produce peace in Sierra Leone.[94] Chief of the Defence Staff at the time of Operation Palliser, General Charles Guthrie, reported that the UK felt some responsibility to stabilize events in Sierra Leone, but would only endorse the UK intervention "as long as the UK did not become part of the UN force".[95] Conveying credibility and maintaining effectiveness required that the UK actually operate independently of the UN. Subordinating British forces under UN command was expected to be counterproductive operationally, because it would signal a status quo approach rather than a committed move to stabilize the country.

## Discussion

One of the unmistakable trends in the empirical record of UN peacekeeping is the spike in missions with Chapter VII authority after the Cold War. Of the 19 Chapter VII operations that the UN has undertaken since its founding – often referred to as peace-enforcement operations – 18 have taken place during the post-Cold War period.[96] This trend is important because peace-enforcement missions are qualitatively different from earlier peacekeeping missions. Earlier Chapter VI UN missions were consensual impartial, and often lightly armed forces that would observe and monitor peace agreements, as they have in the Sinai and Golan Heights.[97]

In contrast, Chapter VII missions are neither consent-based nor impartial, and are often better-armed and employed when the parties of a civil war have not agreed to the nature or terms of peacekeeping activities or are not abiding by those terms. These missions seek to impose order by force, interposing UN forces among warring factions, which necessarily introduces a greater likelihood of danger and requires a greater degree of robustness.[98] As the UN's Supplement to *An Agenda for Peace* cautions, the wars to which peace-enforcement missions go "are often guerrilla wars without clear front lines. Civilians are the main victims and often the main targets", which creates a set of challenges that go well beyond the original peacekeeping missions of self-defense and monitoring interstate borders that were more common during the Cold War.[99]

Though peace-enforcement missions are more difficult and empirically less successful than earlier generations of peacekeeping missions,[100] within this category of peacekeeping there is considerable variation in outcomes. UNAMSIL, for example, had an inauspicious beginning but ultimately produced results that are held up as successes worth emulating. While MONUC has generally maintained the status quo in the DRC, its outcome is not nearly the success that UNAMSIL was, and MONUC troops will remain in the DRC in large numbers for the foreseeable future. Understanding the basis of success on the one hand and status quo on the other may have important implications for peacekeeping effectiveness. This analysis is suggestive of some possible factors that could affect outcomes and the mechanisms by which those factors do much or little to facilitate peace.

### *The effect of capacity*

In comparing the outcomes of UNAMSIL and MONUC, an obvious difference between the two was the constructive insertion of capacity midway through UNAMSIL and the continuing insufficient capacity of MONUC. As with previous studies on peacekeeping capacity, this one finds that it is a necessary condition for peacekeeping success. Where this study diverges is in how it thinks of capacity. Doyle and Sambanis use the term "international capacity" as an overall heading for the amount of economic assistance, the number of troops and type of mandate. James Quinlivan refers to troop numbers vis-à-vis the population.[101]

The former study is helpful in that it sees the effect of international capacity as an interaction between it, the war's hostility and residual domestic capacity, but the model does not explicitly take into account territorial dynamics. The latter study on troop-to-population ratio is methodologically unconvincing because it infers thresholds based on the successes of Kosovo and Bosnia where there are a multitude of factors that might have contributed to successful outcomes, including economic and security incentives to achieve internal peace – such as the lure of joining the EU, receiving EU infrastructure and development funds, and joining the Partnership for Peace and NATO.[102]

Empirically, these accounts also appear to be incomplete. Within-case analysis of UNAMSIL enabled controlling for other factors that may have affected outcomes – territory size, terrain, and the country's history – and isolating the factor that changed over that period. What changed is the insertion of elite British forces in May 2000, which established calm in the country and laid the foundation for the follow-on stability and eventual exit of UNAMSIL. These British forces no doubt brought with them a credible commitment to restore calm in Sierra Leone and had the training, equipment and resolve to confront the RUF that had previously stymied UN forces. These forces, however, did not arrive in overwhelming numbers. Rather, they numbered 700, or 0.14 per thousand of the Sierra Leone population, well below the magic 20 per thousand ratio advanced by previous studies. The difference is that the British were well-equipped, well-trained, and circumvented any UN rules of engagement and chain of command that could constrain their ability to stabilize the country. In this case, capacity was less a function of numbers than training, equipment and operational latitude, capacity that advanced industrialized states may be better able to provide.

The success of British paratroopers also had something to do with geographical advantages, however. Sierra Leone is more densely populated than the DRC, but it also has just one center of gravity. The British forces' ability to stabilize that one center of gravity – the capital of Freetown – had a transformative effect on the rest of the country. The UK paratroopers' first encounter with the RUF near the capital caused the rebel groups to go into hiding, then the British taskforce pushed inland to deter an additional attack. Calm in Freetown percolated radially throughout the country.

Though the Sierra Leone example indicates the importance of capacity – as specified less by numbers than by muscularity – even this case shows an interaction between capacity and territory. The UK brought a well-trained force into Sierra Leone, but the effectiveness was aided by the country's small size and centralized geography. The DRC case offers further support for the importance of capacity, but also suggests that the interaction between capacity and geography may be more pronounced for large states than for small ones.

In the MONUC case, without a peacekeeping force that could patrol a vast territory, rebel groups essentially seep into the many areas that inevitably remain unpatrolled. The insufficient peacekeeping numbers essentially create a balloon effect; peacekeeping forces treat an area of tension or conflict, but the settlement of that area pushes tensions to an area devoid of peacekeepers. Rebel groups migrate to that vacuum, re-convene and create new tensions.

What further complicates peace operations in the DRC is the lack of infrastructure, and the rough and impassable terrain. "Time–distance" factors refer to how quickly government authorities can respond to crises. That factor is extremely high in the DRC. Not only is the DRC a vast country, it is plagued with almost non-existent infrastructure, requiring helicopters for rapid transport. Those require higher amounts of fuel that may not always be available. Tactical vehicles may be possible for traversing the country, but the terrain is so rugged that it takes days to cross, meaning that these vast countries often become self-policing, and within this vacuum rebel forces are able to exploit resources, harm civilians and limit progress toward stabilization.

Larger, impassable states mean inherently greater interstices between areas of government or peacekeeping authority and therefore sufficient forces become more salient for restoring calm across the broad expanse of a state such as the DRC. This is consistent with the research of Jeffrey Herbst, who argues that larger, less densely populated African states are more difficult to govern than smaller states.[103] This argument finds additional support in the context of peacekeeping, which in many cases acts as the de facto authority, with comparable difficulties governing these large spaces.

Within this context, other situational or capacity factors assume more or less importance. Lootable resources, for example, serve as economic engines for the rebel groups, but only because the interaction between capacity and geography creates insecurity around the mines. Lootable resources, then, are the intervening variable between capacity and geography and the success and failure of peacekeeping. Similarly, economic assistance may ultimately affect reconstruction and peacekeeping success, but a state that is insecure is also unable to absorb aid in a productive way. Capacity then affects aid absorption, since aid cannot be delivered adequately if aid workers do not feel secure across the country.

### *Political factors related to peace*

To indicate that peacekeeping success and failure is the result of just one or two factors – capacity and its interaction with geography – oversimplifies what is an overwhelmingly complex phenomenon of civil war and third-party intervention. Close study of these two cases highlights the role of factions and the political process. There are two reasons more factions create challenges for peace. First, more factions mean more interests and grievances that must be addressed. Reconciliation among more parties – including international actors in the DRC case – is more difficult. The dynamic proceeds exactly how the logic of collective action would expect: larger groups face higher costs and greater challenges to acting in their common interest than do smaller groups.[104]

A second point is less obvious and emerged in process tracing the DRC case. More parties mean greater likelihood that one or more will have been left out of the peace process and can act as spoilers. The DRC experience speaks to that problem, as does a parallel case in the Sudan, where the Justice and Equality Movement was left out of the peace process, the party then splintered, and both have subsequently acted as spoilers.[105] These experiences should be a reminder that the purpose of peacekeeping forces is to enforce and facilitate peace agreements. Military power may be important but insufficient if political reconciliation is not on firm ground.

In short, capacity explains some but not all of the variation in outcomes between these two cases. Within case variation in Sierra Leone holds situational factors constant and isolates the effect of capacity in changing the outcome of that operation. A comparison with the DRC points to some situational differences that add to the reasons why MONUC has not been successful. The country's vastness and impassability creates huge ungovernable spaces; adding even 20,000 more forces would still leave interstices unpatrolled. Moreover, the peace process has been uneven and incomplete, leaving out factions and creating spoilers. Unless there is a viable peace agreement to enforce, no reasonable amount of capacity will effectively stabilize the DRC. Capacity, then, is a necessary but not sufficient factor in peacekeeping outcomes.

## Conclusions

The UN's experience with peace-enforcement is not a hopeful one. Perhaps drawing from a realistic understanding of this record, the scholarly and policy literature on the UN's post-Cold War experience is rife with calls for the UN to return to the "business of peacekeeping" rather than peacemaking or peace-enforcement. One scholar has concluded that "the vast majority of interstate and internal conflicts cannot be genuinely solved by imposing a fighting vanguard of Blue Helmets on a country".[106] The supplement to *An Agenda for Peace* made similar concessions, noting that "it would be folly to attempt" additional peace-enforcement missions given that the UN was "hard pressed to handle the less demanding peacemaking and peacekeeping responsibilities entrusted to it".[107]

Though the UN's record of peace-enforcement has been mixed at best, the UN will still likely find itself under pressure to intervene in cases of egregious human-rights abuses, such as those in Darfur. Non-intervention for Bosnia, Somalia or Kosovo was not an option, for example, and the international community felt pressure to "do something".[108] In other words, egregious behavior within or across states will be hard for the UN to ignore.[109] The recommendation to pull back from peace-enforcement missions that lack host-state consent and authorize the full use of force – missions that are considerably more challenging than the Cold War predecessors – is therefore unrealistic given the degree to which human-rights abuses are promulgated in the media. Given that these missions will likely continue, it is wise to understand what factors might produce more favorable outcomes, and this research has been a step in that direction.

Comparing the cases of the DRC and Sierra Leone produces two related insights on peacekeeping. First, the presence of too few, poorly equipped, and too-constrained peacekeepers will cast doubt on the likely success of a particular peacekeeping force. Second, an overly ambitious or ambiguous political objective – one that hopes peacekeepers can make peace amidst spoilers outside the peace process – also creates an environment ripe for failure. Both speak to the utility of force and its interplay with the political process.

This research suggests that several questions should be addressed before crafting and deploying a peace-enforcement operation. Is the size, training and equipment of the intervention force muscular enough to address the instability in the state? Is it a small or large state, and if it is the latter, what can the international community do to minimize the time–distance factors that make governance even more difficult? For example, might construction of infrastructure, helicopters with advanced surveillance technology, and adequate fuel be helpful in improving governance across a large and otherwise inhospitable territory? Have all the relevant factions been incorporated into the peace process and, if not, what can be done to bring them to the table so they do not later act as spoilers? If its goal is to improve the success of the interventions it undertakes, the international community would be wise to orient their intervention strategies around these questions.

## Notes

1 See the UN Mission in Sierra Leone website, www.un.org/Depts/dpko/missions/unamsil/index.html, and its story in 2003 about the successes in Sierra Leone, www.un.org/events/peacekeepers/2003/docs/sierraleone.htm.
2 Press Conference by Under Secretary General for Humanitarian Affairs, 15 September 2006, www.un.org/News/briefings/docs/2006/060915_Egeland.doc.htm.
3 An intervention is successful if the peacekeepers brought war to a close and departed without the war re-starting in two years. It is partially successful if the peacekeepers brought calm but could not leave for fear that the war might restart. It is a failure if the peacekeepers could not bring violence to a close. For definitions of success and failure, see Stedman and Downs, *Ending Civil Wars*, 50–51.
4 Page Fortna, “Does Peacekeeping Keep Peace?”.
5 See discussion on negotiated settlements in Duyvesteyn, “The Effectiveness of Intervention Instruments in Armed Conflict”.
6 In their survey of the peacekeeping literature, Page Fortna and Lise Howard cite three waves, in which the second was more pessimistic and the third has become more auspicious toward the ability of peacekeeping operations to keep the peace. See Page Fortna and Howard, “Pitfalls and Prospects in the Peacekeeping Literature”.
7 Hartzell, Hoddie and Rothchild, “Stabilizing the Peace After Civil War”.
8 See Downs and Stedman, *Ending Civil Wars*, 43.
9 Page Fortna, *Peace Time*.
10 The variations in outcomes may result in part from methodological differences among studies that a) delimit their period of study differently and b) define civil war differently.
11 Doyle and Sambanis, “International Peacebuilding”.
12 She finds that high costs and death tolls actually help keep the peace after interstate wars, in part because they are divided by boundaries, while civil wars keep the citizens in close proximity to each other where those hostilities can play out. Page Fortna, “Does Peacekeeping Keep Peace?”; Page Fortna, “Interstate Peacekeeping”.
13 Walter, “The Critical Barrier to Civil War Settlement”, 347.
14 Hartzell *et al.*, “Stabilizing the Peace After Civil War”.
15 Collier and Hoeffler, “On the Incidence of Civil War in Africa”.

16 Stedman, "Spoiler Problems in Peace Processes". See also the discussion in Doyle and Sambanis, *Making War and Building Peace*, 57–58.
17 George and Bennett, *Case Studies and Theory Development in the Social Sciences*, 155–156; Bennett and Elman, "Case Study Methods in the International Relations Subfield".
18 Robinson, "The Deadliest War in the World", *Time*, 28 May 2006.
19 Rally for Congolese Democracy (RCD) and the Movement for the Liberation of Congo (MLC) were Rwandan- and Ugandan-sponsored groups fighting against the DRC leader and Angola, Namibia and Zimbabwe. For more background on the ethnic factions, see "The Civil War in the DRC", 31 May 2007, www.africaresource.com/content/view/306/68/. As I will discuss later, this situational factor presents the most significant difference between the two cases and may account for some of the divergence in outcomes.
20 The factions included the Revolutionary United Front, Armed Forces Revolutionary Council, West Side Boys and Liberia on one side, and the government of Sierra Leone (and its army), the Kamajors and, eventually, peacekeepers on the other.
21 The cost of war numbers refer to the percentage of the population that was killed at the time UN peacekeepers were inserted into the civil war. At the time of the 1999 MONUC intervention, there were about 1.5 million casualties in the DRC, about 2.3 percent of the population. "Congo Pays the Cost of War", BBC, 26 September 2001.
22 As a proportion, the number of people affected in Sierra Leone was actually higher, as half the population was displaced and mutilations were prevalent, about 100,000 of the population: http://news.bbc.co.uk/onthisday/hi/dates/stories/june/15/newsid_2512000/2512669.stm.
23 This measures GDP per capita (PPP) at the beginning of the conflict.
24 See GDP Data from World Bank Development Indicators for both the DRC and Sierra Leone.
25 www.cia.gov/library/publications/the-world-factbook/geos/cg.html#Govt.
26 Freedom in the World country ratings indicate that the DRC has never registered as "free", while Sierra Leone began registering as partially free in 1999 but would not qualify as having "experience" with democracy. Data available at the Freedom House website.
27 The Transitional Government of the DRC took over in July 2003 and marked the official end of the civil war. "Country Profile: Democratic Republic of Congo", BBC, 30 April 2008.
28 The war began in 1991 and the factions signed a truce in 1999; the war was formally declared over in 2002, but there was some cessation in violence with the Lomé peace agreement in 1999.
29 See CIA Factbook for information about both states' natural resources.
30 Both were ongoing conflicts at the time of intervention.
31 The Lusaka agreement tried to convene the relevant parties and, though it was signed in 1991, not all factions were involved in the talks; continuing violence, particularly in the East, has necessitated subsequent peace talks and accords. The most recent one was concluded in early 2003 in east DRC, but this too has been incomplete in terminating violence. See "Democratic Republic of Congo: Peace Accord Hitch", *Africa Research Bulletin: Political, Social, and Cultural Series*, 45, 2 (2008), 17429B–17430B.
32 Language of peace accord available at www.sierra-leone.org/lomeaccord.html.
33 For an overview of the Congo crisis, see Weissman, *American Foreign Policy in the Congo, 1960–1964*. The toll in terms of UN casualties was about 250.
34 UN Operation in the Congo, facts and figures, www.un.org/Depts/DPKO/Missions/onucF.html.

35 Gondola, *The History of Congo.*
36 Duke, "Congo Begins Process of Rebuilding Nation", *The Washington Post*, 20 May 1997, A10.
37 Edgerton, *The Troubled Heart of Africa*; Gondola, *The History of Congo*, 158–159.
38 "One Year On: Kabila Assassination", BBC, 16 January 2002.
39 The Pretoria Agreement is also known as the Global and All-Inclusive Agreement on the Transition in the Democratic Republic of the Congo, which was signed on 17 December 2002.
40 Those groups included the Rassemblement congolais pour la democratie-Kisangani/Mouvement de liberation (RCD-K/ML), Mouvement de liberation du Congo (MLC) and the RCD-National.
41 UN Security Council, Thirteenth Report of the Secretary-General on the UN Organization Mission in the Democratic Republic of the Congo, 21 February 2003; Lalji, "The Resource Curse Revised".
42 "Congo-Kinshasa: Deploring Casualties, Ban Ki-Moon Appeals for Halt to Fighting in DR Congo", UN News Service, 23 March 2007.
43 Mark Malan and Erin Weir, "DR Congo: Civilian Protection Must Remain MONUC Priority", 6 December 2007, www.refugeesinternational.org/content/article/detail/10334/.
44 Lara Santoro, "Africa leaders Sign Onto Peace", *Christian Science Monitor*, 12 July 1999, 6.
45 See Resolution 1258 of 6 August 1999, www.un.org/News/Press/docs/1999/19990806.sc6711.html and Resolution 1279, www.africa.upenn.edu/Hornet/irin-120899.html.
46 UN Organization Mission in the Democratic Republic of the Congo, www.un.org/Depts/dpko/missions/monuc/background.html.
47 MONUC History background, www.monuc.org/news.aspx?newsID=9564.
48 "Security Council Authorizes Interim Force in Bunia, Democratic Republic of Congo, Until 1 September", 2 June 2003, www.unis.unvienna.org/unis/pressrels/2003/sc7772.html; see Homan, "Operation Artemis in the Democratic Republic of Congo".
49 UN, *Operation Artemis.*
50 International Crisis Group, Democratic Republic of Congo overview, www.crisisgroup.org/home/index.cfm?id=1174&l=1.
51 See MONUC Chronology of Events, www.monuc.org/news.aspx?newsID=884.
52 "DR of Congo: As the Country Moves Boldly Towards Historic Vote, Humanitarian Concerns Continue to Demand Attention", www.un.org/events/tenstories_2006/story.asp?storyID=2300.
53 The UN's map showing headquarters reveals the vast interstices among headquarters: www.un.org/Depts/Cartographic/map/dpko/monuc.pdf. See also the British High Commission's Profile for the Democratic Republic of the Congo, www.britishhighcommission.gov.uk/servlet/Front?pagename=OpenMarket/Xcelerate/ShowPage&c=Page&cid=1007029394365&a=KCountryProfile&aid=1020281605460; Zawadi Mongane from the Eastern Congo remarks to Mike Thomson, BBC, 10 April 2008, http://news.bbc.co.uk/1/hi/help/3681938.htm.
54 Jeffrey Gettleman, "Rape Epidemic Raises Trauma of Congo War", *New York Times*, 7 October 2007.
55 Selah Hennessy, "2007 Sees Renewed Fighting in Congo", *Voice of America*, 20 December 2007. See also UN Report, www.un.org/news/dh/pdf/english/2007/31122007.pdf; "The Kivus: The Forgotten Crucible of the Congo Conflict", *Africa Report*, 56, 24 January 2003; "Democratic Republic of the Congo: Security Council Must Take Steps to Protect Civilians", *Refugees International*, 6 February 2007.

56 "DRC: UN Security Council Must Strengthen MONUC Mandate to Protect Civilians", *Refugees International*, 14 November 2002.
57 "Democratic Republic of Congo: Mission Finding", *Amnesty International*, 1 August 2003, News Service no. 181.
58 "MONUC: Flawed Mandate Limits Success", *Refugees International*, 9.
59 "MONUC: Flawed Mandate", 9; "DRC: UN Security Council Must Strengthen MONUC Mandate to Protect Civilians", *Refugees International*, 14 November 2002.
60 UN, *Operation Artemis*, 9.
61 The largest contingents (based on September 2004 figures) are from Uruguay (1,778 soldiers), Pakistan (1,700), South Africa (1,387), Bangladesh (1,304), India (1,302), Nepal (1,225) and Morocco (801). There are no US personnel serving as peacekeepers or military observers with MONUC. See MONUC, "Military Contributions", 12 September 2004, www.monuc.org/ContribMilit.aspx?lang=en.
62 "Gauteng to Help DRC Infrastructure Development', 3 July 2008, Gauteng Provincial Government, South Africa.
63 MONUC April 2008, www.un.org/depts/Cartographic/map/dpko/monuc.pdf.
64 Author's interview with Colonel Tim Cornett, former strategic planner for MONUC, 6 May 2008.
65 Author's interview with Colonel Tim Cornett, former strategic planner for MONUC, 6 May 2008; see also "Congo Crisis: Military Intervention in Ituri", *Africa Report*, 64, 13 June 2003, International Crisis Group, www.crisisgroup.org/home/index.cfm?l=1&id=1626.
66 "DRC: Who's Who – Parties, Armed Groups and Factions", IRIN, www.irinnews.org/InDepthMain.aspx?InDepthId=33&ReportId=70834.
67 Gondola, *A History of Congo*, 167–176. Nzongola-Ntalaja, *The Congo from Leopold to Kabila*; "Corners of Darkness, Places of Light. Africa: Beyond the Headlines", *The Toronto Star*, 4 June 2000.
68 "DRC: Special Report on Ituri District", *Office for the Coordination of Humanitarian Affairs; Integrated Regional Information Network*, November 2002, www.irinnews.org/pdf/in-depth/Ituri.pdf.
69 "DR Congo: Peace Accord Fails to End Killing of Civilians", *Human Rights Watch Report*, 21 July 2008.
70 Interim report of the Group of Experts on the Democratic Republic of the Congo, pursuant to Security Council Resolution 1698 (2006), 7.
71 Joe Bavier, "Punish Buyers of Rebel Congo Ore, UN Panel Says", Reuters, 29 February 2008.
72 See the CIA Fact Book's numbers on foreign aid, available for 2005 at www.cia.gov/library/publications/the-world-factbook/fields/2064.html. The Congo's record as a recipient of foreign aid is not new. During his reign, the very corrupt President Mobutu took in $20 billion from the international community. See William Easterly, "Can Foreign Aid Save Africa?", 2005 Clemens Lecture, St John's University, December 2005, 16.
73 Bell, *Child Alert*, 2.
74 Author's interview with former UN planning officer, 5 May 2008, Cambridge, MA.
75 Zach-Williams, "Sierra Leone", 143–162; Richards, "The War and Reconstruction in Sierra Leone".
76 www.un.org/Depts/dpko/missions/unamsil/index.html.
77 Reno, "The Failure of Peacekeeping in Sierra Leone", 222.
78 For more detail on the deficiencies of the initial peacekeeping deployment, see Olonisakin, *Peacekeeping in Sierra Leone*.

79 Mark Tran, "Elections in Sierra Leone", *Guardian*, 14 May 2002.
80 William Wallis, "UK Troops Seeks Rapid Result in Sierra Leone", *Financial Times*, 9 May 2000; Kim Sengupta, "Paratroopers Risk Being Drawn into Civil War", *Independent*, 10 May 2000, 14; "Britain Sends Troops to Sierra Leone, Fearing Rebel Attack on Capital", *Deutsche Press-Agentur*, 7 May 2000.
81 Doug Bandow, "The End of UN Peacekeeping", *Washington Times*, 12 June 2000.
82 "Sierra Leone Peacekeepers Are Bailed out by Taskforce", *Scotland on Sunday*, 14 May 2000, 19.
83 Richard Norton-Taylor, "Evacuation Effort Shows How UN Has Failed", *Guardian*, 9 May 2000, 4.
84 "Sierra Leone Fighting Fears Spark Exodus", BBC, 14 June 2000, http://news.bbc.co.uk/2/hi/africa/791062.stm.
85 United Nations, *Lessons Learned from United Nations Peacekeeping Experiences in Sierra Leone*, 13.
86 Ian Bruce, "UN Credibility Stretched as UK Resists Peace-Keeping Pressure", *Glasgow Herald*, 2 October 2000, p. 6.
87 "Freed British Major Heads Home", BBC, 22 July 2000, http://news.bbc.co.uk/1/hi/uk/835004.stm.
88 Ian Bruce, "UN Credibility Stretched as UK Resists Peace-Keeping Pressure", *Glasgow Herald*, 2 October 2000, 6; "Britain Sends in the Troops", *Daily Telegraph*, 8 May 2000, 24.
89 Anton La Guardia, "India to Quit Sierra Leone: Withdrawal Will Leave 'Vacuum' in UN Mission", *Ottawa Citizen*, 22 September 2000, A14.
90 Author's interview with Mr Harry Phillips, UN Department of Peacekeeping Operations military planner for Sudan between 2004–2006, 7 May 2008, Cambridge, MA.
91 Chege, "Sierra Leone".
92 "Secretary-General Praises UN Sierra Leone Peacekeeping Mission as Mandate Concludes", 3 January 2006, www.unis.unvienna.org/unis/pressrels/2006/sgsm10290.html.
93 "Sierra Leone: A Success Story in Peacekeeping", *Year in Review 2005*, www.un.org/Depts/dpko/dpko/pub/year_review05/sierra_leone.htm.
94 "In For The Duration: There Can Be No Quick Fix in Sierra Leone", *Guardian*, 10 May 2000, 21; "A Job Still to Be Done: It is Too Early to Bid Farewell to Freetown", *Guardian*, 23 May 2000, 23.
95 Former Chief of the Defence Staff, General Lord Charles Guthrie, 21 April 2008, international law seminar at Harvard University.
96 The UN intervention in the first Congolese Civil War (1960–1964) was a peace-enforcement mission.
97 Monitoring missions are one step more passive, and are intended to be an interim solution that helps both sides negotiate peace. See Doyle and Sambanis, "International Peacebuilding", 781. The consummate case of first-generation peacekeeping is the UN Emergency Force in which peacekeepers were deployed in a buffer zone with the intention of enforcing an agreed-upon ceasefire, having already obtained the consent of the parties; the peacekeepers, who to a large extent excluded Great Powers, were impartial and were not allowed to use force except in self-defense. See distinction in Utley, *Major Powers and Peacekeeping*, 3.
98 For helpful characterizations of these categories, see Jones and Kennedy-Pipe, *United Nations Peace-Keeping in the Post-Cold War Era*, 120–121. See also Doyle and Sambanis, "International Peacebuilding", 781.
99 Supplement to *An Agenda for Peace*, www.un.org/Docs/SG/agsupp.html.

100 Page Fortna finds that peace-enforcement missions are less likely to be successful than peacekeeping missions. See "Does Peacekeeping Keep Peace?".
101 Doyle and Sambanis, *Making War and Building Peace*; Quinlivan, "Force Requirements in Stability Operations".
102 Serry, "NATO's Balkan Odyssey", *NATO Review*, Winter 2003.
103 This is consistent with the claim of some civil-war-onset scholars who suggest that distance from government authority (peripheral areas) affects the likelihood that individuals, tribes or villages will compete and come into conflict with each other more than where groups are closer to authority. See, for example, Jeffrey Herbst, who discusses the difficulties of exercising control over peripheries of large regions of Africa that are lightly populated because of the region's inhospitable geography. Herbst, *States and Power in Africa*, 11–13.
104 Olson, *The Logic of Collective Action.*
105 "Sudan's Comprehensive Peace Agreement: The Long Road Ahead", *Africa Report*, 106, 31 March 2006, International Crisis Group, www.crisisgroup.org/home/index.cfm?id=4055.
106 Ratner, *The New UN Peacekeeping*, 5. See also Durch, *The Evolution of UN Peacekeeping*, 2; Edward Luttwak argues that, unless the UN is willing to act more aggressively and decisively, the optimal outcome is one in which the warring factions battle and the UN enters at the conclusion. See Luttwak, "Give War a Chance"; "We Are Right to Be There: It is Hazardous, but the British are Needed", *Guardian*, 13 May 2000, 21.
107 See the *Supplement to an Agenda for Peace: Position Paper on the Secretary-General on the Occasion of the Fiftieth Anniversary of the United Nations*, 3 January 1995, www.un.org/Docs/SG/agsupp.html. The debacles of the mid-1990s led to a retraction of UN commitments and an interest in intervening only with consent and cooperation of the parties, and once a ceasefire was in place. Christopher S. Wren, "UN is Cutting Back on its Peacekeeping Ambitions", *New York Times*, 12 November 1995.
108 Jakobsen, "UN Peace Operations in Africa Today and Tomorrow", 171.
109 As James Mayall argues, this is in large part because of the way in which the media mobilizes global public opinion. See Berdal and Economides (eds), *UN Interventionism 1991–2004*, 2.

## Bibliography

Bell, Martin, *Child Alert: Democratic Republic of Congo.* Geneva: UNICEF Publication, 2006.

Bennett, Andrew and Colin Elman, "Case Study Methods in the International Relations Subfield", *Comparative Political Studies*, 40, 2 (2007): 170–195.

Berdal, Mats and Spyros Economides (eds.), *UN Interventionism 1991–2004.* Cambridge: Cambridge University Press, 2007.

Chege, Michael, "Sierra Leone: The State That Came Back from the Dead", *The Washington Quarterly*, 25, 3 (Summer 2002): 147–160.

Collier Paul and Anke Hoeffler, "On the Incidence of Civil War in Africa", *Journal of Conflict Resolution*, 46, 1 (2002): 13–28.

Downs, George and Stephen Stedman, *Ending Civil Wars: The Implementation of Peace Agreements.* Boulder, CO: Lynne Rienner Publishers, 2002.

Doyle, Michael and Nicholas Sambanis, "International Peacebuilding: A Theoretical and Quantitative Analysis", *APSR*, 94, 4 (December 2000): 779–801.

Doyle, Michael and Nicolas Sambanis, *Making War and Building Peace: United Nations Peace Operations.* Princeton, NJ: Princeton University Press, 2006.

Durch, William J., *The Evolution of UN Peacekeeping: Case Studies and Comparative Analysis.* New York: St. Martin's Press, 1993.

Duyvesteyn, Isabelle, "The Effectiveness of Intervention Instruments in Armed Conflict: Conflict Resolution is the Only Solution?", in G. Molier and E. Nieuwenhuys (eds.), *Peace, Security and Development in an Era of Globalization: The Integrated Security Approach Viewed From A Multidisciplinary Perspective.* Leiden, NL: Maetinus Nÿhoff, 2009.

Edgerton, Robert G., *The Troubled Heart of Africa: A History of the Congo.* New York: St. Martin's Press, 2002.

George, Alexander and Andrew Bennett, *Case Studies and Theory Development in the Social Sciences.* Cambridge, MA: MIT Press, 2005.

Gondola, Didier, *The History of Congo.* Westport, CT: Greenwood, 2002.

Hartzell, Caroline, Matthew Hoddie and Donald Rothchild, "Stabilizing the Peace After Civil War: An Investigation of Some Key Variables", *International Organization,* 55, 1 (Winter 2001): 183–208.

Herbst, Jeffrey, *States and Power in Africa.* Princeton, NJ: Princeton University Press, 2000.

Homan, Kees, "Operation Artemis in the Democratic Republic of Congo", in European Commission, *Faster and More United? The Debate about Europe's Crisis Response Capacity.* Brussels: European Commission, 2007.

Human Rights Watch, "DR Congo: Peace Accord Fails to End Killing of Civilians", Human Rights Watch Report, 21 July 2008.

Jakobsen, Peter V., "UN Peace Operations in Africa Today and Tomorrow", in Michael Bothe and Boris Kondoch (eds.), *International Peacekeepers: The Yearbook of International Peace Operations,* vol. 7. The Hague: Kluwer Press, 2003.

Jones, Clive and Caroline Kennedy-Pipe, *United Nations Peace-Keeping in the Post-Cold War Era.* London: Routledge, 2001.

Lalji, Nadira, "The Resource Curse Revised: Conflict and Coltan in the Congo", *Economics of National Security,* 29, 3 (Fall 2007).

Luttwak, Edward, "Give War a Chance", *Foreign Affairs,* 78, 4 (July/August 1999): 36–44.

Nzongola-Ntalaja, Georges, *The Congo From Leopold to Kabila: A People's History.* New York: Zed Books, 2002.

Olonisakin, Funmi, *Peacekeeping in Sierra Leone: The Story of UNAMSIL.* Boulder, CO: Lynne Rienner Publishers, 2008.

Olson, Mancur, *The Logic of Collective Action: Public Goods and the Theory of Groups.* Cambridge, MA: Harvard University Press, 1971.

Page Fortna, Virginia, "Does Peacekeeping Keep Peace? International Intervention and the Duration of Peace after Civil War", *International Studies Quarterly,* 48, 2 (2004): 269–292.

Page Fortna, Virginia, "Interstate Peacekeeping: Causal Mechanisms and Empirical Effects", *World Politics,* 56 (July 2004).

Page Fortna, Virginia, *Peace Time: Cease-Fire Agreements and the Durability of Peace.* Princeton, NJ: Princeton University Press, 2004.

Page Fortna, Virginia and Lise Howard, "Pitfalls and Prospects in the Peacekeeping Literature", *Annual Review of Political Science* (2008): 283–301.

Quinlivan, James T., "Force Requirements in Stability Operations", *Parameters* (Winter 1995).

Ratner, Steven R., *The New UN Peacekeeping: Building Peace in Lands of Conflict After the Cold War.* New York: St. Martin's Press, 1995.

Reno, William, "The Failure of Peacekeeping in Sierra Leone", *Current History* (May 2001): 219–225.

Richards, Paul, "The War and Reconstruction in Sierra Leone", World Bank Development Outreach, www1.worldbank.org/devoutreach/fall99/article.asp?id=23.

Robinson, Simon, "The Deadliest War in the World," *Time*, 29 May 2006, www.time.com/time/magazine/article/0,9171,1198921-1,00.html.

Serry, Robert, "NATO's Balkan Odyssey", *NATO Review* (Winter 2003).

Stedman, Stephen John, "Spoiler Problems in Peace Processes", *International Security*, 22, 2 (Fall 1997): 5–53.

Stedman, Stephen, John Donald Rothschild and Elizabeth Cousens (eds.), *Ending Civil Wars: The Implementation of Peace Agreements.* Boulder, CO: Lynne Rienner, 2002.

United Nations, *Supplement to An Agenda for Peace*, www.un.org/Docs/SG/agsupp.html.

United Nations, *Lessons Learned from United Nations Peacekeeping Experiences in Sierra Leone.* New York: United Nations, 2003.

United Nations DPKO, *Operation Artemis: The Lessons of the Interim Emergency Multinational Force.* New York, DPKO: Peacekeeping Best Practices Unit Military Division, 2004.

Utley, Rachel E., *Major Powers and Peacekeeping: Perspectives, Priorities, and the Challenges of Military Intervention.* Burlington, VT: Ashgate, 2006.

Walter, Barbara, "The Critical Barrier to Civil War Settlement", *International Organization*, 51, 3 (Summer 1997): 335–364.

Weissman, Stephen R., *American Foreign Policy in the Congo, 1960–1964.* Ithaca, NY: Cornell University Press, 1974.

Zach-Williams, Alfred B., "Sierra Leone: The Political Economy of Civil War, 1991–98", *Third World Quarterly*, 20, 1 (February 1999): 143–162.

# 6 Exploring the utility of armed force in peace operations

## German and British approaches in northern Afghanistan

*Kersti Larsdotter*

More than eight years have passed since the Coalition Forces dropped the first bombs over Afghanistan in their campaign to overthrow the Taliban regime. However, the number of insurgent attacks has increased year after year, and a stable and inclusive peace seems more distant than ever.[1] This is not an uncommon trend for military interventions in civil wars. According to Daniel Byman and Taylor Seybolt, military interventions in communal civil wars actually fail or backfire far more often than they succeed in bringing a lasting peace.[2] These failures might suggest that the measures to handle civil wars are insufficient.

Several scholars and practitioners have argued that the conduct of military forces is of key importance to the outcome of peace operations.[3] The robust approach of American forces in, for example, Afghanistan and Iraq is often criticised and compared to the more soft approach of British forces.[4] During operations in Iraq, the British forces were very concerned with the American approach to operations, advising them to adjust to a more 'sensible and sensitive US approach to military operations'.[5] Indeed, several scholars maintain that using too much force – including heavy equipment, numerical preponderance and forceful conduct – in peace operations may be counter-productive. Instead, the so-called British approach, stressing the use of minimum force and a non-threatening posture in order to not intimidate the local population, is endorsed.[6] The soft approach has been repeatedly emphasised in doctrine on peace operations and counter-insurgency (COIN), where the notions of consent, hearts-and-minds and minimum force are considered to be important for successful operations.[7]

However, while there are quite a few studies regarding the relationship between the conduct of forces and the outcome of counter-insurgency campaigns, there are surprisingly few systematic studies on the same relationship in peace operations. According to George Downs and Stephen J. Stedman, this is highly problematic:

> In the absence of systematic evaluation ... our understanding of what does and does not work in peace implementation is prey to any

> number of fallacies, many of which are brought about by a tendency to separate strategy and context. Practitioners frequently attribute successful outcomes to their actions and strategies, and attribute failure to the implementation context. Critics, on the other hand, tend to focus on incorrect strategies as causes of failure and attribute success to context or outside factors.[8]

Moreover, some researchers have argued that military interventions need more military power and a more robust approach in order to control the security problems, rather than less military power.[9] In order to enhance the possibilities of success in peace operations and to avoid fallacies in military doctrine and education, the relationship between the conduct of forces in peace operations and the outcome of these operations needs to be more closely examined.

The aim of the chapter is threefold. First, the theoretical foundations of the relationship between conduct of force and outcome of peace operations will be developed. Second, the empirical knowledge on the use of different kinds of military conduct in peace operations will be enhanced through a closer study of two cases in Northern Afghanistan, namely the Provincial Reconstruction Team (PRT) of Mazar-e Sharif and the PRT of Kunduz. Third, military conduct on the tactical level in these two areas will be compared to the short-term outcome in each area in order to find out if any of these kinds of conduct correlate with a better outcome of operations.

In the first part of the chapter, possible relationships between the conduct of the forces and the outcome of the operations will be examined. Next will be a discussion on how to measure success in peace operations and some methodological considerations. A brief description of the International Security Assistance Force (ISAF) mission in Afghanistan will be followed by a more detailed characterisation of the British and the German way of operations in Northern Afghanistan. Lastly, outcomes of operations in the area of operations (AO) of the British PRT Mazar-e Sharif and of operations in the AO of the German PRT Kunduz will be compared and conclusions drawn.

## Winning hearts and minds

While numerous facets of peace building have been studied, Downs and Stedman claim that the actual implementation of peace agreements in civil wars has received scarce attention by researchers.[10] Apart from quantitative systematic studies with focus on the presence of military forces and on the number of forces deployed, force employment has mostly been ignored in Peace and Conflict Research, as well as in International Relations.[11] Research on the role of the armed forces at the operational and tactical levels in these interventions is surprisingly modest, especially from a theoretical perspective. According to Robert E. Harkavy and Stephanie G. Neuman, these levels have mostly been studied in military history, with a minor emphasis on theory, or as lessons learned by the military.[12] The importance given to force employment is also evident in, for example, military doctrines and training of military forces. General Sir Rupert Smith, in addition, has stressed that it is exactly this level that is of paramount importance in modern 'wars amongst the people'.[13] However, assumptions on the impact of tactics and tactical behaviour on war outcomes are often implicit and, according to Stephen Biddle, nothing more than subjective assessments.[14]

Considering the modest amount of systematic research on the subject, the importance given to the notions of consent, hearts and minds, and minimum force in peace operations seems to be somewhat exaggerated. Further systematic research is needed before these notions can be used as the foundation of operations, doctrines and military education without reservations.

Before turning to the empirical comparison, this alleged relationship needs to be more closely analysed. In research on peace operations, the notion of consent is considered to be one of its cornerstones.[15] According to Charles Dobbie, consent could be obtained at two levels: the strategic and the tactical. Consent on the strategic level concerns, among others, agreements between political elites and peace treaties, while consent on the tactical level concerns the will of the local population. Even if the operation has consent at the strategic level, the intervening forces might have to struggle to get consent from the local population.[16] The notion of consent at the local level in peace operations is strikingly similar to the notion of hearts and minds in doctrines and theories on COIN.[17] According to the US counter-insurgency manual, *FM 3–24*: ' "Hearts" means persuading people that their best interests are served by COIN success. "Minds" means convincing them that the force can protect them and that resisting is pointless.'[18] If there is a failure to win the hearts and minds of the population, the population might turn to the insurgents, and the forces would lose the consent of the population.

Winning hearts and minds, and achieving local consent, is usually thought to be done by interaction. Beatrice Pouligny has studied how local social actors interact with peacekeeping operations. She argues that it is of utmost importance for the military forces in an intervention to communicate and interact with the local population, so as to not marginalise the majority of the people.[19] Although research on this relationship is scarce, Marianne Heiberg suggests that it could be 'convincingly argued that the nature of the relationship a peacekeeping force achieves with the population within its area of control is a decisive element in determining the operation's success or failure'.[20] Furthermore, this relationship must build on communication and confidence in order to be successful: 'a relationship characterised by mounting hostility, suspicion and lack of communication is a sufficient factor for failure.'[21] Interaction and communication could be conducted in several ways; for example, by addressing local powerbrokers. In doctrines, arguably, it is suggested that the local population ought to be the focus of action.[22]

The implicit foundation for interaction with the population seems to be the use of minimum force. In doctrines on peace operations and counter-insurgency, the threat and use of violence is primarily seen as a means of communication, rather than as a way to disarm or overwhelm the antagonists. According to several doctrines, minimum force is often emphasised in order to avoid creating aversion among the population. It is argued that the use of force might make the population unfavourably disposed towards the peacekeeping forces, causing them to turn to the insurgents.[23] Nevertheless, a show of force might, at certain times, be used to make the population perceive the forces in a positive light – as, for example, expressed in the NATO doctrine for Peace Support Operations, AJP 3.4.1:

> In cases of clear breaches of the mandate, the flouting of international law and the abuse of human rights, the use of force, if authorised by the ROE [Rules of Engagement], may serve to enhance the credibility of the force and the consent for the operation nationally and internationally.[24]

Arguably, the use of minimum force is seen as the foundation of interaction. In research on so-called spoilers,[25] as well as on terrorism, the threat and use of violence as a means of communication is prominent. Research on spoilers is one of the few areas where the impact of violence at the tactical and operational level on peace processes has been more closely examined. The use of violence by spoilers is seen as creating or increasing mistrust and decreasing the parties' willingness to continue the peace process.[26] This is similar to the above-mentioned view of the intervening forces' threat and use of violence in military operations – that, by using too much violence, or threat of violence, the forces allegedly create or increase mistrust among the local population.

Military peace operations include many different tasks at the tactical level, for example manning checkpoints, conducting patrols and escorts, collecting weapons, building schools and roads, meeting leaders and training of the domestic military forces. Since some of these tasks are not performed in public – and therefore could not have any (direct) impact on the local populations' perception of the intervention – not all tasks are of interest in this study. In order to distinguish a more threatening kind of conduct, 'show of force', from a more friendly approach, 'minimum force',[27] certain features of the conduct of the forces could be considered: for example, the number of soldiers, the number, and type of, vehicles and weapons – as well as kind of armour. A small number of forces, with non-protected vehicles and no visible weapons, are generally considered not to antagonise the population. More subtle behaviours, like not wearing helmets or sunglasses, are also considered to have a positive effect on the outcome of peace operations.[28] In contrast, it is argued that the use of a high number of soldiers, heavy weapons, many and well-protected vehicles, and visible body armour could frighten or agitate the population, making them turn against the military forces.

Following from the discussion above, it can be argued that the more friendly approach, 'minimum force', would increase the chances of success in peace operations, compared to the more robust approach, 'show of force'. Before moving on to the case studies and the conduct of forces in Northern Afghanistan, it is necessary to discuss how to measure success in peace operations and what methodology will be used.

## Success in peace operations

Definitions of success in peace operations can be divided between inclusive and more limited definitions. In Peace Research, inclusive definitions are common. These entail, for example, to what extent the operation has facilitated conflict resolution, whether the underlying dispute of the conflict has been resolved or whether political, economic and social justice has been improved. Some scholars argue that even the stability of the peace has to be included, namely whether a war has broken out again after several years of peace. However, by using measurements that are too inclusive, the indicators of success risk becoming worthless. If one adopts the definition of success of Roland Paris, one of the advocates of the inclusive approach, who deems the duration of peace essential, then Namibia becomes the only successful peace operation in the 1990s.[29] According to Downs and Stedman, by 'setting such high standards of success ... we are left with a world of undifferentiated failure'; it becomes impossible to evaluate measures that have more limited effect on the outcome. It also becomes difficult to evaluate short-term success.[30]

Due to the problems with inclusive definitions, more limited definitions of success are used extensively by other scholars. According to Downs and Stedman, one of the most commonly used indicators of success in peace operations is the degree of mandate achievement. Some scholars use mission sub-goals, such as demobilisation, respect for human rights or refugee repatriation, which make success even more easily measurable. These kinds of definitions also make it easier to compare different cases, since fewer dimensions have to be compared. Conversely, interactions between different aspects of the outcome might be overlooked. Furthermore, some aspects might be emphasised differently in different missions, making comparisons risky.[31]

Another way to categorise different definitions of outcome in peace operations is by external and internal definitions. In Peace and Conflict research, the focus is almost exclusively on external definitions, such as those described above. However, in War Studies and in military 'lessons learned', internal definitions are also considered. These are often called 'measures of efficiency' or 'measures of performance'. According to Richard K. Bullock, a measure of performance 'describes how well the system utilizes resources'; that is, that 'things are being done right'.[32] These measures could, for example, be to what extent situational awareness is achieved. This way of defining success could complement external definitions but does not say anything about whether 'the right things are being done'.[33]

Since the Afghan conflict in this study will be divided into two specific cases, inclusive definitions are not applicable. A more limited definition is needed. According to Byman and Seybolt, the primary objective of military interventions in civil wars is to 'halt the fighting and preserve peace for a short period of time';[34] one of the main aims of the PRT is to achieve a secure environment. Hence, a good measure of success will be a stable and secure environment in a short timeframe. This does not necessarily imply that the conflict is solved in the long term, or that success at the tactical level of operations can be translated into success at the operational or strategic levels.[35] The indicator for a stable and secure environment will be the number of violent attacks. However, the level of violence is a limited way to measure a stable and secure environment. For the results to be convincing, this indicator needs to be complemented by supplementary indicators that reflect a stable and secure environment in a more comprehensive way. Unfortunately, due to the research design and data availability, other indicators of success have been difficult to identify. Nevertheless, since the level of violence is often used as an indicator for success in military intervention, the results will be easier to compare.[36]

## Methodological considerations

Differences within a conflict are usually not paid systematic attention in most research on peace operations. Rather, the whole conflict is treated as one case. However, when studying the conduct of forces and its effects, comparisons within a conflict are advantageous because of variation in conduct between national contingents.[37] On the one hand, it can also be argued that the conflict environment is more uniform within a conflict than between conflicts. This means that comparisons are made more relevant and it decreases the risk of bias from interfering variables causing differences in outcome.[38] On the other hand, a separation of a conflict in several cases might make comparison more problematic. For example, the strategy and tactics of the insurgents cannot be taken into consideration. The insurgents might direct their attacks against a certain area at a certain time, regardless of whether the local population consents to the intervening forces or not. This can cause differences in the outcome of operations. Different geographical areas, although in the same conflict, might also have different conflict dynamics. For example, different ethnic groupings, specific local warlords or different access to natural resources might influence the outcome of operations for certain areas in the conflict. Despite these problems, because the main focus of this chapter is the conduct of the forces, dividing one conflict into two cases has advantages over other approaches.

The conduct of the British-led PRT of Mazar-e Sharif in the north-western part of Afghanistan and the conduct of the German-led PRT of Kunduz in the north-eastern part of Afghanistan, between 2004 and 2006, will be analysed in this study. According to Sedra, there has been no over-arching strategy for the PRTs in Afghanistan.[39] In addition, the different PRTs were led by different countries.[40] As a result, the armed forces of respective PRTs each developed their own procedures and operated in its own particular way. With respect to the chosen research design of comparing the different approaches to operations within the same conflict, this variation makes the case of Afghanistan ideal for study. Furthermore, in the existing literature on PRTs in Afghanistan, the approaches of British, American and German forces are considered to be different, if not outright opposite.[41]

The case studies have been selected from among the PRTs in the north for several reasons. Most importantly, since two separate missions (ISAF and Operation 'Enduring Freedom' or OEF) are in progress in southern and western Afghanistan, forces from different countries with different modus operandi may operate in the same area. This means that it is difficult to isolate the key variable of force conduct. Furthermore, the relatively lower levels of violence in the northern parts of Afghanistan make these cases more interesting. The security situation is more stable, thus implying that these cases have more in common with other peace operations, such as Kosovo Force (KFOR) in Kosovo. Hence, the results of this study might be more relevant for peace operations in general. Also, the visible conduct of forces might have more impact in a situation where the people are not fighting for their lives. Finally, it might be easier to get reliable information on these cases, considering the lower levels of violence.

Information about the conduct of forces is often secret and difficult to obtain. Therefore, apart from existing research, the main source of information on the conduct of forces has been interviews conducted by the author. Although documentation on, for example, the number of forces and number of patrols probably would have been more accurate considering where and how often the forces were used, the use of interviews contributes to a thick description of how these forces were actually used. For each case, six to eight officers have been interviewed.[42] In order to get information from different perspectives, officers that have been working at different periods between 2004 and 2006, as well as officers on different levels in the organisation, have been interviewed.[43]

## Provincial reconstruction teams in Afghanistan

Following the air campaign against the Taliban regime in Afghanistan in the late autumn of 2001, the Coalition Forces concentrated their efforts on the ground. By 2003, a counter-insurgency force, consisting of around 18,000 troops, had developed, focusing their efforts in the south-east and eastern parts of the country. ISAF, at the time consisting of around 4,500 troops, only had the mandate to operate in and around Kabul. Since the newly constituted Afghan authorities had limited impact outside Kabul, the Coalition Forces established small, joint civil–military teams around the country, so-called 'Provincial Reconstruction Teams'. The first PRTs were established by the Coalition Forces in late 2002, all under US command.[44] A year later, 11 PRTs were established all over Afghanistan.[45]

The ISAF mission in Afghanistan was authorised by the United Nations (UN) through the UN Security Council Resolution 1386 in 2001.[46] At the start, the ISAF mission was restricted to Kabul, but in 2003 the UN Security Council Resolution 1510 extended the mandate.[47] In January 2004, ISAF began to expand its presence outside Kabul in four stages: stage 1 to the north; stage 2 to the west; stage 3 to the south; and stage 4 to the east. In October 2006, the ISAF expansion was completed, at the time consisting of 26 PRTs.[48] Each of the four regions were coordinated by a Regional Command, including one headquarter and one forward support base.[49] According to the mandate, the main aim of the ISAF PRTs was to extend the authority of the Afghan government outside Kabul and to create a safe and secure environment for the Afghan authorities, non-governmental organisations and the UN to conduct reconstruction and humanitarian efforts.[50]

## The British approach

British forces have been engaged in peacekeeping and COIN operations for a long time, and their forces are known for their emphasis on 'minimum force'.[51] According to Huw Bennett, '[v]irtually all writers on the Army since 1945 identify minimum force as a key characteristic of British counter-insurgency'.[52] British forces have considerable experience with policing insurgencies in, for example, Northern Ireland, Kenya and Malaya.[53] In Afghanistan, they are engaged both in the Coalition efforts (OEF) and in the ISAF mission.

The PRT in Mazar-e Sharif was established by British forces in July 2003, at the time as a part of the Coalition Forces. In July 2004 it became part of ISAF and in March 2006 the command was transferred to Swedish forces.[54] From the beginning, the AO of the PRT was the whole of the North West Region of Afghanistan (the five provinces of Balkh, Jawzjan, Faryab, Sar-e Pol and Samangan), and its headquarters was located in Mazar-e Sharif in the province of Balkh. However, since the province of Faryab was remote, and experienced quite a lot of violence, the UK established a small PRT in its capital Meymaneh in June 2004.[55] Since June 2004 falls outside of the time-frame adopted in this study, the province of Faryab will not be included.

The PRT consisted of around 100 personnel covering an AO including almost 2.3 million civilians (23,000 civilians per soldier).[56] The head of the PRT was a military commander, although in close cooperation with two civilian advisers, one from the British Department for International Development and one from the Foreign and Commonwealth Office. Among the usual tasks of the PRT were, for example, escorts, meetings with local governors, chiefs of the police and village elders, as well as advising and monitoring the Afghan National Army and the Afghan National Police, supervising disarmament and attending official ceremonies. One of the most visual displays of PRT Mazar-e Sharif outside the town itself were so-called Military Observer Teams (MOTs). These were small teams, around five to six military personnel, a local translator and two vehicles, patrolling the countryside. Their main responsibility was to gather information and report back to the PRT and the Afghan government in order to coordinate security and humanitarian efforts in the provinces. Most of the time, they drove between villages, talked to governors, warlords, village elders and the people on the ground. In each province, there were two MOTs. They stayed in so-called safe houses, up to two weeks at the time, before returning to the headquarter of the PRT for food, materiel and to report. Using safe houses as their base, each MOT conducted long-range patrols for up to ten days at the time throughout the provinces. During these patrols they often slept between their cars, under the open sky.[57]

On these patrols, the forces did not usually wear visible combat equipment, like helmets, body armour or rifles. Likewise they did not have their guns, as expressed by a civilian adviser at the PRT, 'cocked and loaded, ready for action'.[58] Rather, they wore berets, soft webbing and no visible guns, trying to de-escalate any problem. They often had tea with the village elders. The MOTs usually travelled around their province in two unprotected, white-painted jeeps.[59] When entering a village, each team of six soldiers and the local translator separated into even smaller groups of two soldiers.[60] However, when security deteriorated, they used protected cars, wore helmets, body armour and had their rifles with them when dismounting from the vehicles. If needed, they could also call on the Quick Reaction Force (QRF) of about 25 soldiers, stationed at the forward support base, at the airport of Mazar-e Sharif. The QRF, which was a joint resource for the whole of the Regional Command North, was mostly deployed in groups of 12 soldiers and three vehicles. They did carry more weapons, and, if leaving their weapons in the vehicles, they always had three to four soldiers with full combat equipment nearby for protection. They were always ready to deploy if something happened; for example, in case of a suspected improvised explosive device. When they were not on call, the Quick Reaction Force patrolled the provinces in the same way as the MOTs.[61]

In the town of Mazar-e Sharif, the presence of the forces was somewhat different in comparison to other parts in the AO. The PRT base was in the middle of Mazar-e Sharif. The compound was guarded by soldiers from the Afghan National Army and surrounded by a fence with barbed wire. After an attack or in times of high alert, they patrolled the streets on foot. Since there was more violence in the district of Mazar-e Sharif than in other parts of the AO, three MOTs patrolled the district. There were also more meetings, and therefore probably more escorts, in the town.[62] The QRF and the forward support base were stationed at the airport of Mazar-e Sharif. Therefore, the presence of soldiers in the town of Mazar-e Sharif was probably higher, and more visible, than in the countryside.[63]

When the security situation was good, the British PRT had a minimum display of force. They did not use body armour, visible guns or helmets. They travelled around in small teams, all over the country, in unprotected vehicles. Furthermore, since one of the main aims of the MOTs was to gain confidence from local commanders, warlords and the local population, the British forces had a policy to act in as non-threatening a manner as possible.[64] Being so small, these patrols were not likely to have displayed any 'show of force' in the eyes of the population. Since their presence was more common in the town of Mazar-e Sharif, there might have been more of a 'show of force' there than in the countryside. When confrontations occurred, back-up was at hand, with the possibility of, at times, displays of force of up to 30 soldiers. Nonetheless, after an attack, they did not patrol in larger groups with more than two vehicles at the same time in any specific area.

## The German approach

After the end of the Second World War, Germany did not participate with military forces in international operations until the UN operations in the Balkans in the 1990s. Their recent experience of conducting peace operations, compared to the British forces, is somewhat limited, though they have taken on considerable responsibility in the ISAF mission in Afghanistan. Apart from leading the ISAF mission (III) in 2003, German forces have been in command of the Regional Command North, PRT Kunduz, as well as PRT Feyzabad. They also have forces in several functions in Kabul.[65] However, the pressure from the German government and the population not to lose any soldiers is considerable, and their resulting preoccupation with force protection has been highly criticised.[66]

PRT Kunduz was established in early 2003 by American forces, and in October 2003 it was transferred to German command. When ISAF took command of the PRT in December 2003, it was the first PRT to be included in the ISAF mandate. The AO of PRT Kunduz was the provinces of Kunduz and Takhar, including around 1.66 million civilians. Compared to other PRTs, PRT Kunduz was rather robust, including around 300 to

400 military personnel (around 4,000 civilians per soldier).[67] They also had more civilian personnel than the British PRT in Mazar-e Sharif, with around 15 people. The civilian personnel were separated from the military command, and the PRT had one military and one civilian commander.[68] The main tasks of the PRT were patrols, security tasks and key-leader engagements. Key-leader engagement included, of course, meetings with leaders on different levels, but also attendance at different ceremonies. Pure security tasks included, for example, escorts and airport security. Patrols were tailor-made depending on mission. Sometimes mixed patrols – consisting of both military and civilian personnel – were used where the task demanded it and the security situation allowed it. In other cases, purely military patrols were used to de-escalate potentially threatening situations and to display military presence.[69]

The main exposure of the PRT outside the town of Kunduz was the mixed patrols. These mixed patrols were mission designed and could, for example, include a force-protection element of about ten soldiers, a Civil–Military Cooperation (CIMIC) element, a liaison and monitoring team, a Psychological Operations (PSYOPS) team, a medical element and a reconnaissance team. Civilian advisers were often included in these patrols, which could consist of between 20–30 people and 7–10 vehicles. Normally, the PRT conducted about 20 patrols each day. Most of these were short patrols, about one-to-two hours in the area of Kunduz, or one-day patrols in the surrounding areas.[70] According to several researchers, these were the only patrols conducted by the PRT, since the German forces could not stay outside the base camp overnight, due to their force protection guidelines.[71]

However, in interviews with German officers deployed at PRT Kunduz during the specified time, it appeared that the PRT also conducted longer patrols of two-to-five days, requiring the patrols to stay overnight in safe houses, or guest houses. Safe houses were houses owned by the PRT, guarded by the Afghan National Police. Some of these houses were occupied up to four nights a week. Guest houses were houses where the patrols could stay overnight, but which were not owned by the PRT, such as local police stations. The PRT had guest houses all over the AO. The PRT did some two-day-long patrols every week, and they also carried out some longer patrols, up to four to five days, in the remote parts of the AO. When no safe house or guest house was available, they slept under the open sky in the same way as the forces at PRT Mazar-e Sharif. However, since the German patrols were larger, up to 7–10 vehicles, they could form a ring of vehicles around the camp, providing more protection than the vehicles of the British MOTs.[72]

Apart from deploying larger patrols, another difference from the British PRT was that the German PRT always had to use protected vehicles outside the camp. These were similar to jeeps but with underside Kevlar and mine protection. Although somewhat larger than ordinary jeeps, from the doors up, they were unprotected.[73] According to Peter V. Jakobsen, the German forces also had to include an armoured ambulance and force-protection elements on their patrols outside Kabul, maybe making them appear somewhat more invasive.[74] However, when arriving in a village or a town, they always dismounted from the vehicles into smaller groups of one to three people, coupled with a force-protection element of two soldiers. A further difference was that outside the vehicles the German soldiers had to wear body armour at all times. They also had to bring two weapons when stepping outside the vehicle, one rifle and one handgun, both with ready status – that is, unsecured and with the first bullet ready. This might also have contributed to a difference in appearance between the British and the German PRT. The German soldiers usually wore soft headgear, the same as the British soldiers, but they used helmets when the security situation deteriorated. In such cases, they also used armoured vehicles which, among other things, made it difficult for them to get in contact with the population.[75]

During other tasks, such as patrols, airport security, or cordon-and-search operations, the German soldiers always used armoured vehicles; partly because these tasks demanded more protection, partly in order to signal to the population that things were serious. Cordon-and-search operations were conducted together with the Afghan National Army, and sometimes a whole infantry company was included.[76] Tasks of this kind seem quite similar to the British PRT, where the QRF was used when necessary. One should, however, be careful not to draw too far-reaching conclusions, as data on how often the British and German forces acted in this way is not available.

In comparison to the British PRT, the German patrols were more numerous. Not only did PRT Kunduz have approximately three to four times more soldiers at the PRT and 30 per cent fewer civilians in their AO, they also deployed more numerous patrols. Mostly, the British forces conducted patrols with six to seven soldiers, while the German forces deployed up to 20–30 soldiers on their patrols, displaying a much more robust force than the British soldiers. The PRTs used different kinds of vehicles, as well. When not in a hostile situation, the British used unprotected vehicles, only two at a time. Outside the camp, German forces always used protected vehicles. Protected vehicles, combined with larger patrols, could have been perceived as a greater 'show of force' by the local population. In addition, the German soldiers, when outside their vehicles, always had to wear body armour and carry more weapons than the British soldiers. The British soldiers usually did not use body armour and they left their visible weapons in the vehicles when not in hostile situations.

In a hostile situation, however, the British forces used the same kind of vehicles and armour as the German forces, and they could call in the QRF with around 25 more soldiers. When stopping in a village, the German forces divided into smaller groups, very much like the British approach.

Hence, there might not be such a large difference in conduct as several researchers have argued. When entering a larger town, the individual Afghan citizen might not detect too much difference between the two forces. However, in smaller villages, the presence of up to 30 soldiers, heavily equipped, wearing body armour and weapons, would most certainly be perceived as threatening in comparison to the smaller British teams. In summary, it might be argued that the British used more of a 'minimum force' approach and the Germans used more of a 'show of force' approach.

## Outcome of operations in Northern Afghanistan

The methodological approach of this study requires a comparison between two areas within the conflict in Afghanistan. Unfortunately, most databases do not make a distinction of where attacks and other incidents occur. Rather, the conflict is treated as a whole. As a result, the only indicator of changes in the security situation that could be found in available databases is the number of attacks. As discussed earlier, this is a somewhat limited way of measuring the security situation. Two databases have been used: the RAND-MIPT Terrorism Incident Database (1998–present) and the FAST Event database at the Early Warning Program at Swiss Peace. Both databases are based on open-source material, although the gathering process is somewhat different.[77]

In terms of the reliability of the databases, there are some general problems with open-source material and data-gathering in conflict environments. For example, open-source material has a tendency to become part of the conflict. Both insurgents and the government might contribute to censorship and to the distribution of propaganda.[78] Furthermore, governmental and statistical agencies might not function too well, especially in weak states, making it difficult to obtain reliable data, even on such things as the size of the population. According to Seth Jones, this is especially true in the case of Afghanistan.[79] Additionally, since there are some extremely remote areas in Afghanistan, the data on the number of incidents in some areas might not be as reliable as in other areas of the country.

The RAND-MIPT Terrorism Incident Database serves 'to monitor all terrorism incidents worldwide',[80] and is often used in research on terrorism.[81] According to the website, 'every effort was made to verify the accuracy of the information found in the reports',[82] and, in comparison, the database only 'slightly underestimates the number of attacks and deaths', or closely parallels data collected from sources such as the Afghan Non-Governmental Organisation Security Office or the US State Department.[83] The FAST Event database, meanwhile, includes both incidents and cooperative features of the conflict. However, for this study, only incidents that clearly reflect the number of attacks and are most consistent with the kind of attacks used in the RAND-MIPT database will be used.[84] In comparison to the RAND-MIPT database, the FAST Event database contains a larger number of attacks.

In Afghanistan as a whole, the number of attacks fluctuates according to the seasons, with a decrease in hostilities during the winter months. In order to avoid erroneous analysis caused by this variation, the number of attacks will be measured for a whole 12 months at a time. Since the command of PRT Mazar-e Sharif was transferred to Swedish command in March 2006, the timeframe will be from 1 March 2004 to 28 February 2006. Measuring outcome on an annual basis, however, does have some problems. For example, although PRT Mazar-e Sharif was under British command from when it was established in July 2003, it was not part of ISAF until July 2004 and therefore not part of the ISAF mission during the first four months. This could have had an impact on how the population perceived the British operation in the beginning.[85]

According to the FAST Event database, six attacks took place during the first year (1 March 2004–28 February 2005) in the AO of PRT Mazar-e Sharif. In the second year (1 March 2005–28 February 2006), eight attacks occurred in the same area. This is an increase of two attacks (33 per cent). According to the traditional understanding of peacekeeping theory and the alleged virtues of hearts-and-minds, since the British forces were using a 'minimum force' approach, the security in AO of PRT Mazar-e Sharif should have increased during this time. The results seem to contradict the theory of minimum force.

However, if security became even worse during the same period in the AO of PRT Kunduz, the theory might still have some explanatory power. During the first year, eight attacks occurred in the AO of PRT Kunduz. In the following year, three attacks took place in this area. This is a reduction of five attacks (262 per cent). According to the theory, since the German forces were using more of a 'show of force' approach than the British forces, the security should have deteriorated more in this area. The data show that it did not. Rather, the security in the AO of PRT Kunduz actually improved during this time. The collected data, so far, contradicts the theory.

The corresponding data from the RAND-MIPT database seem to confirm these findings, showing an increase of attacks in the AO of PRT Mazar-e Sharif from one to nine attacks, and a reduction of attacks in the AO of PRT Kunduz from five to zero attacks. Hence, the data of RAND-MIPT database supports the data from the FAST Event database, in both cases presenting evidence in Table 6.1 that contradicts the theory. According to both sets of data, the number of attacks increased in the AO of PRT Mazar-e Sharif during the specified time (þ2 and þ8), at the same time as the number of attacks decreased in the AO of PRT Kunduz (25 in both cases).

According to these results, the statement that minimum force is the better approach in peace operations and counter-insurgency is contradicted.

*Table 6.1* Attacks in Afghanistan and two provinces, 2004–2006

| *FAST Database* | *2004– 2005* | *2005– 2006* | *Change* | *Change %* | *TKB Database* | *2004–2005* | *2005–2006* | *Change* | *Change %* |
|---|---|---|---|---|---|---|---|---|---|
| PRT MeS | 6 | 8 | 2 | 33 | PRT MeS | 1 | 9 | 8 | 800 |
| PRT Kunduz | 8 | 3 | –5 | –62 | PRT Kunduz | 5 | 0 | –5 | –100 |
| Afghanistan | 439 | 571 | 132 | 30 | Afghanistan | 137 | 240 | 103 | 75 |

Rather, it could be argued that the German approach to peace operations and counter-insurgency creates a more secure environment than the British approach. More security measures, such as larger patrols and more weapons, may be preferred when conducting these kinds of operations.[86]

Apart from the methodological problems mentioned above, there are several other problems with this conclusion. First, since the number of attacks in the AO of PRT Kunduz and PRT Mazar-e Sharif are so few, a conclusion regarding a significant change in the security situation is uncertain. As described above (pp. 00–00), it is highly problematic to gather data in a conflict environment, and a small error in the data could cause considerable changes in the result. Second, also as discussed above, the number of attacks might not be the best way to measure the outcome of operations, or even the best way to measure the security situation. Finally, it is important to remember that it is only the visible conduct of the forces that has been considered in this chapter. Nevertheless, there is nothing in the data that implies that minimum force is causing any better security situation than a show of force. Consequently, it is a good idea to conduct further research before the notion of minimum force can be used in operations, doctrine and education, without reservations.

## Conclusion

In order to get a better understanding of how to conduct peace operations, two kinds of military conduct in peace operations have been explored in this contribution. The first could be described as 'minimum force' (the conduct of operations of the British PRT in Mazar-e Sharif) and the second as 'show of force' (the conduct of operations of the German PRT in Kunduz). The existing literature on the differences between the British and the German way of conduct in Northern Afghanistan has been found to be somewhat stereotyped and exaggerated. However, a considerable difference in the visible conduct of the forces can be discerned. Additionally, the relationship between the conduct of military forces in peace operations and the outcome of these operations has been elaborated and compared between the two cases in Northern Afghanistan. The findings, so far, seem to contradict the commonly held idea of the superiority of 'minimum force', since 'show of force' correlated with a better outcome than the 'minimum force' approach.

However, before concluding that 'show of force' is a better way to conduct peace operations, further research is needed. For example, more reliable data on the outcome are needed in order to make the outcome in the two cases more certain. The number of attacks needs to be supplemented with other indicators of a stable and secure environment. Furthermore, in order to establish any causal relations between a certain kind of conduct and the outcome of peace operations, additional comparative case studies are needed.

However, if further research demonstrates that a 'show of force' approach is causing a more stable and secure environment, it could, on the one hand, indicate that the use of minimum force in order to win the hearts and minds of the population, or establishing local consent, has to be abandoned. Rather, more forces and more robust equipment might be needed in peace operations and COIN. On the other hand, it might be the case that the entire idea of winning hearts and minds and establishing local consent by minimum force is overrated. If this is the case, much research and many lessons learned on peace operations and COIN have to be reconsidered. If nothing else, this study has shown that it might be a good idea to be careful before accepting the minimum force approach, without any further consideration.

## Notes

This chapter has been previously published as Kersti Larsdotter, 'Exploring the Utility of Armed Force in Peace Operations: German and British Approaches in Northern Afghanistan', *Small Wars and Insurgencies*, 19, 3 (2008): 352–373. We would like to thank Taylor and Francis for permission to reprint it here.

1 Jones, 'Averting Failure in Afghanistan', 111–116.
2 Byman and Seybolt, 'Humanitarian Intervention and Communal Civil Wars', 33.
3 'Peace operations' is an umbrella term for different kinds of operations; for example, peacekeeping, peace-making, peace-building, peace-enforcement and humanitarian operations (e.g. in the US doctrine *Peace Operations* (2007), 1–7, and the US doctrine *Counterinsurgency* (2006), 1–20). Lately, the similarities between peace-enforcement and counter-insurgency (COIN) have been discussed and COIN is considered part of peace operations in some doctrines (e.g. in the NATO doctrine *Peace Support Operations* (2001), 6–7, and in the UK doctrine on *The Military Contribution to Peace Support Operations* (2004), 2–7). It could also be argued that there are several similarities between the strategic intent of PSO and COIN, as well as between some determinants of success (see below, pp. 00–00). The term 'peace operations' can, in this light, also include COIN. Mockaitis, 'From Counterinsurgency to Peace Enforcement', 40–57; Hills, 'Hearts and Minds Or Search and Destroy?', 2.
4 Thornton, 'The British Army and the Origins of its Minimum Force Philosophy', 83–106.
5 Chin, 'Examining the Application of British Counterinsurgency Doctrine by the American Army in Iraq', 8.
6 Chin, 'Examining the Application', 5–10; Van der Kloet, 'Building Trust in the Mission Area', 433; Jakobsen, *PRTs in Afghanistan*; Bellamy and Williams, 'Introduction', 2; Thornton, 'The British Army and the Origins of its Minimum Force Philosophy', 83–84. However, the true nature of this approach has been extensively discussed. See, for example, Bennett, 'The Mau Mau Emergency as Part of the British Army's Post-War Counter-Insurgency Experience', 143–163.
7 For example, *Counterinsurgency*, 1–25, A-5; *Military Contribution*, A-6; *Peace Operations*, I-1–I-2; *Peace Support Operations*, 3–1–3–6.
8 Downs and Stedman, 'Evaluation Issues in Peace Implementation'.
9 Mueller, 'The Banality of "Ethnic War"', 65; Fearon and Laitin, 'Neotrusteeship and the Problem of Weak States', 23.
10 Downs and Stedman, 'Evaluation Issues in Peace Implementation', 43.

11 Biddle, *Military Power*, 18.
12 Harkavy and Neuman, *Warfare and the Third World*, xi–xii.
13 Smith, *The Utility of Force.*
14 Biddle, *Military Power*,17–19.
15 Bellamy, Williams and Griffin, *Understanding Peacekeeping*, 95, 128–133; Findlay, *The Use of Force in UN Peace Operations*, 15.
16 Dobbie, 'A Concept for Post-Cold War Peacekeeping', 124, 133; Woodhouse, 'The Gentle Hand of Peace?', 30.
17 Hills, 'Hearts and Minds', 2; *Peace Support Operations*, 2–4–2–5.
18 *Counterinsurgency*, A-5.
19 Pouligny, 'Peacekeepers and Local Social Actors', 3; Pouligny, *Peace Operations Seen from Below.*
20 Heiberg, 'Peacekeepers and Local Populations'.
21 Heiberg, 'Peacekeepers and Local Populations'.
22 Gizelis and Kosek, 'Why Humanitarian Interventions Succeed Or Fail', 363–383.
23 *Peace Operations*, I-5–1–6; *Peace Support Operations*, 3–4.
24 *Peace Support Operations*, 3–5.
25 According to Stedman, spoilers are 'leaders and parties who believe that peace emerging from negotiations threatens their power, worldview, and interests, and use violence to undermine attempts to achieve it'. Stedman, 'Spoiler Problems in Peace Processes', 5.
26 Höglund, *Violence in the Midst of Peace Negotiations*; Stedman, 'Spoiler Problems in Peace Processes', 5; Kydd and Walter, 'Sabotaging the Peace', 263–296.
27 In this chapter, minimum force is not the same as the British 'minimum necessary force'. Minimum force is rather about the appearance of the forces, than about how much violence they actually use.
28 Chin, 'Examining the Application', 9.
29 Pushkina, 'A Recipe for Success?', 134; Downs and Stedman, 'Evaluation Issues in Peace Implementation', 47–49.
30 Downs and Stedman, 'Evaluation Issues in Peace Implementation', 49.
31 Downs and Stedman, 'Evaluation Issues in Peace Implementation', 45, 49–50.
32 Bullock, 'Theory of Effectiveness Measurement', 20.
33 Bullock, 'Theory of Effectiveness Measurement', 20.
34 Byman and Seybolt, 'Humanitarian Intervention', 35.
35 Byman and Seybolt, 'Humanitarian Intervention', 35; Brooks, 'Making Military Might', 153.
36 Pushkina, 'A Recipe for Success?', 134; Downs and Stedman, 'Evaluation Issues in Peace Implementation', 47.
37 Single case studies within a conflict are quite common in War Studies. See, for example, Malkasian, 'The Role of Perceptions and Political Reform in Counterinsurgency', 367–394; Hills, 'Fear and Loathing in Falluja', 623–639; Chin, 'British Counter-Insurgency in Afghanistan', 201–225. Unfortunately, comparisons between cases within one and the same conflict are not common. One exception is Malkasian, 'Did the United States Need More Forces in Iraq?', 78–104.
38 As, for example, differences in mandate or implementation strategies, the difficulty of the mission, underlying causes of the conflict, or the salience of a given conflict for international public opinion. Pushkina, 'A Recipe for Success?', 134; Downs and Stedman, 'Evaluation Issues in Peace Implementation', 43, 53.
39 Sedra, 'Civil–Military Relations in Afghanistan'.
40 According to research on military cultures, militaries in different states often have different approaches to the conduct of war. See, for example, the discus-

sion on the 'way of war' in history and on 'strategic culture' in International Relations. Hanson, *Carnage and Culture*, 5–13; McInnes, 'The British Army's New Way in Warfare', 128; Thornton, 'The British Army and the Origins of its Minimum Force Philosophy', 83–106.

41 Sedra, 'Civil–Military Relations'; Perito, *The U.S. Experience with Provincial Reconstruction Teams in Afghanistan*; Jakobsen, *PRTs in Afghanistan*.

42 In addition, several officers working with the Quick Reaction Force in Mazar-e Sharif have also been interviewed.

43 All interviews were conducted on the basis of confidentiality, and the names of interviewees are withheld by mutual agreement. Transcripts are available from the author.

44 Perito, *The U.S. Experience with Provincial Reconstruction Teams in Afghanistan*, 2–3; Lopez, 'Engaging Or Withdrawing, Winning Or Losing?', 246; Maloney, 'Afghanistan Four Years On', 23. These teams were initially called Joint Regional Teams (JRTs).

45 Jakobsen, *PRTs in Afghanistan*, 13.

46 Security Council, Resolution 1386 (2001), 2.

47 Security Council, Resolution 1510 (2003), 2.

48 North Atlantic Treaty Organization, 'Expansion of NATO's Presence in Afghanistan', www.nato.int/isaf/topics/expansion/index.html (accessed 3 March 2008).

49 Forward support bases 'are essential logistic installations, created to provide supply, medical and transport hub in each region to assist the PRTs in their mission to extend the Government of Afghanistan's authority'. North Atlantic Treaty Organization, 'ISAF Regional Command Structure', www.nato.int/isaf/structure/regional_command/index.html (accessed 3 March 2008).

50 Security Council, Resolution 1510, 2.

51 Bennett, 'The Mau Mau Emergency', 145–146; Thornton, 'The British Army and the Origins of its Minimum Force Philosophy', 83–106.

52 Bennett, 'The Mau Mau Emergency', 145.

53 Cassidy, 'The British Army and Counterinsurgency', 53–59; Chin, 'British Counter-Insurgency', 202; *The Military Contribution to Peace Support Operations*, 1–4.

54 Jakobsen, *PRTs in Afghanistan*, 13; Swedish Armed Forces, 'Afghanistan'.

55 Bogland and Bedford, *Sveriges bidrag till PRT i Afghanistan*, 18; interview with officer at PRT Mazar-e Sharif.

56 Central Statistics Office Afghanistan, 'Settled Population by Civil Division (Urban and Rural) and Sex 2005–2006'. For a discussion on the number of forces versus civilians, see Malkasian, 'Did the United States Need More Forces in Iraq?', 80.

57 Jakobsen, *PRTs in Afghanistan*, 21–22; United States Institute of Peace, 'Afghan Experience Project, Interview #26', www.usip.org/library/oh/sops/afghanistan/gov/26.pdf (accessed 5 May 2007); interview with officer at PRT Mazar-e Sharif.

58 United States Institute of Peace, 'Afghan Experience Project, Interview #18', www.usip.org/library/oh/sops/afghanistan/gov/18.pdf (accessed 5 May 2007).

59 Bogland and Bedford, *Sveriges bidrag*, 33–34; Jakobsen, *PRTs in Afghanistan*, 21; United States Institute of Peace, 'Afghan Experience Project, Interview #18'; United States Institute of Peace, 'Afghan Experience Project, Interview #26'.

60 Interview with officer at PRT Mazar-e Sharif.

61 Interview with officer at the QRF.

62 Interview with officer at PRT Mazar-e Sharif.

63 Interview with officer at PRT Mazar-e Sharif.

64 Bogland and Bedford, *Sveriges bidrag*, 35.
65 Merz, *Still on the Way to Afghanistan?*.
66 Merz, *Still on the Way to Afghanistan?*, 9; Jakobsen, *PRTs in Afghanistan*, 25.
67 Central Statistics Office Afghanistan, *Settled Population*.
68 Jakobsen, *PRTs in Afghanistan*, 24; interview with officer at PRT Kunduz.
69 Interview with officer at PRT Kunduz.
70 Interview with officer at PRT Kunduz.
71 Jakobsen, *PRTs in Afghanistan*, 25; Merz, *Still on the Way to Afghanistan?*, 9.
72 Interview with officer at PRT Kunduz.
73 Interview with officer at PRT Kunduz.
74 Jakobsen, *PRTs in Afghanistan*, 25.
75 Interview with officer at PRT Kunduz.
76 Interview with officer at PRT Kunduz.
77 The collection of data for the FAST Event database is conducted by local staff, so-called Local Information Networks, and coded by a Country Coordinator, based on a standardised coding scheme. Swiss Peace, 'About Swiss Peace's Early Warning Program – FAST International', www.swisspeace.ch/typo3/en/peace-conflict-research/early-warning/about/index.html (accessed 31 March 2008).
78 Höglund, *Violence in the Midst of Peace Negotiations*, 58.
79 Jones, 'Averting Failure in Afghanistan', 112.
80 Terrorism is defined

> by the nature of the act, not by the identity of the perpetrators or the nature of the cause. Terrorism is violence, or the threat of violence, calculated to create an atmosphere of fear and alarm. These acts are designed to coerce others into actions they would not otherwise undertake, or refrain from actions they desired to take.
>
> (MIPT Terrorism Knowledge Base, 'Purpose and Description of Information Found in the Incident Databases', www.tkb.org/RandSummary. jsp?page¼method (accessed 31 March 2008))

81 For example, Norris, Kern and Just, 'Lessons From Framing Terrorism', 281–302; Jones, 'Averting Failure in Afghanistan', 111–127.
82 MIPT Terrorism Knowledge Base, 'Purpose and Description'.
83 Norris, Kern and Just, 'Lessons From Framing Terrorism', 281–302; Jones, 'Averting Failure in Afghanistan', 113.
84 The incidents included in the categories are: armed battle, artillery attack, mine explosion, missile attack, other 'bombing', riot or political turmoil, suicide bombing, vehicle bombing, other 'armed actions', other 'force use', give ultimatum, small arms attack and assassination. However, the actual attacks during this period consisted of artillery attacks, mine explosions, other 'bombings', riots or political turmoil, suicide bombings, other 'armed actions', small arms attacks and assassinations.
85 An alternative timeframe would be from when both PRTs transferred to ISAF command, i.e. July 2004. However, if choosing this timeframe, the conduct of the Swedish forces might have influenced the outcome of operations during the last four months in the AO of PRT Mazar-e Sharif. If the conduct of the Swedish forces was different from the British forces, it could have had another influence on the outcome of operation. Since the Regional Command North moved to Mazar-e Sharif, the number of international personnel increased to around 3,000, at the same time as the command of the PRT was transferred to Swedish forces. Since some argue that the Swedish forces acted more like the German forces than the British, the chosen timeframe will be from 1 March 2004 to 28 February 2006. See Interview with officer at PRT Mazar-e Sharif.

86 Even if choosing the timeframe from when NATO received the command over the two PRTs as a starting point, i.e. from 1 July 2004–30 June 2006, the tendency is the same. According to the FAST Event Database there is an increase from 4 to 16 attacks (300 per cent) during this time in the AO of PRT Mazar-e Sharif and an increase from 5 to 6 attacks (20 per cent) during this time in the AO of Kunduz. The corresponding data from the RAND-MIPT database show an increase of attacks in the AO of PRT Mazar-e Sharif from 1 to 16 attacks (1,500 per cent) and a decrease of attacks in the AO of PRT Kunduz from 2 to 1 attacks (250 per cent).

## Bibliography

Bellamy, Alex J., Paul Williams and Stuart Griffin. *Understanding Peacekeeping*. Cambridge: Polity Press, 2004.

Bellamy, Alex J. and Paul Williams. 'Introduction: Thinking Anew About Peace Operations'. *International Peacekeeping* 11, 1 (2004): 1–15.

Bennett, Huw. 'The Mau Mau Emergency As Part of the British Army's Post-War Counter-Insurgency Experience'. *Defense & Security Analysis* 23, 2 (2007): 143–163.

Biddle, Stephen D. *Military Power: Explaining Victory and Defeat in Modern Battle*. Princeton, NJ: Princeton University Press, 2004.

Bogland, Karin and Sofie Bedford. *Sveriges bidrag till PRT i Afghanistan: En utvärdering införövertagandet av ledningen för PRT i Mazar-e-Sharif*. Stockholm: Totalförsvarets forskningsinstitut, 2005.

Brooks, Risa A. 'Making Military Might: Why Do States Fail and Succeed? A Review Essay'. *International Security* 28, 2 (2003): 149–191.

Bullock, Richard K. 'Theory of Effectiveness Measurement'. PhD dissertation, US Department of the Air Force, Air University, 2006.

Byman, Daniel and Taylor Seybolt. 'Humanitarian Intervention and Communal Civil Wars: Problems and Alternative Approaches'. *Security Studies* 13, 1 (2003): 33–78.

Cassidy, Robert M. 'The British Army and Counterinsurgency: The Salience of Military Culture'. *Military Review* 85, 3 (2005): 53–59.

Central Statistics Office Afghanistan. 'Settled Population by Civil Division (Urban and Rural) and Sex 2005–2006'. Available at: www.cso-af.net/cso/documents/estimated% 20population%201384.xls (accessed 1 April 2008).

Chin, Warren. 'British Counter-Insurgency in Afghanistan'. *Defense & Security Analysis* 23, 2 (2007): 201–225.

Chin, Warren. 'Examining the Application of British Counterinsurgency Doctrine by the American Army in Iraq'. *Small Wars & Insurgencies* 18, 1 (2007): 1–26.

Dobbie, Charles. 'A Concept for Post-Cold War Peacekeeping'. *Survival* 36, 3 (1994): 121–148.

Downs, George and Stephen J. Stedman. 'Evaluation Issues in Peace Implementation'. In Stephen J. Stedman, D. Rothchild and E.M. Cousens, eds. *Ending Civil Wars: The Implementation of Peace Agreements*. London: Lynne Rienner, 2002.

Fearon, James D. and David D. Laitin. 'Neotrusteeship and the Problem of Weak States'. *International Security* 28, 4 (2004): 5–43.

Findlay, Trevor. *The Use of Force in UN Peace Operations*. Oxford: Oxford University Press, 2002.

Gizelis, Theodora-Ismene and Kristin E. Kosek. 'Why Humanitarian Interventions Succeed Or Fail: The Role of Local Participation'. *Cooperation & Conflict* 40, 4 (2005): 363–383.

Hanson, Victor D. *Carnage and Culture: Landmark Battles in the Rise of Western Power.* New York: Anchor Books, 2001.

Harkavy, Robert E. and Stephanie G. Neuman. *Warfare and the Third World.* New York: Palgrave Macmillan, 2001.

Heiberg, Marianne. 'Peacekeepers and Local Populations: Some Comments on UNIFIL'. In I.J. Rikhye and K. Skjelsbaek, eds. *The United Nations and Peacekeeping: Results, Limitations and Prospects: The Lessons of 40 Years of Experience.* London: Macmillan, 1990.

Hills, Alice. 'Hearts and Minds Or Search and Destroy? Controlling Civilians in Urban Operations'. *Small Wars & Insurgencies* 13, 1 (Spring 2002): 1–23.

Hills, Alice. 'Fear and Loathing in Falluja'. *Armed Forces & Society* 32, 4 (2006): 623–639.

Höglund, Kristine. *Violence in the Midst of Peace Negotiations: Cases from Guatemala, Northern Ireland, South Africa and Sri Lanka.* Uppsala: Department of Peace and Conflict Research, Uppsala University, 2004.

Jakobsen, Peter V. *PRTs in Afghanistan: Successful But Not Sufficient.* Copenhagen: Danish Institute for International Studies, 2005.

Jones, Seth G. 'Averting Failure in Afghanistan'. *Survival* 48, 1 (Spring 2006): 111–116.

Kydd, Andrew and Barbara F. Walter. 'Sabotaging the Peace: The Politics of Extremist Violence'. *International Organization* 56, 2 (2002): 263–296.

Lopez, Andrea M. 'Engaging Or Withdrawing, Winning Or Losing? The Contradictions of Counterinsurgency Policy in Afghanistan and Iraq'. *Third World Quarterly* 28, 2 (2007): 245–260.

McInnes, Colin. 'The British Army's New Way in Warfare: A Doctrinal Misstep?'. *Defense & Security Analysis* 23, 2 (2007): 127–141.

Malkasian, Carter. 'The Role of Perceptions and Political Reform in Counterinsurgency: The Case of Western Iraq, 2004–05'. *Small Wars & Insurgencies* 17, 3 (2006): 367–394.

Malkasian, Carter. 'Did the United States Need More Forces in Iraq? Evidence from Al Anbar'. *Defence Studies* 8, 1 (March 2008): 78–104.

Maloney, Sean M. 'Afghanistan Four Years On: An Assessment'. *Parameters* 35, 3 (2005): 21–32.

Merz, Sebastian. *Still on the Way to Afghanistan? Germany and its Forces in the Hindu Kush.* Stockholm: Sipri Project Paper, 2007.

MIPT Terrorism Knowledge Base. 'Purpose and Description of Information Found in the Incident Databases'. Available at: www.tkb.org/RandSummary.jsp?page¼method (accessed 31 March 2008).

Mockaitis, Thomas R. 'From Counterinsurgency to Peace Enforcement: New Names for Old Games?'. In Erwin A. Schmidl, ed. *Peace Operations Between War and Peace.* London: Frank Cass, 2001, 40–57.

Mueller, John. 'The Banality of "Ethnic War"'. *International Security* 25, 1 (2000): 42–70.

Norris, Pippa, Montague Kern and Marion Just. 'Lessons From Framing Terrorism'. In Pippa Norris, Montague Kern and Marion, eds. *Just Framing Terrorism: The News Media, the Government and the Public.* London: Routledge, 2003, 281–302.

North Atlantic Treaty Organisation. 'Expansion of NATO's presence in Afghanistan'. Available at: www.nato.int/isaf/topics/expansion/index.html (accessed 3 March 2008).

North Atlantic Treaty Organisation. 'ISAF Regional Command Structure'. Available at: www.nato.int/isaf/structure/regional_command/index.html (accessed 3 March 2008).

Perito, Robert M. *The U.S. Experience with Provincial Reconstruction Teams in Afghanistan.* Washington, DC: United States Institute of Peace, 2005.

Pouligny, Beatrice. 'Peacekeepers and Local Social Actors: The Need for Dynamic, Cross-Cultural Analysis'. *Global Governance* 5, 4 (1999): 403–425.

Pouligny, Beatrice. *Peace Operations Seen from Below: UN Missions and Local People.* London: Hurst, 2006.

Pushkina, Darya. 'A Recipe for Success? Ingredients of a Successful Peacekeeping Mission'. *International Peacekeeping* 13, 2 (2006): 133–149.

Sedra, Mark. 'Civil–Military Relations in Afghanistan: The Provincial Reconstruction Team Debate'. Available at: www.asiapacificresearch.ca/caprn/afghan_project/m_sedra.pdf (accessed 2 April 2008).

Smith, Rupert. *The Utility of Force: The Art of War in the Modern World.* London: Allen Lane, 2005.

Stedman, Stephen J. 'Spoiler Problems in Peace Processes'. *International Security* 22, 2 (1997): 5–53.

Swedish Armed Forces. 'Afghanistan'. Available at: www2.mil.se/sv/Insatser/Afghanistan/ (accessed 31 March 2008).

Swiss Peace. 'About Swiss Peace's Early Warning Program – FAST International'. Available at: www.swisspeace.ch/typo3/en/peace-conflict-research/early-warning/index.html (accessed 31 March 2008).

Thornton, Rod. 'The British Army and the Origins of its Minimum Force Philosophy'. *Small Wars & Insurgencies* 15, 1 (2004): 83–106.

Van der Kloet, Irene. 'Building Trust in the Mission Area: A Weapon Against Terrorism?'. *Small Wars & Insurgencies* 17, 4 (2006): 421–436.

Woodhouse, Tom. 'The Gentle Hand of Peace? British Peacekeeping and Conflict Resolution in Complex Political Emergencies'. *International Peacekeeping* 6, 2 (1999): 24–37.

# 7 Counterinsurgency doctrine

## The US experience

*Thomas Mockaitis*

After almost half a century in the wilderness, counterinsurgency (COIN) has come back into fashion. Two protracted wars in Afghanistan and Iraq, and the prospect of more such conflicts to come, has forced militaries around the world to reconsider how to best to defeat insurgency. This vexing question has drawn attention to the need for good doctrine. Doctrine, however, does not come out of thin air. It must be created by distilling lessons from past campaigns and adapting them to contemporary circumstances. This distillation process always occurs within a context shaped by history. Studying the evolution of COIN doctrine within the military history of a particular nation may provide invaluable lessons, not only about how best to combat insurgency, but also on the important process of doctrine formation.

The United States provides an excellent case study. It has created the most awesome conventional war-fighting establishment in history, but struggles to suppress insurgencies in Iraq and Afghanistan. It also has a long history of fighting unconventional war during the late nineteenth and twentieth centuries. The military captured some of this experience in formal doctrine, but forgot most of it during World War II and the Cold War that followed. Vietnam revived interest in COIN among the Special Forces, but by then most officers believed that even unconventional threats could be defeated with a suitable application of firepower. So traumatic was the defeat in Southeast Asia that the US did not produce a new COIN manual until embroiled in Iraq. That manual has launched a wave of study, doctrine development and analysis likely to continue for the foreseeable future.

### *Insurgency and counterinsurgency*

Few military terms have been more controversial than "insurgency" and "counterinsurgency". Given this controversy, any discussion of counterinsurgency doctrine should begin with a clear definition of terms. Insurgency is an organized campaign to gain control of a country from within, through a combination of information operations, guerrilla warfare and terror. Subversion includes all efforts to persuade disaffected people that the insurgents can do a better job of meeting their needs than the existing government. Since Western nations have usually considered insurgency illegitimate, they prefer negative terms such as "subversion" or "propaganda" to describe these efforts, and use "information operations" to describe their efforts to tell the truth.[1]

Guerrilla warfare consists of using irregular fighters to attack government outposts and ambush small military or police units. Guerrillas seldom wear uniforms, they engage in hit-and-run tactics, and then blend back into the civilian population.[2]

"Terror" is an even more problematic term than insurgency. Terror is a weapon employed by a variety of actors (states, criminal organizations, insurgents and extremist groups) to influence the behavior of an audience. Because it targets those who watch even more than those who die, it has been described as a macabre form of theater. Despite their common desire to spread fear, different groups use terror in different ways. Insurgents and threatened states may both use "enforcement terror" to keep supporters or the general population in line. Insurgents will also use "agitational terror" to disrupt the government they seek to overthrow.[3] "Terrorism" should be confined to describing extremist groups who use terror indiscriminately in the service of broad, ideological, even utopian, goals.[4] The extreme negative connotations associated with the word "terrorism" encourages threatened governments to apply the label loosely to any armed group opposing them, creating considerable confusion. Insurgents make limited use of terror to achieve their political objective, but they are not terrorists in the al-Qaeda sense of the term.

"Counterinsurgency" consists of all measures a threatened state takes to defeat an insurgent movement. The government employs military and police to attack the insurgent guerrillas and disrupt or destroy the insurgent organization. However, armed force alone seldom suffices to defeat an insurgency. Ordinary people rarely support rebellion unless their living conditions become unbearable. By addressing the legitimate grievances of disaffected people, governments separate active insurgents from their supporters. Dubbed "winning hearts and minds", this strategy attacks the root causes of the insurgency. A threatened government also employs its own information operations to counter those of the insurgents.[5]

## The nature of doctrine and the art of war

"Ideas", wrote a prominent historian of imperialism, "fight a grinding battle with circumstances".[6] They shape and are shaped by events. Nowhere is this observation more poignant than in the conduct of war. Strategists study past campaigns to plan and conduct future ones. No plan, observed the great German Field Marshall Helmut von Moltke the Elder, survives contact with the enemy. Plans have to be constantly adjusted to accommodate changing circumstances. Technological change also forces soldiers to adapt. Since, as Carl von Clausewitz noted, war is an instrument of policy, politics also impacts military planning.

Beyond strategy lies the intellectual realm of military activity known as "doctrine". The US Department of Defense defines doctrine as "Fundamental principles by which the military forces or elements thereof guide their actions in support of national objectives. It is authoritative but requires judgment in application".[7] Despite the broad meaning usually given to the term "principles", military doctrine can be very task-specific. It covers the three levels of conventional war-fighting (strategic, operational and tactical), and all the support functions necessary to maintain an army in the field (logistics, intelligence, medical, etc.). Modern military doctrine runs to scores of manuals and thousands of pages. It has a long history dating back at least three centuries.

## Counterinsurgency doctrine

While counterinsurgency has a long history, counterinsurgency (COIN) doctrine began to develop in earnest only in the early twentieth century. Several factors explain this gap between practice and theory. To begin with, armies did not recognize insurgency as a distinct form of warfare. They had dealt with irregular forces since antiquity, but irregulars operating in support of regular troops are not insurgents. The Seven Years War saw the emergence of *petite guerre*, or "small war", as a form of conflict. However, the term referred to small, highly mobile flying columns employed against enemy formations on the march and their supply lines.[8] During his campaign against the Napoleon in the Iberian Peninsula, Arthur Wellesley (later the Duke of Wellington) employed Spanish irregulars to harass French supply lines, attack isolated outposts and ambush small units. These irregulars contributed a new term to the lexicon of war: "guerrilla" (Spanish for "small war").

Because guerrillas operated out of uniform and blended back into the surrounding civilian population after an operation, conventional soldiers did not recognize them as legitimate combatants. They responded to them in much the same manner that royal armies had always dealt with revolt against lawful authority – with brutal repression. They summarily executed captured guerrillas and conducted reprisals against entire communities suspected of harboring them, both as punishment for past attacks and as a deterrent to future violence.

The rapid expansion of European empires during the nineteenth century forced regular armies to devote at least some attention to unconventional conflict, especially when improved communication and growing humanitarian sentiment made scorch-and-burn tactics less acceptable. As conquered people became colonial subjects, dealing with them became a matter of balancing carrots and sticks. Arms continued to suppress revolt with stern measures that included collective punishment, but they also tried to co-opt native populations by removing, as far as possible, causes of their discontent with colonial rule.

Out of this approach grew modern counterinsurgency practice and doctrine.

Of the colonial nations facing unrest, the British developed the most comprehensive approach to countering it. They worked to address the causes of unrest in order to elicit cooperation from their subjects. Cooperation in turn led to information on the activities and whereabouts of the insurgents. Such intelligence enabled the military to use force in a selective manner, targeting rebels without alienating the general population.[9] The British army acquired extensive counterinsurgency experience in the late nineteenth and twentieth centuries. While the War Office transformed some of this experience into formal doctrine, most learning occurred in the regiment, where senior officers and NCOs passed their experience on to each succeeding generation of soldiers.

The most comprehensive study of small wars occurred outside official channels. In 1898, Charles Callwell published his epic study, *Small Wars*, a hefty volume surveying Britain's colonial conflicts and deriving relevant lessons from them.[10] During the first quarter of the twentieth century, the War Office published pamphlets on suppressing revolt: *Duties in Aid of the Civil Power* and *Notes on Imperial Policing*.[11] These publications captured the broad principles of COIN, but much of the practical wisdom from past campaigns remained in the regiments.

## The US military and counterinsurgency, 1865–1941

Contrary to popular belief, the US Army and Marine Corps have a long history of combating unconventional threats. For much of their existence, irregular warfare was their primary activity. Some of this experience occurred on the frontier, and the rest took place in the informal empire the US acquired in the Caribbean, Central America and the Philippines following the Spanish–American War.

Because the framers of the Constitution feared tyranny as much as they did foreign invasion, they preferred a small garrison army supported by state militias. In time of war, the professional officer corps could serve as a leaven to expand the regular army with volunteers and conscripts. The President also had the authority to mobilize and place under federal control the state militias. For almost a century this system served the new republic well, enabling it to fight the War of 1812 and the Mexican War, both relatively small affairs. The Civil War tested the efficacy of the US military system just as it threatened the survival of the nation itself. The President's power to mobilize the militias for a mere 90 days had to be extended, and both the North and the South resorted to a draft once volunteers no longer sufficed to replace the casualties of a modern, industrial-age war. With those adjustments, the system worked well. Following the war, the armed forces contracted to its garrison size, expanding to fight the Spanish–American War and World War I. Following each of these conflicts, the military contracted once more.

America's counterinsurgency campaigns occurred between its major wars. For three decades, cavalry units on the Western frontier pursued native tribes, suppressing renegade bands, forcing groups onto reservations and making sure they remained here. Irregular warfare on the frontier represented an on-going, sporadic resistance to oppression rather than an organized insurgency. Clearing the Great Plains represents one of the darker periods in American history, but the campaign did yield some valuable COIN lessons. The army made little effort to win hearts and minds, but it gained invaluable experience in small-unit tactics and intelligence gathering.

The Philippine insurrection following the Spanish–American War confronted the American army with a full-blown modern insurgency. From 1899–1902, Emilio Aguinaldo and a force of 80,000–100,000 irregular forces organized an effort to drive the US from the Philippines and establish a republic.[12] Following an initial futile effort to engage the Americans in conventional battle, Aguinaldo resorted to guerrilla warfare, dividing his forces into bands of 30–50 men.[13] The US Army deployed at the peak of the campaign 74,000 personnel, but probably no more than 24,000 took the field at any given time.[14] The Americans had too few troops to defeat the insurgents by simply intimidating the Philippine population into submission. They had to separate Aguinaldo's guerrillas from their base of support in the population. To drive the two apart, the occupation government built roads and schools, improved sanitation and inoculated people against smallpox. They gave pro-American Philippine politicians control of patronage jobs, strengthening their influence among the local population.[15] The soldiers did of course pursue the insurgents, and committed some atrocities in the process. Nevertheless, they defeated the insurgency with sound COIN strategy, which included a hearts-and-minds campaign, small-unit tactics and good intelligence work.

The next spate of COIN campaigns came in the aftermath of World War I. The building of the Panama Canal had given the US a strong interest in Central America and the Caribbean. With the exception of Puerto Rico, Cuba and the Virgin Islands, Washington eschewed formal control of territory. It did, however, wish to make sure that the regions' governments remained friendly. This strategy required periodic American intervention to support governments threatened by rebellion. However objectionable such intervention now seems, it nonetheless gave US troops valuable COIN experience. As the primary force deployed to the regions, the Marines benefited most from this experience, and were it not for World War II, they might have become an American imperial police force. During the interwar period, the Marines intervened in Haiti, the Dominican Republic, Mexico and Nicaragua. Of these interventions, the Nicaraguan campaign came closest to a full-scale COIN operation.

Although the Marines intervened in Nicaragua several times between 1910 and 1934, the campaign of 1927–1929 provided the best case of COIN. To combat a rebellion led by the highly competent insurgent leader Augusto Sandino, Marine Captain Merrit "Red Mike" Edson planned to gain control of the Rio Coco River linking Eastern and Western Nicaragua. His plan required planting garrisons along the river, establishing good relations with local populations and using the Rio Coco to attack Sandino's forces in the remote hinterland. Edson understood the importance of living among the indigenous population to gain their trust and garner intelligence on the population. "If we were to succeed in our mission of eradicating the bandit element in Nicaragua", Edson wrote, "we should make every effort to gain the friendliness and cooperation of the peaceful citizenry".[16] Although tactically successful, Edson's plan failed, not due to flawed strategy, but from a decision taken in Washington to withdraw the Marines. The Nicaraguan National Guard murdered Sandino in 1934, bringing his rebellion to an end.

## Early COIN doctrine

By the outbreak of World War II, the US Army had experienced almost a century of irregular warfare. Unfortunately, it preserved little of this experience in formal doctrine. Instead, the army turned "a loose body of broadly defined concepts that blended aggressive military action, punitive measures, and enlightened administration into a carrot and stick approach to the suppression of irregulars and their civil supporters".[17] The last version of its core operational doctrine manual, *FM 100–5*, published on the eve of the American entry into World War II, devoted a scant three pages to combating irregulars, most of it focused on partisans operating in support of regular armies. It referred to insurgency only indirectly as the need to "quell rebellions by semi civilized peoples".[18] The manual did, however, stress the need for small-unit operations, mobility and accurate intelligence, all sound COIN principles.

The Marine Corps did far better in capturing its unconventional warfighting experience in formal doctrine. *The Small Wars* manual was a tour de force in its day and remains a valuable resource down to the present. Published in 1940, this hefty tome runs over 400 pages of practical and theoretical wisdom on insurgency and how to combat it. First and foremost, the Marines recognized that "the application of purely military measures may not, by itself, restore peace and order because the fundamental causes of the condition of unrest may be economic, political, or social".[19] Military force must be used only in support of "peaceful corrective measures".[20] Small-wars strategy had to be adapted to the unique circumstances of each country with sensitivity to social, cultural and religious norms, and with due regard for international law.

## World War II and the early Cold War

Unfortunately, the wisdom garnered during America's early COIN campaigns was lost during the Second World War. The primary mission of the Marine Corps changed from being a small wars constabulary in the Caribbean and Latin America to an over-the-beach invasion force, island hopping through the Pacific. The small garrison army expanded into a massive conventional force. Following the defeat of the Axis powers, neither force reverted to its pre-war size. The Korean Conflict and the Cold War guaranteed that a large military establishment would remain a permanent fixture in American political life.

World War II provided US forces little opportunity to hone their COIN skills. They never faced a concerted guerrilla campaign, although they helped partisans fight the Germans. Despite Hitler's plan to unleash an army of "Werewolves" on the American invaders, very few Nazi stalwarts signed up for a fruitless effort leading to certain death. Both the Japanese and the Germans were too busy surviving after the devastation of their countries to even consider organized resistance to an overwhelming occupation force. The occupation gave American forces experience of military government and civil affairs, both of which have COIN applications, but they did not have to perform these tasks while fighting insurgents.

Insurgencies in the immediate aftermath of the war did arise in record numbers. Most aimed to end colonial rule by severely weakened European powers. The desire of the Soviet Union and China (after 1949) to support these "wars of national liberation" gave the US a stake in their outcome. The US became directly involved in only three of these conflicts during the first decade of the Cold War. American military advisors backed nationalist forces against Communist insurgents during the Greek Civil War (1945–1949) and helped the Philippine army defeat the Hukbalahap Revolt (1946–1954). US forces in South Korea faced Communist guerrillas prior to and during the invasion.

Rather than confirm the lessons of pre-war years, these interventions seemed to contradict them. Unusual circumstances in each country allowed the Americans and their allies to defeat the insurgents using conventional means, especially the application of firepower. The Greek Communists foolishly committed to conventional battle in which the regular forces easily destroyed them. The insularity of the Philippines isolated the Huk, and the Korean insurgents were so desperate for supplies that relocation of civilians combined with search-and-destroy operations almost defeated them. Following the North Korean invasion, the remaining insurgents came out in the open to support their comrades, and the US and South Korean forces destroyed them.[21]

## COIN doctrine, 1945–1960

During the first decade of the Cold War, the army produced a considerable body of doctrine on irregular warfare. The new literature incorporated the lessons of World War II (including German anti-partisan operations), the Korean Conflict, the insurgencies in Greece and the Philippines, and the Communist victory in China. In 1950 Lt. Col. Russell Volckmann wrote *FM 31–20, Operations Against Guerrilla Forces* (1950), the capstone strategic doctrine for the next decade. *FM 31–20* divided guerrillas into two groups: partisans operating in support of regular forces and guerrillas rebelling against the existing government. The manual laid out a sound strategy for COIN, though it eschewed the word and considered guerrillas generally incapable of success without support of regular forces. It also failed to exam Mao Tse-Tung's concept of "people's war" and appeared just as the Korean Conflict was beginning.[22]

*FM 31–15: Operations Against Irregular Forces* followed in 1953, which reflected the experience of Korea and an understanding of Maoist people's war as well as the other campaigns touched on in *FM 31–20.* The manual underscored the need for unity of effort, intelligence gathering and separation of guerrillas from their base of support in the population. It stressed the importance of respecting law and civil liberties. Above all, the manual stressed the importance of removing the causes of unrest that fed the insurgency. "A stronger force, such as a military unit, can destroy the irregular force, but the resistance will, when convinced that it is militarily feasible to do so, reconstitute the irregular force", *FM 31–15* warned, "unless the original causative factors are also removed or alleviated".[23] *FM 31–20* and *FM 31–15* were representative of a new body of sound COIN doctrine merging in the 1950s. This body of doctrine did, however, have one serious weakness that would haunt US forces in Southeast Asia. It presumed that the Americans would be able to make key decisions in a counter-guerrilla campaign and failed to consider what would happen if the US backed an inept government.

The spate of so many publications reflecting such a clear understanding of COIN raises a puzzling question: why was the US military so unprepared for the Vietnam War? The answer lies in a combination of doctrine, mission and institutional culture. In 1958, the army consolidated its doctrine in a way that "virtually eliminated counterguerrilla theory from U.S. Army doctrine".[24] This decision was probably caused by the Pentagon's understanding of its primary mission, which had crystallized after the Korean Conflict. From the mid-1950s until the end of the Cold War the US military would focus on two key tasks: first and foremost, the defense of Western Europe, and second, defense of South Korea, and with it Japan and Taiwan. Given the Warsaw Pact's numerical superiority in armor, artillery, aircraft and infantry, NATO relied on firepower to compensate. The US extended its strategic nuclear umbrella over its European allies to deter Soviet aggression and, if need be, stop an actual attack. They developed a similar strategy for the Korean Peninsula. The new doctrine of "massive retaliation" left little room for unconventional war. The omission seems to have troubled few officers. The experience of Greece and Korea reinforced faith in firepower and mobility to solve any military problem. This faith coupled with a conviction that counter-guerrilla action could be handled by light infantry with little additional training kept a generation from taking the challenge of COIN seriously. A young Marine officer at the Command and Staff College in 1962 captured the American attitude when he dismissed the applicability in Vietnam of Britain's long-haul, low-cost approach in Malaya: "We will work them over with so much steel that six months will see the end of it."[25]

## Vietnam and US COIN doctrine

The US took over the war in South East Asia from the French, who withdrew following their disastrous defeat at Dien Bien Phu in 1954, surrendering the North to the Communists and setting up a republic in the South. While the US considered the war an anti-Communist crusade, the Vietnamese and many outside observers viewed it as a war of independence from colonialism. Fearing invasion from the North, the US helped to reorganize and equip the Republic of Vietnam army as a large conventional fighting force based on heavy divisions.[26] However, the Vietnamese disbanded their four light divisions, the most mobile units that could be adapted for COIN operations, by 1962, even though the threat of invasion had declined.[27]

Failure of the corrupt Diem regime to counter a growing threat from Viet Cong insurgents backed by the North led the US to shift from an advisory to a direct action role. Unfortunately, the Pentagon did not restructure its forces accordingly. Military Assistance Command Vietnam (MACV) reached a peak of 543,000, but only 80,000 of them were combat troops.[28] Growing American involvement in Vietnam began as doctrinal controversy over COIN wracked the army. President John F. Kennedy vowed to aid countries threatened by Soviet-backed "wars of national liberation", and replaced "massive retaliation" with a new strategy of "flexible response", equipping the military to fight a broad range of conflicts. Special Forces played an important role in the new doctrine. Army Chief of Staff General George H. Decker accepted the need for COIN doctrine but insisted the military remain focused on NATO. "We could lose in Asia without losing everything, but to lose in Europe would be fatal", he warned.[29]

This divergence of views between the White House and the Pentagon meant that the US took on the COIN campaign but did so with a conventional approach to operations. To make matters worse, Vietnam was a hybrid war with conventional and unconventional elements. Unfortunately, MACV concentrated too much on the conventional threat. Not until late in the war did it begin to correct this imbalance. Conventional thinking also seeped into COIN doctrine. *FM 3–16, Counterguerrilla Operations* advocated "fire flush" operations in which troops would contain a 1,000-square-meter area to be pounded with artillery and aerial bombardment, driving surviving guerrillas into the concentrated fire of the infantry cordon.[30] Throughout the Vietnam War, the US Army would "move inextricably towards a more expansive use of firepower".[31]

The Marine Corps developed a somewhat better approach. Marine Major General H. Krulak challenged MACV's wasteful search-and-destroy missions through the Central Highlands and questioned its concentration on interdicting the Ho Chi Minh trail when Viet Cong seemed largely self-sustaining. He proposed pacifying the densely populated Mekong Delta and coastal plane, where 90 percent of South Vietnamese lived. Krulak's strategy called for securing villages under Viet Cong control to separate the insurgents from the rural population. Because of the Pentagon's commitment to conventional war-fighting, Krulak's ideas had little impact on operations.[32]

While they could not change overall US strategy, enlightened officers did achieve some notable results. The Strategic Hamlet program borrowed the Malayan Emergency strategy of relocating entire villages to break the link between insurgent guerrillas and their supporters. Rather than apply the program gradually, however, the Saigon government built the hamlets rapidly, locating many where they could not be defended.[33] Combined Action Platoons (CAPs) of the US Marine Corps achieved far more impressive results. Drawing from their experience in Latin America in the 1920s and 1930s, they wedded a rifle squad (12–15 men) with a 30-man South Vietnamese militia. The CAP lived in the village, building trust, gathering intelligence, aggressively patrolling at night and defending the community from attack.[34] The Phoenix Program also yielded impressive results. The Central Intelligence Agency operatives worked with the South Vietnamese to target the insurgent organization. All told, Phoenix agents killed 26,000 Viet Cong, captured 33,000 and persuaded 22,000 of the captured to work with the government against their former comrades. They considered the insurgents they turned a greater indicator of success than those they killed.[35]

Despite the effectiveness of individual programs, the US campaign lacked unity of effort. The office of Civil Operations and Rural Development (CORDS), set up under Robert Komer in 1967, sought to correct this problem. CORDS brought CAPS, Phoenix and other elements of pacification under the same umbrella. General Creighton Abrams, who replaced William Westmoreland as commander of MACV in June 1968, shifted strategic emphasis from conventional operations to pacification and its corollary, Vietnamization, transferring responsibility for security to local forces. By then, however, it was too late. The TET offensive had already soured the American public on the war.

## Post-Vietnam COIN doctrine

It would be difficult to exaggerate the degree to which the Vietnam War damaged the US military. As they recovered and transitioned to an all-volunteer force, the Army and Marines wished to avoid direct involvement in COIN. However, the Pentagon realized that unconventional threats could not be wished away, so it developed a new approach to handling them. At a 1969 press conference, President Richard Nixon articulated a new doctrine that would bear his name. While the US would continue to support threatened states, it would not do so through direct action. "We shall furnish military and economic assistance when requested in accordance with our treaty commitments", he insisted. "But we shall look to the nation directly threatened to assume the primary responsibility of providing the manpower for its defense."[36]

Military doctrine began to reflect the change. *FM 31–23: Stability and Support Operations* (1972) dictated that "assistance is predicated upon the requirement that any country whose security is threatened and who requests assistance must assume primary responsibility for providing the manpower needed for its own defense."[37] COIN began to disappear from the military lexicon. *FM 100–5: Field Operations* (1976) omitted any mention of it.[38] "Low-Intensity Conflict" became the new operational term for unconventional war. *FM 100–20: Low-Intensity Conflict* (1981) mandated that a Military Advisory Group, consisting of a Special Forces mobile training team, would provide assistance to a threatened state under a new strategic concept, "Foreign Aid for Internal Defense".[39] Until Afghanistan and Iraq, the US military assiduously avoided COIN. "Only in exceptional cases where major US interests are at high risk will US military forces be committed to combat", *Joint Publication 3–07: Doctrine for Joint Operations in Low-intensity Conflict* (1994) asserted.[40]

The new doctrine necessitated an expansion of Special Operations Forces (SOF) to implement it. The Goldwater-Nichols Defense Reorganization Act of 1986 created a Special Operations Command (SOCOM) to oversee all US SOF. By the mid-1990s, SOCOM managed 46,000 troops, including 30,000 Green Berets, Rangers, psychological warfare groups, a civil affairs battalion and DELTA Force, the top-secret counterterrorism unit. Navy Sea, Air, Land Teams (5,500) and the Air Force Air Commandos (9,500) made up the rest.[41] The Marines maintained separate Force Reconnaissance units until 2006, when Secretary of Defense Donald Rumsfeld created the Marine Special Forces Command within SOCOM.

SOCOM relieved US conventional forces from having to worry about COIN. SOF soldiers advised the Salvadoran government during its long civil war and aided the Nicaraguan Contras. The results were mixed. Advisors could encourage but not compel the host military to respect human rights, use limited force and develop a hearts-and-minds campaign. The Salvadoran government fought the insurgents to a standstill and then reached a negotiated settlement with them, and the Contras helped bring down the Sandinista government of Nicaragua. Both US-backed groups, however, committed horrible atrocities. Relegating COIN to the SOF community also had a long-term negative consequence. Relieving conventional soldiers from the need to study COIN left them woefully unprepared for the conflicts in Afghanistan and Iraq.

## COIN doctrine in immediate post-Cold War era

The end of the Cold War and the stunning victory against Iraq in 1991 seemed to vindicate the US focus on conventional war. For much of the 1990s, the American armed forces faced no serious threat. It did, however, encounter a new, more complex security environment. While it responded to crises as they arose on an ad hoc basis, the military did not adjust its thinking or its force structure to new realities. However, the Army and Marines had to adjust to at least one new reality.

The collapse of Communism created a power vacuum and civil conflict that led to failed states and humanitarian crises. These crises cried out for intervention, and an end to the deadlock in the UN Security Council meant that the US could participate in a new generation of peace missions. These missions required the use of military force to safeguard delivery of humanitarian aid. US forces entered the war-ravaged country of Somalia in December 1992 only to leave in disgust a year later. Initially safeguarding food deliveries, they ended up pursuing Somali warlord Mohammed Farah Aided. The failed operation to capture Aided left 18 Americans dead and an angry public asking why American soldiers were dying half a world away in a conflict of no interest to the US. The Pentagon remembered why it hated unconventional war. Nonetheless, the US found itself compelled to intervene in Bosnia (1995) and Kosovo (1999). These missions required at least some reconsideration of doctrine.

The US Army embraced a new doctrinal concept for unconventional conflict: Operations Other Than War (OOTWA). By defining nontraditional activities in operational terms, OOTWA provided a far better strategic framework than "low-intensity conflict", which focused on a single variable: use of force. The 1993 version of the Army's basic manual, *FM 100–5: Operations*, demonstrated a clear understanding of the new security environment. "Army forces face complex and sensitive situations in a variety of operations", the manual proclaimed. "These range from support to US, state, and local governments, disaster relief, nation assistance, and drug interdiction to peacekeeping, support for insurgencies and counterinsurgencies, noncombatant evacuation, and peace enforcement."[42] Coming out of the unhappy experience of Somalia, the military clearly had its eye on humanitarian intervention, which contemporary literature had dubbed "peace enforcement". COIN received scant attention. Regular forces still considered it a SOF function to be exercised in support of a threatened state. The missions to Somalia, Bosnia and Kosovo did provide valuable experience to a generation of soldiers and Marines who participated in them, but they had little impact on US doctrine, which remained focused on maneuver warfare.

## COIN in the aftermath of 9/11

The terrorist attacks of September 11, 2201, caught the army at the beginning of a long period of transition from land forces organized in divisions for conventional war to brigade combat teams designed to meet a range of threats. Few soldiers had any significant combat experience. Senior officers and NCOs had participated in the 1991 Gulf War, but that conflict ended in less than a week with fewer than 100 American casualties. The US still had the most formidable fighting force in history, but it was designed for a single purpose: decisive defeat of a conventional force through a combination of air, sea and land operations. However, America's armed forces had little training and even less experience of COIN.

The strengths and weaknesses of the American war machine would become obvious during the first year of the Afghan invasion and the Iraq War. CIA and SOF teams entered Afghanistan to assist the Northern Alliance of Afghan tribes to oust the Taliban and pursue al-Qaeda. Operating as forward air controllers, these teams directed American airpower in a lightning campaign that reached Kabul in weeks. Bin Laden escaped with many of his leaders, and operations to corner him in the Tora Bora caves failed. Rather than hunker down for a protracted COIN campaign, the Pentagon decreed the war nearly finished and moved on to its next objective, Iraq.

A US-led coalition invaded Iraq in the spring of 2003 with a skillful combination of surprise and firepower. The "shock and awe" air campaign that preceded the ground offensive focused on Iraqi communications and command and control, leaving most of the country's critical infrastructure intact and sparing civilians as much as possible. They reached Baghdad in a matter of weeks, but then faced a host of unanticipated problems. Assuming the Iraqi administration, particularly the police, would remain in place, the Pentagon made no plans for a protracted COIN campaign and sent in too few troops to implement a plan if they had one. Widespread looting and lawlessness wrecked what was left of an infrastructure weakened by war, sanctions and neglect. With too few troops to occupy the country and no adequate COIN doctrine, soldiers and Marines reverted to default mode, employing firepower to kill and capture bad guys. For three years the situation deteriorated as foreign terrorists entered the country to back Sunni insurgents, and Shi'a militias waged sectarian warfare. In the midst of this firestorm, officers, NCOs and ordinary soldiers adapted. The Army and Marine Corps began to incorporate the hard lessons of Iraq and Afghanistan into new COIN doctrine. Application of this doctrine and the decision to work with former insurgents (the Anbar awakening) backed by an increase of forces (the "surge") began to produce results by the fall of 2007.

## *FM 3–24*: counterinsurgency

In 2006, the Army and Marine Corps jointly produced the first COIN manual since the Vietnam War. *FM 3–24* candidly admits the failures in Iraq and presents a healthy balance of academic analysis and practical wisdom. It begins by clearly recognizing the political nature of COIN: "Political power is the central issue in insurgencies and counterinsurgencies. Each side aims to get the people to accept its governance or authority as legitimate."[43] Legitimacy in turn rests on the ability of government to provide security under the rule of law, meet the basic needs of its people, and promote economic and social development. Armed forces in COIN provide the security necessary for these other activities to occur. They must separate the insurgents from the people who support or tolerate them, build trust, garner intelligence and use that intelligence to bring focused but decisive force to bear on the insurgents. Success requires unity of effort between military units and civilian agencies, police, etc. Operations against insurgent guerrillas require considerable decentralization of decision-making and flexibility. *FM 3–24* also cautions soldiers to "Prepare for a Long-Term Commitment".[44]

The process that produced *FM 3–24* is even more encouraging than the manual itself. The military employed soldiers and academics to write it. They drew lessons from past campaigns while recognizing new challenges in the contemporary world. Most importantly, the authors do not consider the manual a definitive work. They council vigilance, adaptation and ongoing revision of existing doctrine.

## Conclusion

Armed forces reflect the societies that create them. Culture and experience shape their thinking as they do that of any other institution. American culture and history combined to create armed forces that, in the twentieth century, became overwhelmingly committed to a single type of conflict: large-scale conventional war. The experience of unconventional conflicts on the Western frontier, in the Philippines and in Latin America had a limited impact on military doctrine. However, the Army and Marine Corps forgot during World War II and the Cold War. They entered Vietnam convinced that firepower would achieve the same result it had against Germany and Japan. That long, bitter conflict led some officers to rediscover old COIN lessons, but the defeat in Southeast Asia created a strong aversion to unconventional conflict. The regular forces relegated COIN to the SOF community and forgot about it until Iraq and

Afghanistan forced them to develop COIN doctrine as part of a new approach to winning those wars. It remains to be seen whether this trend will continue or whether the military will once again turn its back on the most persistent type of warfare.

## Notes

1 *Department of Defense Dictionary of Military and Related Terms*, Washington, DC: GPO, 2008, defines insurgency as "An organized movement aimed at the overthrow of a constituted government through use of subversion and armed conflict." Available at www.dtic.mil/doctrine/jel/doddict/data/i/02719.html, accessed 27 August 2008.
2 For a more detailed discussion of insurgency, see Mockaitis, *Iraq and the Challenge of Counterinsurgency*, 16–17.
3 Thornton, "Terror as a Weapon of Political Agitation", 73.
4 Mockaitis, *The New Terrorism*, 6–10.
5 Mockaitis, *Iraq and the Challenge of Counterinsurgency*, 19–20.
6 Thornton, *The Imperial Idea and its Enemies*, 1.
7 *Department of Defense Dictionary of Military and Related Terms*, www.dtic.mil/doctrine/jel/doddict/data/d/01757.html, accessed 27 August 2008.
8 Szabo, *The Seven Years War*, 65.
9 For a discussion of the evolution of British COIN, see Mockaitis, *British Counterinsurgency*.
10 Callwell, *Small Wars*.
11 *Duties in Aid of the Civil Power*, *Notes on Imperial Policing*.
12 Deady, "Lessons from a Successful Counterinsurgency", 55.
13 Gates, "The Pacification of the Philippines".
14 Deady, "Lessons from a Successful Counterinsurgency", 55, suggests as many as 44,000 may have been in the field at one time; Boot, *Savage Wars of Peace*, 109–110, sets the figure at 24,000. The difference may hinge on what constitutes "in the field". Troops posted to static defense duties in quiet areas may contribute to the larger number.
15 Deady, "Lessons from a Successful Counterinsurgency", 59.
16 Edson, "The Coco Patrol".
17 Birtle, *U.S. Army Counterinsurgency and Contingency Operations Doctrine, 1942–1976*, 9.
18 *FM 100–5*, 238.
19 *Small Wars Manual United States Marine Corps*, I-9, 15.
20 *Small Wars Manual United States Marine Corps*, I-9, 16.
21 Birtle, *U.S. Army Counterinsurgency and Contingency Operations Doctrine*, 52, 97; Cable, *Conflict of Myths*, 38–40.
22 Details on and discussion of *FM 31–20* based on Birtle, *U.S. Army Counterinsurgency and Contingency Operations Doctrine*, 134–142.
23 *FM 31–15*, 4.
24 Birtle, *U.S. Army Counterinsurgency and Contingency Operations Doctrine*, 151.
25 Mockaitis, *British Counterinsurgency, 1919–1960*, 56.
26 Herring, *America's Longest War*, 58–59.
27 Krepinevich, *The Army and Vietnam*, 23–24.
28 Krepinevich, *The Army and Vietnam*, 197.
29 Quoted in Bittle, *U.S. Army Counterinsurgency*, 226.
30 Quoted in Bittle, *U.S. Army Counterinsurgency*, 244.
31 Krepinevich, *The Army and Vietnam*, 194–197.

32 Krepinevich, *The Army and Vietnam*, 297.
33 Thompson, *Make for the Hills*, 127–131; see also his classic study, *Defeating COIN.*
34 Boot, *Savage Wars of Peace*, 304–308.
35 Boot, *Savage Wars of Peace*, 310.
36 President Nixon's Speech on Vietnamization, 3 November 1969.
37 *FM 31–23*, 4–1.
38 Paschal, "Low-Intensity Conflict Doctrine?", 42.
39 Taw and Leicht, *The New World Order and Army Doctrine*, 58.
40 *Joint Publication 3–07*, II-5.
41 Waller, *The Commandos*, 33.
42 *FM 100–5*, 13.
43 *FM 3–24 COIN*, 1.1.
44 *FM 3–24 COIN*, 1–24.

## Bibliography


Birtle, Andrew J., *U.S. Army Counterinsurgency and Contingency Operations Doctrine, 1942–1976.* Washington, DC: GPO, 2006.

Boot, Max, *Savage Wars of Peace: Small Wars and the Rise of American Power.* New York: Basic Books, 2002.

Cable, Larry, *Conflict of Myths: The Development of American Counterinsurgency Doctrine and the Vietnam War.* New York: New York University Press, 1986.

Callwell, Charles, *Small Wars: Their Principles and Practice.* Lincoln: University of Nebraska Press, 1996.

Deady, Timothy K., "Lessons from a Successful Counterinsurgency: The Philippines, 1899–902", *Parameters* (Spring 2005), www.carlisle.army.mil/usawc/Parameters/05spring/deady.pdf, accessed 31 August 2008.

*Department of Defense Dictionary of Military and Related Terms.* Washington, DC: GPO, 2008, www.dtic.mil/doctrine/jel/doddict/data/i/02719.html, accessed 27 August 2008.

*Duties in Aid of the Civil Power.* London: His Majesty's Stationery Office, 1923, 1927.

Edson, Merritt, "The Coco Patrol", *Marine Corps Gazette*, August 1936, http://pqasb.pqarchiver.com, accessed 14 May 2007.

*FM 3–24 COIN.* Washington, DC: Headquarters, Department of the Army, 2006.

*FM 31–15: Operations Against Irregular Forces.* Washington, DC: GPO, 1953.

*FM 31–23: Stability and Support Operations.* Washington, DC: GPO, 1972.

*FM 100–5: Field Regulations.* Washington, DC: GPO, 1941, http://cgsc.cdmhost.com/cgi-bin/showfile.exe?CISOROOT=/p4013coll9&CISOPTR=23, accessed September 2, 2008.

*FM 100–5: Operations.* Washington, DC, 1993.

Gates, John M., "The Pacification of the Philippines", in Joe E. Dixon ed., *The American Military in the Far East: Proceedings of the 9th Military History Symposium, USAF Academy.* Washington, DC: GPO 1982, www.wooster.edu/history/jgates/book-ch3.html, accessed 31 August 2008.

Herring, George, *America's Longest War: The United States and Vietnam, 1950–1975.* New York: Alfred Knopf, 1986.

*Joint Publication 3–07: Doctrine for Joint Operations in Low-intensity Conflict.* Washington, DC: Joint Staff, 1994.

Krepinevich, Andrew, *The Army and Vietnam.* Baltimore: Johns Hopkins University Press, 1986.

Mockaitis, Thomas R., *British Counterinsurgency: 1919–1960.* Basingstoke: Macmillan, 1990.

Mockaitis, Thomas R., *The New Terrorism: Myths and Reality.* Westport: Praeger, 2007.

Mockaitis, Thomas R., *Iraq and the Challenge of Counterinsurgency.* Westport: Praeger, 2008.

Nixon's Speech on Vietnamization, 3 November 1969, http://vietnam.vassar.edu/doc14.html, accessed May 25, 2007.

*Notes on Imperial Policing.* London: HMSO, 1934.

Paschal, Rod, "Low-Intensity Conflict Doctrine: Who Needs It?", *Parameters,* 15, 3 (Autumn 1985).

*Small Wars Manual United States Marine Corps.* Washington, DC: GPO, 1940.

Szabo, Franz A.J., *The Seven Years War: 1756–1763.* New York: Longman, 2007.

Taw, Jennifer and Robert C. Leicht, *The New World Order and Army Doctrine: The Doctrinal Renaissance of Operations Short of War?* Santa Monica: RAND Corporation, 1992.

Thompson, Robert, *Defeating COIN: Lessons from Malaya and Vietnam.* London: Chatto and Windus, 1966.

Thompson, Robert, *Make for the Hills.* London: Leo Cooper, 1989.

Thornton, Archibald Pater, *The Imperial Idea and its Enemies: A Study in British Power.* New York: St. Martin's Press, 1959.

Thornton, Thomas P., "Terror as a Weapon of Political Agitation", in Harry Eckstein ed., *International War: Problems and Approaches.* Westport: Greenwood, 1964.

Waller, Douglas, *The Commandos: The Making of America's Secret Soldiers From Training to Desert Storm.* New York: Simon and Schuster, 1994.

# 8 No simple formula

## The use of military force in counterinsurgency

*James S. Corum*

Since 2001 the US military has gone through a painful process of relearning the art of counterinsurgency. Fighting non-state forces – whether they are insurgents, terrorists or criminals – is to engage in a fundamentally different type of war from the state-on-state conventional war that the US military has long been trained for. Success in counterinsurgency, or any conflict against irregular enemies, requires that one masters an environment that is far more complex than conventional war, and requires the employment of a wide assortment of organizations, tools and methods.

In a 2008 speech to the Army's cadets at West Point, American Secretary of Defense Robert Gates noted the complex leadership challenges facing the US Army officers of the future in fighting non-state, or "irregular", conflicts:

> We will see yet more irregular and difficult conflicts, of varying types, in the years ahead; conflicts where the traditional duties of an officer are accompanied by real dilemmas – dilemmas posed by a non-linear environment made up of civilian detainees, contractors, embedded media and an adversary that does not wear uniforms or obey the laws of war; an adversary that could be your enemy on one day or, as we've seen in Iraq's Anbar province, your partner the next.[1]

The aim of this chapter is to identify and discuss some of the implications for employing force in counterinsurgency based on an analysis centered on the special conditions of such a conflict. One of the most complex questions in conducting counterinsurgency, certainly the most common form of irregular warfare today, is the degree of force that the counterinsurgent should employ to attain the strategic goal. In conventional war – that is, a conflict that pits a conventional state with conventional armed forces against another state with a similar force – there is little confusion for the military commander about the level of force to use. Any sensible commander will use all the military instruments readily available and try to bring overwhelming force against the enemy and strike at the enemy's most vital and sensitive points. The use of overwhelming force maximizes enemy casualties while minimizing one's own, and also has an enormous effect on the enemy morale. A well-conducted campaign employing overwhelming force and firepower will often compel the enemy to quickly surrender or seek terms rather than throw away his forces in a futile struggle. The doctrines oriented toward employing overwhelming force ratios and superior firepower work very well in conventional war because conventional armed forces have centers of gravity that can be readily targeted and destroyed by superior force and firepower. Although there have been cases in conventional war when victory went to the nation with inferior forces and less firepower, the sound money usually bets on the big battalions and heavy firepower.

Yet superior forces and overwhelming firepower do not provide much assurance of victory when one is fighting non-state irregular forces waging an insurgent war. Many insurgent groups have started out with weak forces and little firepower and faced overwhelming government troop numbers and firepower, yet they have still gone on to win the final victory. In many cases in counterinsurgency warfare, the insurgents possessed very minimal military capabilities right up to the end of the conflict – but they won anyway. One can look at many cases of great nations equipped with the most modern weapons defeated by small numbers of poorly armed rebels. The Greek Cypriot rebels of the 1950s never had more than 400 active fighters in the field, and they faced British military and security forces of 40,000. Yet the Greek Cypriots forced Britain to grant independence. Fidel Castro had only a handful of fighters and a few weapons opposing the overwhelming military force of the Cuban government forces of President Fulgencio Batista. Yet, after one small battle, the government forces collapsed and Castro seized power. In October 1993, irregular Somali clansmen engaged a US force in Mogadishu and killed 18 and wounded another 80 Americans. The Americans could have brought overwhelming military force to bear, but chose not to and withdrew from Somalia instead.

If military force does not give one a decisive military advantage in fighting insurgents, irregulars and other non-state forces, it still confers some advantages. The fundamental cause of the breakdown of order in Iraq after the fall of Saddam Hussein, and the inability of the Coalition forces to restore order and provide stability to the Iraqi population in 2003, was the failure of American and Coalition forces to have enough troops on the ground in Iraq at the end of the conventional campaign that toppled Saddam Hussein's regime with minimum battle casualties. The principle of personnel on the ground was understood by some. In late 2002, US Army planners planned to have at least 250,000 US troops on the ground in Iraq by the time Saddam Hussein's regime collapsed. The US Army planners did not request the 250,000-man force to defeat Saddam Hussein, as US intelligence accurately predicted that few Iraqis would fight for the hated dictator. Instead, the large ground force was to be deployed to establish order and control in the country. Creating a secure environment for the Iraqis would be necessary to a new political order and restore the economy. Indeed, the decision by the Bush administration to reject their own military planners and to send inadequate forces into Iraq in 2003 allowed the post-war chaos to flourish and enabled the swift growth of many armed factions and insurgent movements that plunged Iraq into a frenzy of violence between 2003 and 2009.

Even worse, only a small US force was committed to securing the city of Baghdad, a vast city of more than five million people, and there was no plan at all for the US forces to secure the city and reestablish order.[2] Indeed, even a large force would have been ineffective because of the lack of a plan to secure the city. One can rightly blame the civilian politicians for overriding the military's request to commit adequate forces to Iraq, but senior military officers failed in their fundamental duty to plan for the post-battle phase of operations. Rarely in US history has there been such a comprehensive failure at the strategic level by both civilian and military leaders.

## Strategic and doctrinal views on using force

The fundamental question of how much military force to employ in counterinsurgency, and the development of policies to guide the employment of force when a conflict is fought in the midst of a civilian population, provides modern military planners with some major dilemmas. The first issue at hand is to understand the fundamentally different role that force plays in developing a strategy to address different conflicts.

Strategy is defined as the allocation of military, political, economic and other resources to attain a political goal. The strategic level of war refers to applying these national or coalition resources in a coordinated manner to meet national or theater objectives.[3] In a conventional conflict, the military normally has the paramount role in strategy. In counterinsurgency, this is often not the case. A counterinsurgency strategy that relies overwhelmingly upon military force and military operations, and largely ignores the social, political and economic aspects of the insurgency, will not get one to the desired endstate – or even close to it. Indeed, while fighting an insurgency, the political, informational and economic aspects of the strategy are just as important as the military contribution.

Secretary Gates alluded to political complications of irregular warfare and counterinsurgency in his West Point address when he noted that, in Iraq, the people the Americans fight today might become allies tomorrow. Such role changing and transfer of allegiances is exceptionally rare in conventional, state-on-state warfare, but it is a very common phenomenon of counterinsurgency. The fact that whole groups and factions can be induced to change allegiance is one of the major factors that governs the use of force.

As an attempt to address some of these questions, the US Army and Marine Corps in December 2006 published the first truly comprehensive counterinsurgency doctrine since the Vietnam War. With insurgencies raging in Iraq since 2003, and Afghanistan since 2001, and little apparent progress being made, both the US Army and Marines decided that it was time to revisit the basic assumptions of counterinsurgency in light of recent experience. The result was a document of over 200 pages that reads more like a book than a manual. The doctrine begins with a theory of counterinsurgency and then covers the major strategic and operational issues of counterinsurgency to include campaign planning, intelligence, training host nation forces, logistics, social environment and the use of airpower. One of the characteristics of the doctrine is the heavy use of historical studies and precedents. The doctrine looks at various approaches to counterinsurgency and gives as many examples of unsuccessful operations as successful ones. Two of the themes that are constantly repeated are how military force is used in counterinsurgency, and the degree of force that is appropriate.

At the strategic level, which is the level of much of the Army and Marine doctrine, one lesson that is emphasized throughout is that the solution may not be a military one.[4] A military approach may kill a lot of the insurgents – but, unlike conventional war and its focus on fielded forces, killing insurgents is not a sure path to success in counterinsurgency. Chapter 1 of *FM 3–24* lays out a doctrine with a set of principles beginning with the need for the counterinsurgency strategy to adequately address the social, political and economic issues in order to reach the political solution – which is usually the way that a successful counterinsurgency campaigns are concluded.

The emphasis on the non-military factors of counterinsurgency strategy means that the military is often a supporting force, and not the main effort. This goes very much against the military culture of the US and most Western nations. This is not to say that the military effort is unimportant, but it does mean that we have to consider military force within a very broad and complex political context.

The complexity of counterinsurgency strategy has to take into account that insurgency is local, and often driven by local, rather than national, conditions. This means that, in addition to a national counterinsurgency strategy, one will likely have to craft comprehensive local strategies. One can take the case of Bangladesh as a current example of a nation combating local insurgencies. Today there are three small insurgencies in three different regions of Bangladesh. In organizing to combat the insurgencies, the Bangladesh government employs essentially the Malaya committee model, in which the provincial governor chairs the committee that includes the senior military, intelligence and police officers as well as the civilian directors of the social services, economic development and government information service. With the civilian governor in charge and the military serving in the role as support to the civilian authorities, each province crafts its own local strategy to combat the insurgents and provide security to the local population. The strategy in each province and district – especially the policies about employing force – are likely to be very different and dependent upon the local conditions. One Bangladeshi officer put it too me, "In one province the strategy is to take a hard line against the insurgents, in another province they have taken a soft line and the third province is somewhere in the middle."

In fact, this is not much different from the tone of the US Army and Marine doctrine and the actual practice of conducting counterinsurgency campaigns in Iraq and Afghanistan. In those countries the Coalition forces have found that both national and local strategies are necessary. In addition, each strategy has to be comprehensive and address the social, economic and political dimensions of the conflict as well as the military side. One of the most constant refrains from those who have served in Iraq is just how different the conditions of the conflict were from province to province. The Iraqi case is different from the insurgencies of the Cold War era that usually featured a strong, centrally commanded, single insurgent movement. In Iraq and Afghanistan, Coalition and government forces end up fighting a variety of factions, groups and tribes. Moreover, factions and tribes will change allegiances and might fight as your enemy one month and then become your ally, and vice-versa.

## Human intelligence in guiding operations

The role of intelligence in counterinsurgency is fundamentally different from intelligence in conventional war. Conventional military intelligence is about looking for things you can see and count. Thanks to modern technology, with its signals intelligence, and ability to monitor the battlefield by space and aerial surveillance, the primary mission of intelligence in conventional war – locating the enemy's main conventional forces – is relatively easy. The high-tech intelligence assets are featured in conventional war operations: the space, reconnaissance and signal assets. In counterinsurgency, the first mission of the intelligence agencies is to try to understand the context of the conflict. This means collecting information about the whole society, understanding local conditions, monitoring public opinion, analyzing social and political relationships and networks. And that is just the start. The next step is to find the insurgents and try to understand their organization. This is difficult because the insurgents are likely to wear civilian clothes and hide among the population. The insurgents will have a local, perhaps national, organization – and it is all underground. If you are lucky, the insurgents will stand and fight, and give the counterinsurgent the chance to use military force and airpower against them. But even if the insurgent forces are decimated, if the underground support network is not broken, the insurgent combatant forces will quickly revive.

The kind of intelligence one needs to understand the insurgent social context and to understand the insurgent organization is human intelligence (HUMINT). Again, this goes against the US and Western military bias of a high-tech approach to war. Of course, high-tech intelligence also has its place. Space surveillance and other reconnaissance assets can give one great data. For example, high-tech surveillance might tell one that the people are all leaving a particular village. But it does not say *why* they are leaving. You need really competent analysts to answer questions like that.

To wage effective counterinsurgency, one needs personnel who understand the local culture and traditions, personnel who can help the commander predict and assess the impact of counterinsurgent actions. At times holding fire and pulling forces back might send the right political signal to the insurgent and pave the way for a peaceful settlement. At other times such an action might be interpreted as weakness and encourage the insurgents to increase their acts of violence. As in the Bangladesh examples, the degree of force used in an area ought to be dictated by local political and cultural conditions. Sometimes employing very large forces and a lot of firepower might be exactly the right response to set the conditions for strategic success, as also argued by Kersti Larsdotter in this volume.

The kind of intelligence analysis needed in counterinsurgency is essentially a foreign area officer, someone who speaks the language fluently, has studied the country and the region in depth, and who understands the societal context of official and unofficial networks. In fighting insurgents, a mature, competent specialist intelligence officer is far more useful than a B-2 bomber. The good news is that a human intelligence specialist is a lot cheaper than a B-2 bomber. The bad news is that it takes about as long to develop a competent country and regional expert as it does to develop the bomber.

One of the primary problems that our forces in Iraq and Afghanistan have faced is the lack of fully qualified human intelligence specialists. Unlike logistics, which requires some general skills, you cannot easily contract out for such people when you need them. From 2003 to 2006, many of the American unit commanders in Iraq often commented that their greatest operational limitation was the lack of solid intelligence. Only human intelligence could provide the vital information needed by the commanders to seek out and destroy the terrorists. Information such as how the local insurgent cells were organized, the personalities of the insurgent leaders, the networks developed between different insurgent groups, the location of explosive and arms caches and details of the financial support system of the insurgents could only be obtained by human intelligence methods such as interviews, interrogations and document translation. Even when good data were obtained, one needed a highly skilled human intelligence specialist to sift through the mass of data and put it all together into intelligence that would be operationally useful to the commander.[5] Having a vast array of firepower at one's disposal is of little use if no one knows how to find a hidden enemy. As one officer in Iraq said, "We have plenty of trigger pullers, what I need are more intelligence people so I can find the enemy. If I can find him, I can eliminate him."

Democratic nations must rely on intelligence and try to separate the insurgents from the mass of the civilian population as much as possible. Democratic nations must adhere to the norms of international law and behavior because otherwise public and political support for the armed forces will decline – and democratic nations require public support to wage war. However, limiting collateral damage to the population is not only a moral imperative, it is good strategy. Keeping civilian casualties low reduces the natural resentment that civilians might have against any and all combatant forces that disrupt their lives.

The lesson is that if a nation is going to have adequate human intelligence support in a conflict, it needs to build up its human intelligence capabilities long in advance. Unfortunately, at the end of the Cold War the US largely gutted the human intelligence specialist force. Although the Americans are making progress in rebuilding the human intelligence system, the lack of adequately trained human intelligence specialists remains one of the greatest factors to limit the American employment of force.

## Operational leadership in counterinsurgency

The quality of leadership has always been one of the primary factors that determine success or failure in war. In fact, in irregular warfare, leadership becomes an even more critical factor than in conventional war. In conventional war, where the focus is to find and destroy the enemy's armed forces, senior officers do not have to grapple with the thorny question of how much force to use to the same extent.

Some of the requirements of effective leadership in counterinsurgency are the same as in conventional war. For example, an officer must still be highly skilled at coordinating military operations of several arms. However, there are several requirements of leadership needed in counterinsurgency that are not common to successful leadership in conventional operations. The radically different context of counterinsurgency requires a high degree of mental flexibility, not necessarily a common trait of military leaders. An officer can be a great soldier in a conventional war, and fail badly in counterinsurgency. For a senior officer in counterinsurgency, the specific requirement is political "situational awareness" – a thorough understanding of the political context of the insurgency. A senior leader, and often a junior leader as well, needs to understand the various combatant factions operating in a region. A leader needs to understand their motivations, organization and their goals, as well as their military potential. A senior leader in counterinsurgency needs to recognize the political dimension of policies and carefully employ the resources available to further the political objectives. In short, the effective leader always links the use of force to a clear political objective to be gained. Violence is therefore inherently strategic at all levels of war.

A good example of an effective senior leader in counterinsurgency is General Sir Gerald Templer, who served as military commander and Governor General of Malaya at the height of the insurgency, from early 1952 to 1955. Templer inherited a bad situation. When he took command, the insurgency had been going on for four years and was steadily increasing in the level of violence. The British expanded the Malay Police from 10,000 personnel to almost 50,000. The British military garrison in Malaya was expanded from 10,000 personnel to almost 40,000. The RAF in the country was dramatically increased. The British applied massive force against the rebels, with Lincoln heavy bombers blasting away large areas of jungle in an attempt to destroy rebel hideouts, while the army carried out artillery-supported brigade sweeps of Malay districts. The level of violence on both sides rapidly increased. Although the British and Malay forces had killed many of the insurgents, insurgent strength still steadily increased and rebel organization likewise continued to grow. The massive British force applied against the insurgents between 1948 and 1951 failed to impede the advance of insurgent strength.

However, the situation turned around with the arrival of General Templer as Governor General and military commander in early 1952. Templer brought a new leadership team and a new strategy to Malaya with a new approach for employing force against the rebels. Templer's leadership over three years was a key element in turning the situation around, and his approach to counterinsurgency is worthy of study.

Gerald Templer was a highly intelligent officer, but no genius. He had an outstanding record as a conventional soldier, having commanded a division and later a corps during World War II. But his success as leader in Malaya was less a function of his conventional war skills than an intrinsic talent to study a situation and grasp its essentials. Templer's knowledge of Malay language and culture was minimal. However, in the weeks before he went out to Malaya, he corresponded and talked with several senior officials and other experts on Malaya. Even before he arrived, he had a good understanding that the solution to the insurgency was going to be as much political as it was military.

The insurgents in Malaya were overwhelmingly ethnic Chinese, who were actually a plurality of the population of the Malayan states, but excluded from government power by the ethnic Malays. Templer understood that one of the core causes of the insurgency was this systematic exclusion of the Chinese from the government, civil service and security forces. Templer immediately initiated a series of reforms to help close the deep social and political divide between the Malays and the ethnic Chinese. Against considerable resistance from the Malay state governments, and many in the British military and civil service, Templer pushed the Federation government to admit Chinese to the civil service and into the security forces with the goal of making the Malayan army and police forces representative of all of Malaya's groups. Since Chinese were excluded from the Malaya Regiment, he created new Malayan regiments that would be multi-ethnic and he personally recruited officers from among the Chinese community. Sir Arthur Young, who came out with Templer to take over and reform the Malayan Police, worked with him to weed out the most corrupt elements from the police force and to recruit as many Chinese as possible into the force. Rather than worry about increasing the forces at his disposal, Templer placed the priority on using his forces more efficiently. A smaller but better-trained and better-led Malay Police produced much better intelligence than before. Better intelligence enabled the government to focus their operations more carefully and with better effect. The massive army/police sweep operations supported by heavy firepower were out. Instead, smaller operations guided by improved intelligence, focused on eliminating specific insurgent units and clearing small districts, one by one, became the modus operandi. As a result, rebel forces began to decline and the security for the majority of the population was quickly improved.

Along with a more carefully focused approach to military and police operations, Templer carried out a clever political campaign. The combination of force, backed up by the offer of political representation and economic opportunity to groups that had been shut out of the Malayan political system, proved to be a highly effective strategy. Neither the political approach nor the employment of force would have worked alone. Together, the strategy succeeded. Some of the most prominent Malay political leaders had also concluded that an independent Malaya required a partnership between the Malays and ethnic Chinese and began to cooperate with Templer in promoting reforms. While the literature about the Malayan insurgency concentrates overwhelmingly on the military side of the insurgency, it was arguably the political reforms, coupled with the improvement in the leadership and training of the Malaya security forces, that did more to defeat the insurgency than overt military operations. Templer's role as commander and governor general, and his focus on achieving realistic political objectives, were keys to ending the insurgency and setting Malaya up as a stable independent nation.[6]

A prime example of poor situational awareness, as well as a bad strategy for using military force, is a contemporary of Templer's, Field Marshal Sir John Harding. Like Templer, Harding had a brilliant record as a conventional soldier. He had served as a division and corps commander in World War II and was serving as Chief of the Imperial General Staff in 1955 when the insurgency in Cyprus broke out. Although he was preparing to retire, he accepted Anthony Eden's plea to take over as commander and Governor General of Cyprus. However, unlike Templer, Harding had no talent at all for understanding the political context of the insurgency and the use of force against insurgents.

The motivation of the insurgency was simple: 80 percent of the Cypriots were ethnic Greek and wanted an end to colonial status, and were seeking union with mainland Greece. British policy was that Cyprus would remain a colony. Harding saw his job as a simple one: crush the insurgency with overwhelming force. Cyprus had only a small police force and military garrison at the start of the insurgency in 1955. In short order Harding brought in massive reinforcements and rapidly increased the size of the police force. The island was saturated with troops. Indeed, with 40,000 soldiers and police at his disposal by 1956, Harding had the largest force-to-population ratio in the history of counterinsurgency – 40,000 military/police personnel versus a total Greek Cypriot population of 400,000. Yet, despite such overwhelming force at his disposal, the British failed to break the Greek Cypriot insurgent organization or inhibit the Cypriots' ability to increase the level of violence. The British weakness was not the lack of forces but the lack of an effective intelligence system. With little intelligence or means to locate the Greek Cypriot insurgents, the massive British forces were employed on large operations that yielded no tangible results.

Field Marshal Harding made a series of bad decisions based on a terrifically bad understanding of Cypriot politics and conditions. Each bad decision made the insurgency and violence on the island worse. First, Harding was convinced that the Cypriots really did not want independence or *enosis*, but had been pushed to violence by Archbishop Makarios, the political and spiritual leader of the Greek Cypriots. Harding believed that if Makarios were removed, then he could sit down and discuss some form of minimal home rule with the moderate Cypriots, who he believed were in the majority. So he had Makarios arrested and sent to exile in the Seychelles Islands.[7]

Harding's assessment was incredibly wrong. Makarios, in fact, was one of the more moderate Greeks, and had insisted that the insurgent military force limit its attacks on the British. With Makarios gone, more-radical leaders were left in charge and the violence escalated dramatically. Harding's next step was to arm a large force of ethnic Turkish Cypriots who, with minimal training and little in the way of leadership, were unleashed on the Greek community. As was characteristic of most of his decisions, Harding did this against the advice of Britain's Cyprus experts. What they had most feared – open communal warfare between the island's two ethnic groups – came to pass. Harding's strong-arm counterinsurgency policies also caused a wave of blatant human-rights violations that were documented by the international press, which was present in force in Cyprus. Harding publicly denounced all allegations against his troops as Greek propaganda, even while the journalists reported very public collective indiscipline of British troops and the routine abuse of the civilian population.

If there had been any Greek moderates on Cyprus before Harding arrived, his actions pushed them into supporting the insurgents simply as an act of self-defense. Even with overwhelming forces at his command, Harding failed to bring the island under control and failed to capture the leader of the Cypriot insurgents. Harding left the island in 1957, believing that the tide had turned and that outright victory was around the corner. The next year the British government agreed to grant Cyprus independence, and also granted full amnesty for the insurgents.

The contrast between the two soldiers faced with an insurgency illustrates some basic principles of leadership and the employment of force within a counterinsurgency strategy. One soldier, Templer, adjusted to the conditions of counterinsurgency and always kept the political issues in mind. Under his administration in Malaya, the political ramifications of every decision were carefully considered. He listened and took advice from some top experts and always linked the use of force to the achievement of some political objective. Harding saw the Cyprus insurgency as a military operation with a military solution – and that was his downfall. He had a special talent for ignoring good advice and his clumsy strategy of overwhelming the insurgents by force actually increased the violence against the British as well as the amount of public support for the insurgents. His strong-arm tactics put a bad face on British imperialism and left behind a legacy of communal violence that persists today.[8]

## Modern dilemmas – the media's negative portrayal of the counterinsurgent force

One of the most common critiques made by officers involved in counterinsurgency operations around the world is that the counterinsurgent forces are doing very poorly in employing the media to convey government message to the population – while the insurgent, terrorist and radical groups are using the media very effectively.[9] For one thing, insurgents, radical groups and the states that support them are not hindered by any requirement to stick to the truth. Disinformation campaigns and deliberate falsifications are standard methods of attacking the legitimacy of counterinsurgency operations and of whipping up local and world opinion against the US and Coalition allies.

The effective use of the mass media by revolutionaries and insurgents is nothing new. Richard McKenna's insightful 1962 novel *The Sand Pebbles* depicts the confusion and frustration of US military officers on a gunboat in Chinese waters as they are confronted with the mass propaganda campaign of the Chinese revolutionaries in the 1920s. McKenna had been a US Navy "China sailor" in that era, and had lived many of the events described in his novel. His description of the revolutionaries' use of newspapers and pamphlets rings true for any US officer confronting irregular warfare today. In the novel, Lieutenant Collins USN, Captain of the *USS*

*San Pablo,* a gunboat stationed on the Yangtze River in 1926, tells his crew about new orders restricting their operations:

> For instance, our little sortie against the river pirates last month, they fired first, and we killed only one pirate. But the consul has a clipping from a local newspaper stating that we killed 30 unarmed people, including women and children. They have suddenly begun making fantastic charges against gunboats on the main river. That is why we have new orders not to fire back blindly against ambushers, because if we do the students will make a big lie of each occasion. We are up against lying as a matter of planned strategy.... We are fighting lies now, not armed men.[10]

Insurgents and non-state forces confronting the regular military forces, especially of Western states, will commonly focus their propaganda efforts against the technological advantage of the counterinsurgent forces. America and Western nations are portrayed as using their asymmetric technological advantage as a means to bully and repress the civilian population. In China in the 1920s, the gunboat was the symbol of Western technology and oppression. Today, airpower is the force that is especially singled out as the primary symbol of repression. It is easy to make fantastic charges against air forces, and accuse them of deliberately bombing civilians, because the insurgent still controls the ground at the end of the day. This means that the insurgent also controls the story – and accusations of brutality through airpower make great material for sensational news stories. Insurgents and non-state forces are also assisted by a news media, very often the Western media, because they will print the insurgent and radical casualty claims without disclaimer or comment, often repeating ludicrously high figures of civilian casualties and damage to civilian homes. Indeed, insurgents and non-state groups get so much propaganda value from civilian casualties that they readily use the civilian population as human shields. The tactic of placing heavy weapons in highly populated areas in the hope that the air force will attack them and inflict collateral damage has become a common insurgent strategy.

A case that aptly illustrates the counterinsurgent's dilemmas and difficulties to manage the image of the conflict is Israel. During Israel's 1982 invasion of Lebanon, the Palestine Liberation Organization (PLO) placed artillery pieces and anti-aircraft guns in civilian neighborhoods, on the roofs of apartment buildings and even on hospital grounds.[11] They hoped to provoke the Israelis to attack targets with the assurance of heavy civilian casualties. If the Israelis refrained from attacking, the PLO preserved its forces and equipment; if Israel attacked, the resulting dead civilians could be displayed to the world as victims of Israeli aggression. For the PLO, it was a win/win situation. Using civilians as human shields is a gross violation of international law. The same problem was apparent again in the Lebanon campaign in 2006. The case of Israel is not unique in terms of non-state forces using the media to their advantage.

Insurgents also used this win/win media strategy in Iraq. In Fallujah in 2004, insurgents placed munitions and weapons in 20 mosques, and used mosques as fighting positions. Of course, targeting a mosque that is being used as a military installation is a perfectly acceptable act under the laws of war. Although the US employed precision weapons, and tried to keep damage to mosques to a minimum, there was just enough damage in Fallujah to ensure that insurgents could portray the conflict as Americans attacking the Islamic faith – a theme that resonated throughout the Arab nations and helps further radicalized Islamic opinion.

Because the use of military forces, and especially air attack, is automatically viewed in the Third World as cruel and heavy-handed, it creates a paradox for policy-makers. While airpower might be the most effective means to strike at insurgents and terrorists, its use will provoke outcry in many quarters of Western society and throughout the Third World. In short, there is a heavy political price to pay when military force, and especially airpower in the form of air-strikes, is employed against terrorist groups or insurgent factions.

The US and other Western nations must do much better in presenting their side of the conflict to the world media. Western nations need to be ready to counter a large-scale disinformation campaign mounted by insurgent and radical groups against their military operations. A senior commander today, operating against irregular enemies, needs a highly trained specialist cadre who can handle media and information operations.[12] This shows that it is necessary to anticipate the opponent's media campaign and to proactively develop responses that use themes, words and images that appeal not only to one's own public, usually the audience for government media campaigns, but also to the people of the region. When the supporters of insurgents and terrorists employ lies and disinformation, or when the Western media uncritically repeats such disinformation, the Western nations need to quickly and effectively counter such propaganda. Counterinsurgency is still about winning hearts and minds, and effective media operations are one of the main weapons we have. As the US Army and Marine Corps counterinsurgency doctrine argues: "Some of the best weapons for the counterinsurgent do not shoot."[13]

In the winter of 2008–2009, the Hamas faction in the Gaza Strip adjoining Israel began a campaign of violence when they fired hundreds of rockets into nearby Israeli settlements. Provoked by the attack on its civilians, Israel responded and, in December 2008, began a ground invasion of the Gaza Strip with the intention of inflicting maximum damage upon the Hamas fighters. Israeli intelligence was apparently very good, and the use of ground forces was very targeted. In four weeks of ground operations carried out before a ceasefire was signed, only ten Israeli soldiers were killed. In contrast, hundreds of Hamas fighters were killed. Israelis successfully used air-strikes to kill some of Hamas's top leaders.[14]

One policy that Western nations might adopt is to directly confront the insurgent use of human shields as a means of inhibiting military action against insurgent fighters as they try to win a propaganda advantage by increasing the likelihood of civilian casualties. A step to deter this practice would be to aggressively prosecute leaders of radical and insurgent groups as war criminals whenever they use civilians as human shields. The precedent of the Nuremberg Trials is clear – leaders can be held responsible for the systematic policy of war crimes committed by their subordinates. This is becoming common practice in many conflicts. Hamas and Hezbollah have routinely conducted operations to ensure that civilians were endangered by Israeli operations. Indicting and prosecuting insurgent leaders for such acts that endanger civilians will serve to change the legal and political dynamics in which insurgency and terrorism are viewed.

## Positive and negative lessons from the American experience

The United States has considerable experience in counterinsurgency, and this experience forms the basis for some important lessons for developing an effective counterinsurgency strategy and doctrine. Two of the most important conflicts for establishing current US doctrine for counterinsurgency are the Vietnam War (US involvement from 1961–1973) and the civil war in El Salvador (1981–1992).

The American military experience in the Vietnam War provides some very useful lessons in how not to apply force against irregular forces. When the US committed conventional forces to South Vietnam in 1965, it was still largely a war against the Viet Cong insurgents. The nature of the conflict soon changed to a conventional war waged against the North Vietnamese Army. Essentially, two wars ran in parallel: a counterinsurgency war fought against the Viet Cong and a conventional war fought against the North Vietnamese regular army. Much of the war, especially the war against the Viet Cong, was fought in heavily populated areas of South Vietnam, and the civilians were constantly in the middle.

The US military found two means to fight the enemy in South Vietnam. One means was developed by the Marine Corps and consisted of Civic Action Platoons (CAPs), small detachments of Marines who lived in the South Vietnamese villages and organized and trained local South Vietnamese self-defense forces. The Civic Action Platoons were employed in the northern part of South Vietnam. The CAP program relied on a maximum of human interaction and a minimum of heavy firepower, and was highly successful in meeting its objective in keeping hundreds of villages secure from the Viet Cong while using only a few thousand Marines.

However, the highly successful CAP program was an anomaly in terms of the American approach to the Vietnam War. General William

Westmoreland, the US commander in South Vietnam from 1964 to 1969, had little use for programs such as the Marine's CAP platoons and viewed the war solely in conventional terms. For Westmoreland, victory was to be achieved through firepower. Following Westmoreland's conventional war strategy, a strategy approved by the US Joint Chiefs and political leaders, the US military expended vast amounts of ordnance in South Vietnam. Per Westmoreland's directives, mid-level and junior leaders had the leeway to declare whole areas "free fire zones" in which any person could be immediately engaged as the enemy and in which there were few restrictions on the amount of firepower that could be used. The heavy reliance upon firepower by the American forces in Vietnam resulted in many thousands of civilian casualties.[15] Westmoreland's conventional war approach emphasized the "body count". If a commander reported a lot of enemy kills, he was considered successful and rewarded. It also meant that some commanders regarded every dead Vietnamese, even those who were obviously civilians, as an "enemy fighter" for the body-count reports.[16]

By any reckoning, the US approach to force and firepower in Vietnam was a colossal failure. The relatively indiscriminate use of firepower in the South Vietnamese countryside helped alienate many South Vietnamese from their government. Civilians viewed soldiers not as protectors, but as a danger to themselves and their homes. The US takeover of the war from 1965 to 1970, coupled with the American policy of shunting the South Vietnamese forces to the side, worked to erode the legitimacy of the South Vietnamese government in the eyes of the people. In the course of the war, a great many Viet Cong and North Vietnamese were killed, but South Vietnam never developed a capable armed forces and, after the US pullout in 1973, were unable to mount an effective defense of their country.

Most of the American military leadership reacted to Vietnam by focusing their attention on high-tech conventional warfare and simply rejected the notion of fighting insurgencies and irregular forces in the future. As a result, most of the lessons of Vietnam, both positive and negative, were quickly lost in the mainstream US military. However, there were a dedicated few, largely within the Special Operations community, who did study the lessons of Vietnam and were determined to get it right in the future.

When a civil war organized by Marxist insurgents broke out in El Salvador in 1980, the US offered financial assistance to the Salvadoran government, as well as military training, advice and equipment to the Salvadoran military. After the recent experience in Vietnam, Congress was adamant that no American combat troops were to be committed to the El Salvador conflict and the number of US advisors sent to the country were to be strictly limited.

The limits on US combat involvement in El Salvador worked in America's favor because they forced the US military to eschew the high-tech, high-firepower war it preferred and return to basic counterinsurgency theory, which emphasized the comprehensive nature of a counterinsurgency war and the need to avoid actions and strategies that would alienate the population. Because the US military was not allowed to "Americanize" the war, they could only succeed by training and supporting the Salvadorans. The first and most essential task for the US military was a program to systematically build, train and equip the Salvadoran military. In 1981 the Salvadoran forces were poorly trained and equipped, and were responsible for widespread human-rights abuses against the civilian population. The American commanders in Latin America and in the US State Department established a policy of first reforming the largely dysfunctional Salvadoran armed forces. The key to this effort would be training a cadre of competent officers who, over time, would form the majority of the officer corps. Over a 12-year period the US military advisors and trainers worked patiently to transform El Salvador's dysfunctional military into a capable force.[17]

A key aspect of the US program to rebuild the Salvadoran military was to send younger officers, the ones least influenced by the dysfunctional military culture of El Salvador, to extended training courses in the United States and Panama Canal Zone. Over time the US mentors expected that the Salvadoran officers would not only learn basic military and leadership skills, but would be part of a professional military environment where corruption and a politicized officer corps were not part of daily life.[18] Respect for the civilian population was a central point of the US training program, and the US government brought constant pressure upon the Salvadorans to carry out military reforms and to remove officers who were known human-rights violators. While the US provided a considerable amount of new armament to the Salvadoran army and airforce, the US Southcom advisors also insisted that the Salvadorans take care to minimize civilian casualties.

By 1984, the US efforts to train and equip the Salvadoran military began to pay off as the Salvadorans demonstrated an improved capability to conduct true joint operations. Up to this point, the government forces had been losing, but now the conflict turned in its favor. The army had been expanded and whole battalions of specially trained light infantry had been trained in the US military training center in Honduras. Even while the army was expanded, the training levels also improved and, by mid-decade, the complaints of human-rights abuses by the Salvadoran military and police, which had been epidemic early in the decade, had dropped significantly. In 1985 and 1986 the tide had clearly turned as the military scored several notable successes in clearing several districts which had formerly been rebel strongholds.[19] The Farabundo Martí National Liberation Front (FMLN) tried one last time by mounting a major offensive in the capital city of San Salvador in 1989. Although the initial attack took the government forces by surprise, the FMLN was defeated in an urban battle in which the population refrained from supporting them. After the failure of the FMLN offensive, the opposing groups entered into peace talks. In 1992 a peace agreement that was very much in the government's favor was signed and the war ended.

Max Manwaring, one of the US strategists who helped guide the US strategy in El Salvador, developed an analysis of the essential tasks of counterinsurgency. Manwaring argued that legitimacy is the central strategic problem and the primary factor that determines the success or failure of a counterinsurgency campaign.[20] If the military commonly abused the population, it would lose legitimacy. Concurrently, Manwaring argued that the discipline and training of the armed forces was one of the key requirements for success in counterinsurgency. It was fine that the military could employ force, but they needed to be able to employ it in an effective manner that effectively targeted the insurgents and minimized civilian casualties and suffering.

Manwaring's argument was no more than classic counterinsurgency theory that had proven its worth in successful counterinsurgency campaigns in the Philippines in the 1950s and in Thailand in the 1970s. As a core part of any successful counterinsurgency strategy, military force had to be used with great care and alienation of the civilian population avoided. However, little of this study had much impact in the US military that increasingly focused on the high-tech conventional war. The dramatic victory of US forces in the 1991 war against Iraq, a war that featured precision strategic strikes, rapid and decisive ground campaigns and an imposing array of high tech weapons, pushed many in the US military to conclude that this kind of campaign was the future of warfare. A cadre of specialists in irregular warfare and counterinsurgency remained in the US armed forces, but they were definitely not in the mainstream of US military thinking.[21]

Thus, when the US blundered into an insurgency in Iraq in 2003, few of the senior US military leaders were mentally prepared to fight an irregular enemy. Even as the insurgency grew in size and scope, senior US commanders hoped that an American military campaign would quickly defeat them. By 2006 the violence was continuing to escalate and Iraq was sliding into a full-scale ethnic civil war. The US Army and Marines finally responded by developing a new, comprehensive counterinsurgency doctrine under the direction of Lieutenant General David Petreaus, who had served as division commander in Iraq and had proven to be one of the most mentally flexible of the US military's senior officers.

The new doctrine, the first such doctrine since the Vietnam War, was released in December 2006. US Army *FM 3–24 Counterinsurgency* was over 260 pages long and examined counterinsurgency in all its aspects. In many respects, it directly challenged the conventional-war mainstream thinking of the US military's senior leadership. *FM 3–24* begins with a discussion of classic counterinsurgency. Notably, the concept of legitimacy as an essential strategic requirement is an important part of the new American doctrine. Throughout the Army and Marine doctrine, the counterinsurgent forces are cautioned not to take actions or impose policies that will undermine the long-term legitimacy of the host nation government.[22]

Especially important is the understanding that legitimacy refers primarily to the host nation government, and Chapter 6 of *FM 3–24* is devoted to discussing the training and advising of the host nation security forces, which stresses that "Developing effective host nation security forces is one of the highest priority counterinsurgency tasks".[23] Thus, it is not only the use of force by American military units that is discussed, but how the Americans train the host nation to employ force. This is emphasized in the US doctrine with the understanding that, in the end, the US military cannot win another nation's internal war for them. The United States or other democratic powers might provide aid, equipment, training and advice. Outside powers can buy time for the defending government so that they can build up their own forces and infrastructure. But, in the end, to defeat insurgents the threatened nation has to field its own forces, develop its own strategy, and find its own political solution.

Under the principle of legitimacy, the use of American force in another country's internal conflict will always be viewed suspiciously by the host nation's populace – even those in sympathy with American aims and friendly to the American effort. In Afghanistan and Iraq, the US military has faced a stream of constant complaints about the American use of force from national governments dependent upon the support of American forces. Most of the Iraqi and Afghani government complaints of excessive use of force by US troops seem to have little grounding in hard fact or reliable evidence – but are more an act to demonstrate the independence of the government from the American outsiders, who are also infidels as well. If the host nation governments wish to establish their legitimacy in the eyes of the people, they might even feel compelled to constantly condemn American counterinsurgency operations in their own country, even while they quietly ask for US assistance behind the scenes.

## Conclusion

In an examination of a subject as complex as the use of force in counterinsurgency, a few principles can be laid out. First of all, having overwhelming military force is no assurance of success. By any reckoning, the British in Cyprus and the Soviets in Afghanistan had overwhelming military force at their disposal. The US had vast firepower resources available in Vietnam. In all these cases, overwhelming force could not prevent ultimate defeat. Second, for force to be effective, it has to be carefully coupled with realistic political objectives. Third, in an effective counterinsurgency strategy, military force might not be the main focus of the effort, but only a supporting player. Fourth, the effective use of force in irregular warfare requires a commander of very high mental flexibility and political understanding. Conventional mindset military leaders will usually fail in counterinsurgency operations. Fifth, the media will be likely to portray the use of force by the counterinsurgent as excessive and cruel. The counterinsurgent must be prepared to counter a very hostile media. Finally, the use of force must always be considered in the light of the principle of legitimacy: how will the host nation population perceive the use of force, especially the use of force by foreign troops? One should remember that in counterinsurgency you can have military success, but still fail politically – and the political effort is often more important than the military.

The argument in this chapter also raises important questions of force and capability generation. Given the reality of American and international politics, the most reasonable solution in helping an allied nation to combat an insurgency is to refrain from the temptation to take over the war effort of the host nation, and to instead emphasize the training and support of the host nation's security forces. In order to assure the host nation's success, the Western training and advisory programs must emphasize the careful use of force within a comprehensive strategy, as well as respect for the civilian population. Such an approach will require major adjustment in the American military mindset, where a great part of the senior officer corps still thinks only in terms of the high-tech conventional war, with decisive operational maneuvers winning a conflict in days. Effective counterinsurgency is the kind of war that American soldiers intrinsically dislike. Even the most successful counterinsurgency campaigns last for years, and much of the most important work is the unglamorous training and advisory mission which is measured by small steps. Training Third World forces, and developing a leadership corps that will respect the population and use force effectively, is an arduous task that normally takes years.

In the coming decade the US military will likely see another debate about the nature of modern conflict similar to the one that occurred after the Vietnam War. After the experience of Iraq and Afghanistan, the US military, and the armed forces of its Western allies, will be tempted to reject the study of counterinsurgency and instead seek out high-tech conventional wars while avoiding conflicts with irregular forces. This would, in fact, be the height of irresponsibility. Various movements and factions throughout the world will continue to mount rebellions against governments allied with the West, and hostile factions will take advantage of power vacuums in unstable states to offer sanctuary to international terrorists. It is difficult to imagine a world in which insurgency and conflict with irregular groups is not a major feature of conflict. The political realities of our age require that the Western armed forces continue to deepen our study of counterinsurgency and all facets of irregular warfare. Among these core requirements of future strategy is the continuous refinement of our understanding of how to use force against non-state enemies.

## Notes

1 Secretary of Defense Robert Gates, Address to the cadets of the US Military Academy, West Point, 21 April 2008.
2 An excellent account of the problems facing the US military in Baghdad in April 2003 is found in DePue, *Patrolling Baghdad*. The 223rd Military Police Company, a National Guard unit, was the first MP company to arrive in the city of five million. The unit can attest to the complete absence of any plan or directives for the occupation and administration of the city.
3 *Joint Publication 1–02, Department of Defense Dictionary of Military and Associated Terms*, August 2006. Definition:

> The strategic level of war: The level of war at which a nation, often as a member of a group of nations, determines national or multinational (alliance or coalition) strategic security objectives and guidance, and develops and uses national resources to achieve these objectives. Activities at this level establish national and multinational military objectives; sequence initiatives; define limits and assess risks for the use of military and other instruments of national power; develop global plans or theater war plans to achieve those objectives; and provide military forces and other capabilities in accordance with strategic plans.

4 *FM 3–24 Counterinsurgency* (December 2006).
5 On the problems of the intelligence effort in Iraq and Afghanistan and the lack of human intelligence specialists, see Corum, *Fighting the War on Terror*, 121–159.
6 See Corum, "Building the Malayan Army and Police".
7 Good overviews of Britain's relationship with Cyprus and the origins of the insurgency are found in Holland, *Britain and the Revolt in Cyprus 1954–1959*. For a good general history of the insurgency, see Cranshaw, *The Cyprus Revolt*; Foley and Scobie, *The Struggle for Cyprus*.
8 For a study of Templer's and Harding's leadership in counterinsurgency operations, see Corum, *Training Indigenous Forces in Counterinsurgency*.
9 Between 2006 and 2007 I polled students at the US Army Command and General Staff College, talked with Philippine officers fighting insurgents on Luzon and the southern Philippines, and talked with NATO officers dealing with insurgents in Afghanistan. All point out that a major weakness of counterinsurgent forces is the failure to use the media to get across the government's message to the population.
10 McKenna, *The Sand Pebbles*, 343.
11 MacBride, *Israel in Lebanon*, 147. The international commission noted that the PLO had placed flak guns on top of civilian building, but said that attacks on such targets were illegal as the guns posed no threat. In fact, the Israeli Air Force lost several aircraft to anti-aircraft fire.
12 In 2004 the US Army changed its views on the importance of media operations. The interim counterinsurgency doctrine published that year stated that, in counterinsurgency, the psychological operations officer is now to be considered as a primary staff officer. See *FM 3.07.22* (2004), chapter 5, pages 5–1 to 5–5, which deal with psychological operations. These are described as "an integral part' of the commander's plan and operations.
13 *FM 3–24 Counterinsurgency*, chapter 1, para. 153.
14 See Boot, "The Gaza Aftermath".
15 A thorough study of the US use of force in Vietnam is found in Greiner, *Krieg Ohne Fronten*, see 74–75, 136–138.

16 Michael MacLean in *The Ten Thousand Day War* refers to American units shooting up areas simply to increase the body count. See MacLean, *The Ten Thousand Day War*, 270–279. On the body-count mentality and the resulting civilian casualties, see also Spector, *After Tet*, 202–221.
17 A good discussion of the difficulties of the training mission in the context of El Salvador's largely dysfunctional military leadership is found in Ramsey, *Advising Indigenous Forces*, 83–103.
18 The basic military strategy for the US support for the Salvadorans was worked out in early 1981 when Brigadier General Fred Woerner and a small staff met for ten days with the Salvadoran general staff and developed a five-year plan to support, retrain and reorganize the Salvadoran armed forces – then facing likely defeat at the hands of the rebels. General Woerner later became the US Commander in Chief of Southcom in the late 1980s.
19 Ramsey, *Advising Indigenous Forces*, 87–90.
20 See Manwaring, "The Threat in the Contemporary Peace Environment", 52–54.
21 See Corum, *Fighting the War on Terror*, chapter 2.
22 US Army/US Marine Corps, *FM 3–24 Counterinsurgency*. See especially chapter 1.
23 *FM 3–24 Counterinsurgency*, para. 6–22.

## Bibliography

Boot, Max, "The Gaza Aftermath", *The Weekly Standard Online*, 14, 19 (2 February 2009).

Corum, James S., *Fighting the War on Terror*. St. Paul: Zenith Press, 2007.

Corum, James S., "Building the Malayan Army and Police – Britain's Experience During the Malayan Emergency 1948–1960", in Randall Gott (ed.) *Security Assistance: U.S. and International Historical Perspectives*. Leavenworth: Combat Studies Institute Press, 2006, 291–314.

Corum, James S., *Training Indigenous Forces in Counterinsurgency: A Tale of Two Insurgencies*. US Army War College, Strategic Studies Institute, March 2006.

Cranshaw, Nancy, *The Cyprus Revolt*. London: Allen and Unwin, 1978.

DePue, Mark, *Patrolling Baghdad: A Military Police Company and the War in Iraq*. Lawrence: University Press of Kansas, 2007.

Foley, Charles and W.I. Scobie, *The Struggle for Cyprus*. Stanford: Hoover Institution Press, 1975.

Greiner, Bernd, *Krieg Ohne Fronten: Die USA in Vietnam*. Hamburg: Hamburger Edition, 2007.

Holland, Robert, *Britain and the Revolt in Cyprus 1954–1959*. Oxford: Clarendon Press, 1998.

*Joint Publication 1–02, Department of Defense Dictionary of Military and Associated Terms*, August 2006.

McKenna, Richard, *The Sand Pebbles*. 1962; reprint: Annapolis: Naval Institute Press, 2000.

MacBride, Seán, *Israel in Lebanon: Report of the International Commissions to Enquire into Reported Violations of International Law by Israel During its Invasion of Lebanon*. London: The International Commission, 1983.

MacLean, Michael, *The Ten Thousand Day War*. New York: St. Martin's Press, 1981.

Manwaring, Max, "The Threat in the Contemporary Peace Environment: The Challenge to Change Perspectives", in Edwin Corr and Stephen Sloan (eds.) *Low Intensity Conflict: Old Threats in a New World.* Boulder: Westview Press, 1992.
Ramsey, Robert, *Advising Indigenous Forces: American Advisors in Korea, Vietnam and El Salvador.* Global War on Terror Occasional Paper 18, Fort Leavenworth: Combat Studies Institute, 2006.
Spector, Ronald, *After Tet.* New York: The Free Press, 1993.
US Army/US Marine Corps, *FM 3–24 Counterinsurgency,* December 2006.

# 9 Adapting to counterinsurgency

## The US Army in Iraq, July 2004–December 2006

*Brian Burton and John Nagl*

> This is a difficult and unprecedented task – and we're learning as we go.
>
> President George W. Bush, Commencement speech at the US Air Force Academy, Colorado Springs, Colorado, 28 May 2008[1]

When the US military went into Iraq in March 2003, it did so without any of the doctrine, training or other preparations necessary for countering an insurgency. The civilian leadership's flawed planning for the war and reconstruction were certainly a primary cause for what was to come. However, the lack of counterinsurgency preparation in the US military, particularly the Army, helped to make the problems of post-Saddam Iraq immeasurably worse. General Barry McCaffrey noted that the Army 'was not prepared to train the Iraqi security forces. It wasn't prepared to conduct area security or counterinsurgency.'[2] As former Army Vice Chief of Staff General John Keane described it:

> we put an army on the battlefield that I had been a part of for 37 years. The truth of the matter is: It doesn't have any doctrine, nor was it educated and trained, to deal with an insurgency ... After the Vietnam War, we purged ourselves of everything that dealt with irregular warfare or insurgency, because it had to do with how we lost that war. In hindsight, that was a bad decision.... We have responsibility.[3]

### Sanchez sets the stage

The top-to-bottom lack of understanding of insurgency and counterinsurgency initially resulted in incoherent strategy in Iraq, exacerbated by counterproductive operations and tactics. In the aftermath of Major Combat Operations, political and military leaders clung to the belief that they were facing the last gasp of a Ba'athist remnant rather than organized and well-armed enemies. The delayed recognition of the significance of the ongoing violence hindered development of a military strategy in Iraq based on local realities. Unity of command was lacking as the civilian

Coalition Provisional Authority and the military's Combined Joint Task Force – Seven (CTJF-7) operated in almost completely separate spheres. The military's security effort was therefore not harnessed to the achievement of a set of political ends.

Furthermore, Lieutenant General Ricardo Sanchez, the CTJF-7 commander, does not seem to have provided overarching guidance of any kind to his own military subordinates, the divisional commanders. Instead, he left them largely to their own devices, resulting in varied and often ineffective operational and tactical methods. Although US troops did attempt to avoid excessive civilian casualties and deliberate targeting of non-combatants, they often did not use the minimum force necessary to accomplish their mission, sometimes employing indirect artillery fire and air-strikes in close proximity to civilian areas or returning fire in less-than-discriminate fashion when attacked. Some force restructuring occurred on the ground, but large-unit sweeps seemed to prevail as the operation of choice as the military's priorities focused on killing or rounding up as many suspected insurgents as possible. US tactics for detaining suspects similarly employed excessive force, and were largely ineffective in producing the intelligence necessary to combat the insurgency.[4] An Army report acknowledged that the 'military means employed' by the United States 'were not sufficient' to achieve the ultimate objective of replacing the Iraqi regime and securing a new political order in the country.[5] The military tools were not optimized for the job at hand, and Sanchez did not lead any effective effort to adapt the force to the demands of counterinsurgency during his command.

## Casey at the bat

General George Casey's command in Iraq, from summer of 2004 to the end of 2006, saw considerable improvements in US military counterinsurgency operations, but was hampered by a faulty strategy. This time period was characterized by increasing awareness and employment of counterinsurgency on the strategic, operational and tactical levels, though substantial inconsistencies remained. The military in Iraq adopted a campaign plan and recognized the necessity of linking military means to the wider political objective of building up state capacity. On the operational and tactical levels, American forces altered and refined their methods to conform to accepted counterinsurgency principles, making more coordinated efforts at using the minimum amount of force necessary to accomplish tasks and reorganizing to deal with an unconventional threat. The development of effective Iraqi security forces became a central priority for American military leaders and the effort to train them received more attention and resources.

Yet strategic success proved elusive. Although Casey's leadership proved to be better than Sanchez's, the campaign plan Casey developed did not emphasize counterinsurgency principles as much as 'Iraqization', a transition from American to Iraqi responsibility for security. Despite clear evidence that the Iraqi security forces were not ready to perform this mission, US military leaders continued to emphasize transition. Violence in Iraq spiked to new heights during this time, as the insurgency continued to develop and communal sectarian violence rose. An indigenous security apparatus was built that was plagued by incompetence and mixed loyalties. At worst, the Iraqi security forces contributed to the disorder with sectarian-motivated killings. Given this situation, maintaining the emphasis on transition contributed to the American inability to stop the escalating insurgent and sectarian violence. Only the deadly and dire circumstances of late 2006 finally prompted the much-needed strategic review that produced the effective counterinsurgency strategy and troop surge of 2007.

If the failure of Sanchez's command lay in an over-emphasis on the use of force, Casey's was in underestimating the utility of force in counterinsurgency. In *The Utility of Force*, General Sir Rupert Smith theorizes that force is less useable and useful in 'wars amongst the people'. The US fought according to Smith's principles: fighting 'so as not to lose the force, rather than fighting by using the force at any cost to achieve the aim'.[6] The transition strategy was premised on the view that US troops could not achieve the objective of establishing security in Iraq and that their very presence was counterproductive, provoking popular resistance. The 'main preoccupation', therefore, was 'how to leave the territory', rather than to establish the security conditions necessary for 'a desired political outcome of stability'. The transition approach focused on minimizing American losses in an attempt to avoid the heavy but essential costs of counterinsurgency and skip ahead to the achievement of political objectives. Rather predictably, this strategy failed. America's adversaries maintained their violent strategies of insurgency without effective opposition from US or Iraqi forces. The resulting violence precluded any meaningful strategic progress.

The experience of American forces serving under Casey's command in Iraq demonstrated an enduring truth of counterinsurgency: that the lack of utility of conventional combat power should not be confused with a lack of utility of force in counterinsurgency. As the most successful operations and tactics by US forces in Iraq demonstrated, the use of force to secure and control the population and separate it from the insurgents is crucial to success in 'war amongst the people'. It was a lesson that was learned fairly rapidly among operational and tactical leaders, but comparatively slowly by strategic commanders.

## Shifting to transition

US strategy under Casey placed more emphasis on population defense, developing effective Iraqi security forces, and political and economic reconstruction than had Sanchez. However, the overriding emphasis was to shift the burden of providing security to Iraqi forces as quickly as possible. As a result, US military strategy in Iraq continued to lack a fully coherent counterinsurgency approach. It was only late in 2006 as increasing violence continued to drag on with no end in sight that an effective counterinsurgency strategy was finally developed.

The first strategic shift began soon after the transfer of sovereignty from the CPA to the interim Iraqi government at the end of June 2004, when General George Casey replaced Ricardo Sanchez as the top US commander in Iraq. Casey and his staff immediately began to develop a campaign plan, which had been stunningly lacking throughout Sanchez's tenure. As Thomas Ricks describes, the campaign plan that Casey issued on 5 August 2004 in conjunction with the US Embassy

> essentially called for containing the insurgent violence, building up Iraqi security forces, rebuilding economically, and reaching out to the Sunni community through both coercion and cooptation, in an effort to persuade them of the inevitability of success for the US-led side.[7]

The military mission in Iraq was intended primarily to neutralize the insurgency and to train Iraqi security forces to eventually take full responsibility for the country; it also included 'lines of operation' such as helping Iraqis build democratic institutions, assisting in the provision of essential services such as water and sanitation, aiding economic reconstruction, strengthening the rule of law and civil rights, and promoting a free and responsible Iraqi media.[8]

Not only was this the first time US military objectives had been laid out in any clear form, but it was also the first real attempt to link the military strategy to political goals as prescribed in classic counterinsurgency literature. Many of the plan's elements were the products of Casey's strategy office, which was directed by Colonel William Hix and included experienced military professionals tasked with developing improvements to the US approach to Iraq. One of the strategy office's members, retired Special Forces Lieutenant Colonel Kalev Sepp, was assigned to produce a paper detailing the best practices for counterinsurgency which became the only unclassified part of the campaign plan. This 'best practices' paper, which emphasized the necessity of population control, political development, training indigenous security forces, and a single executive authority for successful counterinsurgency, was quite influential in the initial strategic shift to a better counterinsurgency strategy.[9]

Casey was committed to working more closely with non-military agencies, particularly the State Department, to ensure that the military was properly supporting civilian reconstruction efforts. He found an effective counterpart in career diplomat John Negroponte, the US Ambassador to Iraq, who arrived in the country on the day sovereignty was transferred to the Iraqis. Both Casey and Negroponte wanted to avoid the conflicts and communication problems that had plagued Sanchez and Bremer. Negroponte shifted $3.3 billion in funding to military commanders for security and local-level reconstruction projects which helped provide jobs for Iraqis. Casey set up an office in the US Embassy to ensure that he would be in constant contact with the ambassador.[10] The willingness to personally coordinate military and political efforts was one of the biggest improvements that Casey and Negroponte offered over their predecessors.

The US military leadership recognized at some level that the achievement of political and economic reconstruction was dependent on protecting the population, but there was no commitment to population security as a centerpiece of a counterinsurgency strategy. This cognitive dissonance was clearly demonstrated by the surging of troops for the major public elections of 2005. The insurgents had stepped up attacks to new heights in June 2004 as the transfer of sovereignty was being prepared, and it was assumed that future elections and other events signifying political progress would be targeted. Therefore, Central Command (CENTCOM) increased the number of US brigades in Iraq from 17 to 20 (150,000 US soldiers) to 'help assure security for the important January 2005 national elections', which went off smoothly despite another surge in insurgent attacks. However, the increased troop presence was only temporary and did not reverse the overall trend of increasing attacks, especially those directed against Iraqi civilians.[11] US troop levels were boosted again for the October 2005 constitutional referendum and the December 2005 elections (to a high of 160,000 US troops) with similar results: no major incidents occurred during the voting, but the overall level of violence was not tamped down for any sustained period.[12]

Securing polling sites and conducting elections did not lead to victory. The continuing violence and the military's apparent inability to stop it led to some internal questioning within the Bush administration and the military, but no fundamental shift of direction on the ground in Iraq. In September 2005, Jim Jeffrey, the White House's coordinator of Iraq policy, argued to Secretary of State Condoleezza Rice that the US military was not adequately executing a plan for victory; the insurgency was barely being contained, not neutralized, and the political and economic reconstruction 'was not going well'. The 'National Strategy for Supporting Iraq', a classified report outlining the plans for spending $21 billion in US funding for reconstruction, was effectively a blueprint by which the State Department and the US military were attempting to operate, but infrastructure security was a major problem hindering its effective implementation.[13] General John Abizaid, the CENTCOM commander, privately displayed a lack of confidence in the American campaign plan. He reportedly told then-Deputy Secretary of State Richard Armitage in the summer of 2004, 'We've got a really bad situation here. Can't win it militarily', and said to retired Army friends in the fall of 2005, 'We've got to get the [expletive] out' and have the Iraqis step up to defeat the insurgency.[14]

Abizaid's attitude reflected the quiet understanding between him and Casey that America could not 'win' the war; rather, the goal was to hand over responsibility to the Iraqi government and security forces. American forces, in their view, were a foreign presence in Iraqi society that was generating resistance, necessitating a rapid handover to an Iraqi government to restore order to the country. The American military was not going to defeat the insurgency; instead, the US military's mission was to focus increasingly on training Iraqi security forces that could stand on their own. The Bush administration sought to outline this strategy in the public domain, but demonstrated its own strategic confusion by simultaneously continuing to emphasize 'victory' and promote counterinsurgency strategies. For example, when testifying before the Senate Foreign Relations Committee on 19 October 2005, Secretary of State Condoleezza Rice said of US policy, 'In 2005, we emphasized transition: a security transition to greater reliance on Iraqi forces and a political transition to a permanent, constitutional democracy.' However, Secretary Rice went on to describe a population-centric counterinsurgency approach by using the term 'clear, hold, and build' to explain the US effort planned for 2006: 'In short, with the Iraqi Government, our political–military strategy has to be to clear, hold, and build: to clear areas from insurgent control, to hold them securely and to build durable Iraqi institutions.'[15]

The release of the National Strategy for Victory in Iraq on 30 November 2005 set 'clear, hold, and build' as the centerpiece of the 'security track' for victory in Iraq. The security track was summarized as follows:

> **Objective**: To develop the Iraqis' capacity to secure their country while carrying out a campaign to defeat the terrorists and neutralize the insurgency.
>
> To achieve this objective, we are helping the Iraqi government:
>
> **Clear** areas of enemy control by remaining on the offensive, killing and capturing enemy fighters and denying them safe-haven.
>
> **Hold** areas freed from enemy control by ensuring that they remain under the control of a peaceful Iraqi government with an adequate Iraqi security force presence.
>
> **Build** Iraqi Security Forces and the capacity of local institutions to deliver services, advance the rule of law, and nurture civil society.[16]

The security track was accompanied by a 'political track' centered on building functional institutions of democratic government and an economic track focused on restoring Iraq's infrastructure and reforming economic regulations. The document's appendix detailed eight 'strategic pillars' in Iraq. The first strategic pillar was 'Defeat the Terrorists and Neutralize the Insurgency', and the second was 'Transition Iraq to Security Self-Reliance'.[17]

On the surface, the principles of counterinsurgency were being addressed, reflecting the growing understanding that military means alone could not win the war in Iraq. This understanding was partially informed by the military's operational and tactical experiences there, as more officers concluded that providing jobs for civilians and rebuilding infrastructure were necessary to obtain the support of Iraqi citizens in hunting down hidden enemies. However, at the same time there was resistance to fully embracing the effectiveness of 'clear, hold, and build'-style population security in a strategy emphasizing the slow grind of counterinsurgency rather than rapid transition to Iraqi control. 'Clear, hold, and build', like all counterinsurgency strategies, takes time and patience to implement effectively, particularly as it requires improving host-nation security forces and institutions to the point that they can operate independently; the military leadership, on the other hand, sought to transition to Iraqi security responsibility more quickly. The two approaches were incompatible, and the focus on transition hindered strategic coherence and the operational and tactical effectiveness of counterinsurgency efforts on the ground.

## Developments in military operations, tactics and organization

By May 2005, Tall Afar, a predominantly Sunni town of approximately 250,000 people near the Syrian border in northwestern Iraq, had developed into an insurgent stronghold that served as a base for attacks on Mosul and other population centers in the area. Then the 3rd Armored Cavalry Regiment (ACR) under the command of Colonel H.R. McMaster was deployed to the area. The 3rd ACR had previously served in western Iraq through spring 2004 and, according to media reports, had been responsible for several instances of civilian and detainee abuse. McMaster assumed command of the unit in June 2004 and began intensive retraining of the unit in counterinsurgency methods. In preparation for the deployment, he ordered one out of every ten soldiers to attend a three-week conversational Arabic course, gave his subordinate officers a reading list consisting of Arab and Iraqi history and counterinsurgency texts, and told his soldiers: 'Every time you treat an Iraqi disrespectfully, you are working for the enemy.'[18]

Once in Iraq, the 3rd ACR did not attack the insurgency in Tall Afar head-on. First, it acted in conjunction with Iraqi forces to weaken the insurgency's infrastructure by securing the nearby border and seizing an outlying town that served as a way station for fighters and supplies coming from Syria. McMaster also sought out local Sunni leaders to admit previous American mistakes and express a commitment to understanding and working with Iraqis. At the suggestion of some Iraqi allies, the 3rd ACR constructed a nine-foot-high sand berm to encircle the city and enable all movements in and out of Tall Afar to be easily monitored. Intelligence was used to trace tribal links of known fighters and track their movements to possible locations outside the city, thus facilitating the capture of some 120 insurgents. Civilians were encouraged to leave Tall Afar for a temporary encampment before US troops moved to clear the city in September. American and Iraqi troops cleared the city block-by-block and established 29 patrol bases within the city.

McMaster had the officers and soldiers responsible for Tall Afar's security live at these bases inside the city, rather than at a larger facility outside, to enable constant surveillance. US forces helped build up a 1,400-man police force that included both local Sunni Arabs and others from different parts of Iraq. With the creation of a functioning civic government, US forces received increasing support from Iraqis in the form of 'actionable intelligence', such as warnings about when and where insurgents were planting roadside bombs. Detainees were treated with respect, and no instances of abuse were reported during the deployment. The progress in security was measurable: daily insurgent attacks decreased from six to one during the 3rd ACR's deployment to the area.[19]

The 3rd ACR's rotation in Iraq concluded in the spring of 2006, and the 1st Armored Division took its place. Almost immediately, insurgent activity in the area began to increase. Units of 1st Armored Division responded by rounding up all men aged 18 to 65 in parts of Tall Afar.[20]

The Tall Afar experience is an instructive example of the results that could be achieved through the appropriate application of counterinsurgency principles, and of the opportunities lost due to inconsistent operational and tactical methods across units and the flawed transition strategy. In spite of the continuing inconsistencies in strategic guidance, lower-level leaders increasingly found operational and tactical success through population-centric counterinsurgency measures. Military operations and tactics began to conform to counterinsurgency principles more consistently than in the first year of the US occupation. The lower-level focus had shifted with the strategy to combat the insurgency through methods aimed at engaging with and securing the Iraqi population. Many of the basic operational concepts remained largely unchanged but were carried out with less force and more patience for local sensitivities. Organizationally, the US forces maintained a system of isolated forward operating bases that were not ideal for counterinsurgency, but some commanders occasionally strayed from this approach to push their forces closer to the population. Units and leaders who had made mistakes on their first tours learned and adopted better methods on their next tours. Overall, the operational, tactical and organizational shifts toward counterinsurgency principles were uneven, but demonstrated improved awareness of the delicate nature of the unconventional tasks required.[21] Yet because there was no overarching strategic framework supporting counterinsurgency, the efforts of those units that were embracing effective population-centric operations and tactics added up to less than the sum of their parts.

Initial deployments of American forces in 2003 and 2004 came more prepared for invasion and conventional combat than counterinsurgency. Over a year into the struggle, US troops started to conceptualize the struggle they were in differently than they had in the first year of the occupation. By 2005, 'American soldiers began to think about the war in ways that would have been unrecognizable two years earlier', viewing their primary duties as population protection and support for Iraqi forces rather than actively rounding up insurgents. Less time was spent on combat operations, with one officer estimating that 'he spent half his time on engagement of the population' and another quarter of it 'working with Iraqi troops'. Rules of engagement were tightened in an effort to avoid alienating the population through excessive use of force.[22]

The concept of 'full-spectrum operations', a term used by then-Major General Peter Chiarelli to denote a simultaneous fusion of combat operations and training of Iraqi security forces, with the restoration of basic services and political and economic reconstruction through military-sponsored public works projects to employ local people, gained credibility as an effective approach to counterinsurgency in the field (see Figure 9.1). In Chiarelli's view:

> It is no longer sufficient to think in purely kinetic terms. Executing traditionally focused combat operations and concentrating on training local security forces works, but only for the short term. In the long term, doing so hinders true progress and, in reality, promotes the growth of insurgent forces working against campaign objectives.... The reality is that there are cultural mechanisms at play that demand a more integrated plan. No longer is it acceptable to think sequentially through stability operations and support operations by believing that if you first establish the security environment, you can work sequentially toward establishing critical infrastructure and governmental legitimacy then drive toward economic independence.[23]

Much as the French officer David Galula called on soldiers involved in counterinsurgency operations to be prepared to handle civilian reconstruction tasks, Chiarelli advises that those civilian reconstruction tasks be considered as integral to the military's counterinsurgency effort as combat operations. He writes, 'We should consider paraphrasing Clausewitz: full-spectrum operations are the continuation of major combat operations by other means.'[24]

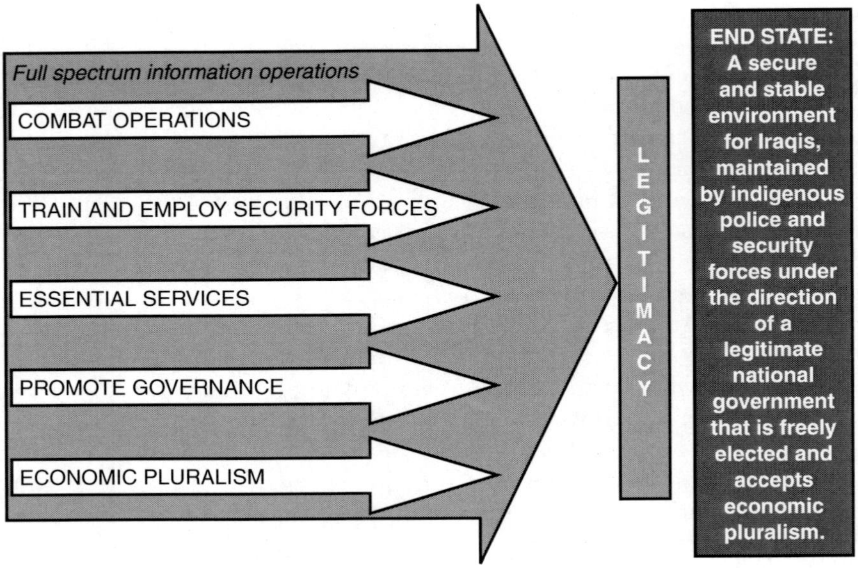


*Figure 9.1* Chiarelli's concept of full spectrum operations in counterinsurgency.

Although there remains substantial debate among military officers about the effectiveness of reconstruction projects, particularly in areas where security issues are more severe, the concept of full-spectrum operations has been increasingly employed. Counterinsurgency principles and the effort to 'clear, hold, and build' were being translated into operational planning. As Lieutenant Colonel Chris Gibson, a battalion commander in 82nd Airborne Division with two tours in Iraq between December 2004 and January 2006, writes:

> Our COIN tactics, techniques and procedures (TTP) were consistent with the current administration's direction to 'clear, hold, build'. We addressed –
>
> - Establishing security by gaining situational awareness, developing intelligence, and dominating the battlespace.
> - Affiliating with the local populace.
> - Developing the ISF [Iraqi Security Forces].
> - Strengthening civilian institutions.
>
> We pursued these aims concurrently, although security was the first among equals.[25]

Gibson's battalion conducted frequent reconnaissance patrols, both mounted and dismounted, to 'saturate' zones of the city with troops to deter attacks and 'actively pursued interaction with the people to make them feel safe and help us achieve security dominance'. When homes were damaged during security operations, soldiers distributed claims cards to those affected and paid for repairs. The claims cards were also used as a basis for monitoring who lived in the area. With the population feeling more secure, intelligence tips began to flow in. Gibson and other commanders also learned to be more careful about acting on intelligence tips, using multiple sources to corroborate information to lessen the chances of detaining the wrong people.[26]

As another lieutenant colonel with command experience in southern Baghdad elaborated:

> We had four lines of operation [LOO] in our area: combat operations, ISF operations, information operations, and civil–military operations. To be effective, we couldn't just focus on one or two LOOs; we had to integrate all four lines into each of our operations and the overall campaign, and we had to apply them simultaneously. As an example, when we executed a raid, we included our ISF partners, used tactical psychological operations and our own soldiers to ensure the public knew what our intent was, and then followed up the raid the next day by making goodwill gestures to the population, such as distributing soccer balls, repairing roads, or providing clothing and food.[27]

Company commanders found a variety of ways to bring more security to their areas of responsibility. Most were simply refinements of previous methods adjusted for more discriminating use of force, such as conducting 'cordon-and-knock' missions and talking politely to families in their homes rather than kicking in doors and breaking things during searches. Some units found that they received more intelligence during 'cordon-and-knock' missions covering specific blocks of a city or town, because it enabled citizens to quietly pass information to the Americans anonymously. Interpreters' cell phones could serve as tip lines for citizens to report illegal activities in their neighborhoods. Information operations, such as informing civilians of the intentions of military actions in their towns or providing positive news, assumed a more prominent role in many units' operations.[28] Commanders displayed increasing flexibility in adjusting their methods to local realities. The way units behaved in the field varied considerably with the local environment and attitude of the commander, but on the whole counterinsurgency principles such as the minimization of force were employed more uniformly.

The concentration of US forces in large and isolated forward operating bases, a contradiction of counterinsurgency principles, continued during this period, although instances of commanders moving their troops to smaller bases among the population were the norm in the more effective operations. The force posture of big bases enabled more soldiers to enjoy a substantially improved quality of life, with air-conditioned rooms, Internet access, abundant food and other amenities.[29] Yet the bases also drove a wedge between US soldiers and the Iraqi populace and made it more difficult to project the American security presence into populated areas. Some commanders felt it necessary to move their troops to outposts in urban zones in order to provide better security. On his unit's first assignment, which was to help secure the Baghdad airport road in late 2004, Lieutenant Colonel Gibson planned for half of his unit to live in outposts within his area of responsibility, which at the time was rare and unpopular among the troops. Yet living among the population enabled better monitoring and had the effect of dramatically reducing attacks in the area to close to zero in a short period of time.[30] The 3rd ACR's concurrent experience in Tall Afar was that operating from small patrol bases scattered around the city allowed them a 'degree of observation [that] made it very difficult [for insurgents] to plant bombs', and added an element of flexibility in patrol and raiding operations.[31] Another officer believed that living among the populace was necessary because 'the people in our AO [Area of Operations] would allow the insurgents to move freely through them unless we or the ISF were physically present 24 hours a day'.[32]

Yet, while counterinsurgency learning did occur at the unit level, the persistent problem of this period in Iraq was the fact that each unit had to learn its own lessons the hard way. Without effective counterinsurgency doctrine in place in the US Army, there was no common understanding across the military of the problems it faced in Iraq and how to deal with them. As a result, individual commanders continued to be left to their own devices. At the tactical and operational levels in the field, it was as if American football teams were working without playbooks, having instead to draw up impromptu and unpracticed plays in the middle of the game. Without a shared operational understanding of how to deal with the situation, it was unsurprising that every unit was doing its own thing. Increasing adherence to counterinsurgency principles was due largely to individual leaders experiencing success with population-centric methods and occasionally learning from their peers. While operational and tactical flexibility are important to successful counterinsurgency, the continuing failure to translate strategy into operations and tactics, or integrate effective operations and tactics into strategy, meant that battlefield gains did not lead to any sustainable progress toward either transition or security in Iraq. Tall Afar is a prime example: effectively secured by 3rd ACR, it deteriorated under a different unit using less-effective methods. By unintentionally working at cross-purposes, strategic and operational goals undermined one another. The strategic goal of transition, in fact, was a major hindrance to US military cooperation with the Iraqi population to enhance security.

## Iraqi security force development

From the summer of 2004 to the present day, the development of effective security forces was the top declared US priority in Iraq. President Bush's 2005 statement that American forces will stand down as Iraqis stand up is illustrative of US civilian and military policy-makers' view of the centrality of this task to American strategic success in Iraq. With the dissolution of the CPA, the Iraqi military training program became much better organized and generated Iraqi units of improved quality as the resources allotted to the program increased. The Iraqi Police, however, remained largely ineffective due to incompetence, corruption, enemy infiltration and a less well-organized development program. Iraqi security forces were increasingly expected to take the lead in security operations, with mixed results.

In the summer of 2004, Lieutenant General David Petraeus took charge of the Multinational Security Transition Command–Iraq (MNSTC–I), the Coalition military organization created to oversee training operations. Under his command, MNSTC–I was reorganized into three sections: the Coalition Military Assistance Training Team (CMATT) for building up the Iraqi armed forces, the Joint Headquarters Advisory Support Team (JHQ) for advising Iraqi leaders on military command and control, and the Civilian Police Assistance Training Team (CPATT) for developing police and internal security forces.[33]

Core Iraqi military strength at the time was small: six battalions comprising 6,323 Iraqi troops (from both regular army and elite forces) were reportedly trained and equipped as of that fall, and high levels of desertion meant that even that figure was an overestimate. The Iraqi National Guard, which would merge into the regular army in January 2005, theoretically maintained large numbers of local troops in different areas of the country, but those units possessed virtually no training and could not be counted on. Improved training emphasized leadership and unit cohesion to make the few Iraqi combat units available more effective. Some 2,000 Iraqi soldiers fought alongside US forces in Fallujah in November 2004, but they still lacked the experience and heavy weapons to take the lead in security operations without US support. As of mid-November 2004, for example, the regular Iraqi Army had only 4,507 fully trained and equipped troops from a nominal strength of 12,364. Nevertheless, the Iraqi performance in Fallujah generated some cautious optimism, and plans for substantially increasing the size of the Iraqi Army were being implemented by the end of 2004.[34] The notable aspect of this security force structure was how it deviated from the smaller, lighter forces originally planned by the Americans. Petraeus and the rest of the US military leadership in Iraq, recognizing the seriousness of insurgent violence, effectively redrew the plans in order to create forces that would be more capable of standing up against a heavily armed insurgency.

However, the improvements in the security force structure and training could not obscure significant problems. Incompetence and corruption were prevalent, particularly among the police forces, but those responsible often could not be fired for reasons of manpower shortages and tribal political favoritism in local employment. Many security force recruits were of very low quality and only joined because they had no other employment prospects; they were often illiterate and possessed little experience with modern technology or a disciplined work environment. The relatively limited available training alone could not transform them into effective soldiers and police. When combined with the lack of adequate facilities and equipment, the result was still a security force that was in many ways a paper tiger. Operational readiness and unit cohesiveness were generally low and mass desertions in the face of insurgent intimidation were not uncommon. While the battle in Fallujah raged in November 2004, for example, insurgents launched a significant assault on Mosul that caused the desertion of approximately 3,000 Iraqi police, soldiers and guardsmen out of a total of 4,000 assigned to the city.[35] Casting further doubt on the competence of MNSTC–I was the US Government Accountability Office's finding that, in the rush to equip the Iraqi forces, over 50 percent of the weapons delivered between June 2004 and September 2005, approximately 190,000 firearms, went unaccounted for;[36] some likely ended up in the hands of insurgents and militia fighters.

Mixed progress on Iraqi force development continued in 2005 and 2006. Troop numbers expanded significantly; by July 2005, the Iraqi Army reportedly had 78,200 trained and equipped soldiers and the Iraqi Ministry of the Interior had 64,100 police and 30,700 other internal security personnel. In August 2006, the Iraqi Army had increased to 127,200 troops and there were 115,500 police and 49,600 other Ministry of Interior security personnel. Yet those reported numbers were tainted by the fact of continued absences and desertions from the ranks.[37] Iraqi forces participated in securing several thousand polling sites for the January 2005 elections, and, according to a CENTCOM briefing given to President Bush, by the end of the year were performing more independent operations than US and other Coalition forces.[38]

Increasing numbers of Iraqi units were deemed capable of operating independently and taking lead security responsibility for designated areas of operations: by August 2006, lead security responsibilities were transferred to five divisions, 25 brigades, and 85 battalions, and one province (Muthanna in southern Iraq) had been entirely transferred from Coalition forces to the Iraqis.[39] Despite this progress, Iraqi security forces were notably unsuccessful in tamping down insurgent and rising sectarian violence. Operation 'Together Forward', a highly publicized crackdown led by Iraqi military and police units intended to suppress violence in Baghdad, began during the summer of 2006 and drifted on with no substantial results other than an increase in the number of daily attacks (from

24 in June to 34 in July). Iraqi forces were not effectively holding neighborhoods that had been cordoned and cleared of insurgent and militia fighters.[40]

Sectarian loyalties ran deep within the mostly-Shi'ite security forces, with an estimated 70 percent of police officers having some militia affiliation. Elite police commando units were frequently accused of perpetrating extrajudicial 'death squad' killings.[41] Other Iraqi units refused to deploy beyond their home areas.[42] US military officers interviewed on this point saw the Iraqi police forces as an obstacle to peace and to security. Captain Forrest Cook described the Iraqi police as a 'shady organization' with more loyalty to 'their friends' than the Iraqi government.[43] Captain Chad Caldwell described them as 'highly corrupt' and 'unwilling to do their job', and found that his unit regularly had to confront and arrest police officers who were setting up illegal checkpoints and threatening civilians.[44] The advances of the previous years had not produced Iraqi military or police units that were really capable of upholding law and order, rendering a policy of transitioning more responsibility to them dangerously untenable.

## Descent into chaos

The overall result during this period was continuing and deepening insecurity in Iraq. The Sunni-dominated insurgency grew more virulent and more sectarian in nature. The Jordanian terrorist Abu Musab al-Zarqawi emerged in 2004 as the leader of al-Qaeda in Iraq, one of the extreme Sunni religious resistance factions notable for their vitriolic anti-Shi'ite rhetoric and indiscriminate attacks on Shi'ite civilians as well as US forces. This faction gained more prominence following the major US assault to retake Fallujah in November 2004; a significant number of militants escaped Fallujah to spread more violence in cities like Baghdad, Ramadi and Mosul, and were responsible for an increased number of suicide bombings in the lead-up to the January 2005 elections.[45] The tempo of attacks increased throughout the year; there were 26,496 reported attacks in 2004 and 34,131 in 2005. Improvised explosive device and other bomb attacks increased to approximately 1,800 per month in the autumn of 2005, and insurgents were building more sophisticated and powerful bombs, which caused increasing numbers of casualties (see Table 9.4).[46]

Iraqi security forces were targeted, with 2,545 military and police personnel killed in 2005. The attacks against security forces were linked to a wider campaign against all Shi'ites, including civilians, by the most extreme jihadist faction of the Sunni insurgency seeking to provoke sectarian warfare.[47] Sunni civilians were also victims of brutal intimidation campaigns by the insurgents, particularly in northwestern areas of the country, such as Tall Afar, where militants were setting up way stations close to the Syrian border. Additionally, the Shi'ite-dominated security forces, particularly police commandos, were responsible for numerous incidents of torture and extrajudicial killing of Sunni civilians as 'suspected insurgents'.[48] These incidents were a troubling harbinger of what was to come.

The year 2006 was dominated by sectarian violence, most of it against both Sunni and Shi'ite civilians. The bombing of a Shi'ite shrine in Samarra on 22 February 2006, which killed no one but destroyed the revered mosque's golden dome, helped to transform what had formerly been a considered a secondary facet of the violence in Iraq into the largest cause of instability, particularly in Baghdad. In the days following the mosque bombing, Shi'ite militias swept through Sunni areas and killed hundreds in retaliation. The conflict degenerated into a dirty war of Sunni and Shi'ite death squads kidnapping, torturing and murdering innocent Iraqis in droves. Sometimes they came by night, but mass abductions also occurred in broad daylight in crowded areas of Baghdad, and were often carried out by men wearing police uniforms. Bodies were dumped all over the capital as well as in other cities that had mixed sect populations. Most of the deaths in Baghdad were caused by execution-style shootings, a death squad hallmark.[49] The Iraq Study Group aptly described the nature of the conflict following the mosque bombing as a 'deadly cycle' in which 'Sunni insurgent attacks spark large-scale Shi'ite reprisals, and vice versa', and also led to a kind of sectarian cleansing as civilians fled areas where their sect was in the minority for the relative safety of areas dominated and protected by a militia of the same sect.[50]

So many deaths were occurring in unreported death squad attacks that the creators of the Brookings Institution's 'Iraq Index' stopped differentiating between civilians killed by confirmed 'acts of war' versus 'acts of crime', which was the former classification the 'Iraq Index' used for sectarian murders, because 'it is no longer practical'. That is the reason for the substantially different numbers in Table 9.1, which employed the older counting method, and Table 9.2, which includes all violent deaths including 'crime' or 'sectarian violence' as tallied by the UN and the Iraqi Ministry of Health. Over 34,000 Iraqi civilian deaths were counted in 2006.[51]

Particularly notable is the fact that civilian deaths and the average number of daily attacks in Baghdad both began to increase significantly in the late summer and early fall, during the same time period as Operation 'Together Forward', the major Baghdad security effort by Iraqi and US forces (Table 9.3).[54] This operation demanded the deployment of more troops to Baghdad from other areas of the country, contributing to a reduction in manpower available for security in the other provinces. At this point, the Americans and the Iraqi government were getting the worst of both worlds: the influx of troops to Baghdad failed to slow the rate of violence there, but left the rest of the country undermanned and more vulnerable to insurgent and militia attacks.

*Table 9.1* Iraqi civilians killed as a result of war-related violence, July 2004–December 2005[52]

| *Month* | *Civilians killed* |
| --- | --- |
| July 2004 | 375 |
| August 2004 | 746 |
| September 2004 | 663 |
| October 2004 | 558 |
| November 2004 | 1,867 |
| December 2004 | 562 |
| January 2005 | 677 |
| February 2005 | 828 |
| March 2005 | 562 |
| April 2005 | 429 |
| May 2005 | 1,006 |
| June 2005 | 746 |
| July 2005 | 844 |
| August 2005 | 2,489 |
| September 2005 | 1,150 |
| October 2005 | 565 |
| November 2005 | 826 |
| December 2005 | 534 |
| Total | 14,599 |

*Table 9.2* Iraqi civilians killed as a result of all violent causes, January 2006–December 2006[53]

| *Month* | *Civilians killed* |
| --- | --- |
| January 2006 | 1,778 |
| February 2006 | 2,165 |
| March 2006 | 2,378 |
| April 2006 | 2,284 |
| May 2006 | 2,269 |
| June 2006 | 3,149 |
| July 2006 | 3,590 |
| August 2006 | 3,009 |
| September 2006 | 3,345 |
| October 2006 | 3,709 |
| November 2006 | 3,462 |
| December 2006 | 2,914 |
| Total | 340,529 |

*Table 9.3* Average number of daily attacks by province, August 2005–February 2007[55]

| *Province* | *February–June 2005* | *August–January 2006* | *February–May 2006* | *May–August 2006* |
|---|---|---|---|---|
| Baghdad | 20.3 | 21 | 28.9 | 30.3 |
| Anbar | 12.3 | 23.3 | 22 | 31.1 |
| Salah ad Din | 8 | 13.8 | 13.7 | 15.5 |
| Diyala | 3.1 | 5.4 | 8.3 | 14.5 |
| Ninawa | 10.4 | 8.5 | 7.6 | 10.3 |
| Tamim | 3.1 | 4.7 | 4.3 | 4.8 |
| Basra | 1.2 | 1.1 | 2 | 2.4 |
| Babil | 1.5 | 1.8 | 1.2 | 1.9 |
| Maysan | 0.6 | 0.5 | 0.8 | 0.6 |
| Al Qadisyah | 0.1 | 0.2 | 0.1 | 0.8 |
| Dhi Qar | 0.2 | 0.2 | 0.5 | 0.4 |
| Wasit | 0.2 | 0.2 | 0 | 0.4 |
| Karbala | 0.2 | 0.2 | 0.2 | 0.1 |
| Muthanna | 0.2 | 0.1 | 0.2 | 0.2 |
| Najaf | 0.1 | 0.1 | 0.1 | 0.1 |
| Irbil | 0.1 | 0 | 0 | 0 |
| Sulaymaniyah | 0.1 | 0 | 0 | 0 |
| Dahuk | 0.1 | 0 | 0 | 0 |
| Total | 61.8 | 81.1 | 89.9 | 113.4 |

## A surge in strategy

As the situation on the ground in Iraq forced units to adapt on the fly, the military began working to better prepare its units for the irregular tasks of counterinsurgency by revamping its training and doctrine. In late 2005, General Casey established a counterinsurgency academy at the major US military base at Taji and began requiring all incoming company, battalion and brigade combat commanders to attend the educational regimen upon arrival in Iraq. The five-day course emphasized the classic counterinsurgency principles developed by French counterinsurgent Lieutenant Colonel David Galula: analyzing the population and thinking ahead about potentially harmful second- and third-order effects stemming from excessive force or other abuses during US operations.[56] The Defense Department's 2005 National Defense Strategy and 2006 Quadrennial Defense Review Report as well as the White House's 2006 National Security Strategy of the United States asserted the importance of building up capabilities for counterinsurgency and 'irregular warfare', including improved cultural awareness and advising skills and the development of more Special Operations units.[57] The National Training Center, 'the Army's most important training facility', moved away from its conventional combat exercises in favor of 'a complex insurgency scenario complete with civilians and all of the other things a unit could expect to find in Iraq'.[58]

*Table 9.4* Multiple-fatality bombings in Iraq, July 2004–December 2006[61]

| *Month* | *Attacks* |
|---|---|
| July 2004 | 11 |
| August 2004 | 13 |
| September 2004 | 16 |
| October 2004 | 17 |
| November 2004 | 11 |
| December 2004 | 17 |
| January 2005 | 28 |
| February 2005 | 18 |
| March 2005 | 13 |
| April 2005 | 21 |
| May 2005 | 36 |
| June 2005 | 34 |
| July 2005 | 26 |
| August 2005 | 27 |
| September 2005 | 46 |
| October 2005 | 39 |
| November 2005 | 41 |
| December 2005 | 21 |
| January 2006 | 30 |
| February 2006 | 39 |
| March 2006 | 37 |
| April 2006 | 40 |
| May 2006 | 56 |
| June 2006 | 57 |
| July 2006 | 53 |
| August 2006 | 52 |
| September 2006 | 57 |
| October 2006 | 56 |
| November 2006 | 65 |
| December 2006 | 69 |
| Total | 1,046 |

The Army issued an interim counterinsurgency manual, its first since the 1980s, in October 2004 in response to demand from the field. This manual was focused on tactical specifics, such as the use of vehicles in crowd control, and lacked a strategic orientation toward security for the population.[59]

As events in Iraq were at perhaps their nadir in 2006, the Army and Marine Corps released *Field Manual 3–24,* a completely new operational-level manual that sought to institutionalize the population security model as the centerpiece of the military's counterinsurgency doctrine. Coordinated by Army Lieutenant General David Petraeus and Marine Lieutenant General James F. Amos, the new manual marks a return to the basics of counterinsurgency. The manual's foreword, co-authored by

Petraeus, recalls Galula's warning that soldiers must be prepared to be more than just soldiers when fighting an insurgency:

> [Counterinsurgency] requires Soldiers and Marines to employ a mix of familiar combat tasks and skills more often associated with nonmilitary agencies. The balance between them depends on the local situation. Achieving this balance is not easy. It requires leaders at all levels to adjust their approach constantly. They must ensure that their Soldiers and Marines are ready to be greeted with a handshake or a hand grenade while taking on missions only infrequently practiced until recently at our combat training centers. Soldiers and Marines are expected to be nation builders as well as warriors. They must be prepared to help reestablish institutions and local security forces and assist in rebuilding infrastructure and basic services. They must be able to facilitate establishing local governance and the rule of law.... Conducting a successful counterinsurgency campaign requires a flexible, adaptive force led by agile, well-informed, culturally astute leaders.[60]

*Field Manual 3–24* establishes general principles for counterinsurgency that mirror the older conceptual works on the subject. Rather than detailing specific tactics, the manual outlines in general terms aspects of counterinsurgency that have increased in prominence in US strategy as the Iraq war has progressed. It includes a chapter devoted to 'developing host-nation security forces' and appendices on 'linguistic support' and 'social network analysis' for locating and tracking insurgents. Its first chapter lists 'paradoxes of counterinsurgency', such as 'the more force is used, the less effective it is' and 'the best weapons for COIN do not shoot', that strongly echo Galula's writings.[62]

In spite of the increasing commitment to counterinsurgency principles, 2006 was marked not by a triumph for Iraqi and US forces employing Galula's methods but by armed sectarian extremists slaughtering civilians with impunity and questions of whether Iraq had entered a period of civil war between Sunnis and Shi'ites. 'Clear, build, and hold' did not happen, and President Bush 'ended the year closeted with his war cabinet on his ranch trying to devise a new strategy, because the existing one had collapsed'.[63] Casey's approach, which had been endorsed by the President and the National Strategy for Victory in Iraq, had envisioned the gradual pull-back of US troops to support roles as Iraqi security forces took on the bulk of security tasks. It fell into disfavor during a year that had begun with high hopes for improved security and saw the death of al-Qaeda's leader in Iraq Abu Musab al-Zarqawi, but was dominated by rampant sectarian killing in Baghdad and unrelenting insurgent attacks in Anbar province. Amid the continuing deterioration of security in 2006, in the words of one of Lieutenant General Raymond Odierno's aides, 'This race to victory based on a timeline did not pass the common-sense test.'[64] The President was unwilling to accept the conclusions of the congressionally commissioned Iraq Study Group, a bipartisan blue-ribbon committee that recommended in December 2006 that US forces should accelerate security handovers to the Iraqis, shift more quickly to a support-only role and begin to withdraw sooner.[65]

Instead, the administration focused on a new strategy that relied on a more direct application of counterinsurgency principles and put more support behind 'clear, hold, and build'. As the violence in Iraq escalated alarmingly, American political and military leaders eventually came to accept the idea of a 'surge', an influx of US troop reinforcements to be deployed for a limited duration, mainly to Baghdad, to flood the streets and prevent more insurgent and militia attacks on civilians. Generals Abizaid and Casey had previously opposed the addition of more US troops because they felt that doing so would only make the US presence appear more like an occupation and exacerbate Iraqi dependence on Americans for security, but they eventually went along with the Bush administration's view that more troops were needed to tamp down the violence in the capital before the Iraqi government could make progress on political reconciliation.[66] The idea of adding more troops and engaging more directly in counterinsurgency-type population security operations in addition to supporting the Iraqi security forces was supported by the then-Lieutenant General Petraeus, the counterinsurgency manual's co-author and the officer who would succeed Casey in command of the effort in Iraq.[67] On 10 January 2007, President Bush announced the new strategy, marking the beginning of a new phase of the US effort in Iraq.

## Results and conclusions

The US military's approach to Iraq changed substantially on the operational and tactical levels during General Casey's command, but still failed strategically in emphasizing transition as a priority over counterinsurgency. Casey, unlike his predecessor, acknowledged the significance of the threat posed by the insurgency and proved open to advice in developing a campaign plan. The US military accepted that it could not win by trying to hunt down every last insurgent, but rather than adopting a focus on population security and gradually developing local government and security force capacity, its preference was to minimize the American role and transition rapidly to Iraqi responsibility even as security continued to deteriorate.

Although still not uniform, the application of counterinsurgency principles had improved across the board. In the operational and tactical realms, more officers saw a need to employ methods beyond combat operations to achieve security; the concept of full-spectrum operations was employed in the field to help make military means serve the political goal of luring the population into supporting the authorities. Military units tended to use more discriminating methods than they had before, seeking to use as little force as necessary and employing information operations to connect with the population rather than alienate them with more violence. Force posture was problematic in its emphasis on large, isolated bases, but this practice shifted in some areas in favor of small outposts in populated areas to keep closer watch on areas of responsibility. The US military improved its ability to train Iraqi forces by creating specialized teams. The US military's approach to Iraq since the summer of 2004 has demonstrated that it has learned aspects of effective counterinsurgency. Yet battlefield progress was not effectively translated into lasting strategic success due to the stubborn focus on an exit strategy of rapid transition. This emphasis on transition caused the military to fail to exploit successes like Tall Afar for strategic advantage and brought the United States to the brink of completely losing Iraq to civil war and chaos in late 2006.

It is evident that the American approach to Iraq, however much improved in its adherence to counterinsurgency principles, was ineffective in securing Iraq through the end of 2006. It was a mistake to focus too much on transition too soon. Major Niel Smith and Colonel Sean MacFarland, commander of the 1st Brigade of the 1st Armored Division, whose cooperation with tribal leaders in Ramadi in the fall of 2006 helped spark the Anbar Awakening that severely damaged al-Qaeda in Iraq, describe the alternative:

> Instead of telling [the Sunni sheikhs] that we would leave soon and they must assume responsibility for their own security, we told them that we would stay as long as necessary to defeat the terrorists. That was the message they had been waiting to hear. As long as they perceived us as mere interlopers, they dared not throw in their lot with ours. When they began to think of us as reliable partners, their attitudes began to change.[68]

Transition of responsibility to, and exit from, a host nation, in most cases, is and should be the ultimate objective when the United States intervenes on foreign soil. However, the American experience in Iraq from 2004 through 2006 illustrates the dangerous consequences when transition is the objective. Counterinsurgents must demonstrate staying power if they are to break the grip of militants over the population. Without the promise of protection, civilians will bandwagon with insurgents in order to survive. Effective host-nation security forces cannot be expected to take on the responsibility of securing the population for at least a few years, particularly when, as in Iraq, they have to be created from scratch. The aim of transitioning out of Iraq rapidly was based on valid concerns about the utility of force and the military's ability to achieve much success in a complex counterinsurgency environment in which it was widely perceived as an occupying force. However, transition pursued too quickly invites strategic failure.

Armed force alone does not achieve the political goals that determine 'victory' in counterinsurgency, or in any war. That does not mean that it is ineffective or has no role to play in the achievement of incremental objectives, such as separating the insurgents from the population, that are crucial to ultimate success. While 'war amongst the people' is political, it is still a kind of war. In counterinsurgency, military force to 'capture' the will of the people through security has an essential utility. The failure of Iraqi forces to perform this function required US troops to perform this role, which they proved capable of fulfilling as the post-2006 strategy shift took hold. Militaries confronting insurgency must therefore seek to use force appropriately on the operational and tactical levels, in concert with an integrated political–military approach, rather than simply focusing on exit strategies. There is no shortcut to success in counterinsurgency.

## Notes

This chapter has been previously published as Brian Burton and John Nagl, 'Learning As We Go: The US Adapts to Counterinsurgency in Iraq, July 2004–December 2006', *Small Wars and Insurgencies*, 19, 3 (2008): 303–327. We would like to thank Taylor and Francis for permission to reprint it here.

1 The White House, 'President Bush Delivers Commencement Address at United States Air Force Academy'.
2 General (ret.) Barry McCaffrey, US Army, *NewsHour with Jim Lehrer*, 30 June 2008.
3 General (ret.) John Keane, US Army, *NewsHour with Jim Lehrer*, 18 April 2006.
4 For detailed accounts of the flawed approach during Sanchez's command, see Ricks, *Fiasco*, and Hashim, *Insurgency and Counterinsurgency in Iraq*, passim.
5 Wright and Reese, *On Point II*, 569.
6 Smith, *The Utility of Force*, 19.
7 Ricks, *Fiasco*, 392–393.
8 Woodward, *State of Denial*, 410.
9 Ricks, *Fiasco*, 392–4; Burton interview with Kalev Sepp, Monterey, CA, 2 March 2007; Sepp, 'Best Practices in Counterinsurgency', 8–12. Sepp now serves as the Deputy Assistant Secretary of Defense for Special Operations.
10 Woodward, *State of Denial*, 315–316.
11 Metz, *Learning from Iraq*, 51–52.
12 O'Hanlon and Campbell, 'Iraq Index: Tracking Reconstruction and Security in Post-Saddam Iraq', 21.
13 Woodward, *State of Denial*, 410–411.
14 Woodward, *State of Denial*, 319, 426.
15 Woodward, *State of Denial*, 417–448; Secretary of State Condoleezza Rice, 'Iraq and US Policy'.
16 National Security Council, 'National Strategy for Victory in Iraq', 8.
17 National Security Council, 'National Strategy for Victory in Iraq', 28–29.
18 Ricks, 'The Lessons of Counterinsurgency'.
19 The 3rd ACR's performance in Tall Afar was so impressive that it was used as the case study illustrating 'Clear, Hold, and Build' in *Field Manual 3–24*. See pp. 182–184.

20 Ricks, 'The Lessons of Counterinsurgency'; Ricks, *Fiasco*, 419–424; Packer, 'The Lesson of Tal Afar'.
21 A similar learning process in the direction of population-centric counterinsurgency was occurring concurrently among US forces deployed to Afghanistan. See Phillips, 'To Woo Afghan Locals, US Troops Settle In'.
22 Ricks, *Fiasco*, 414–416.
23 Chiarelli and Michaelis, 'Winning the Peace', 15.
24 Chiarelli and Michaelis, 'Winning the Peace', 16.
25 Gibson, 'Battlefield Victories and Strategic Success', 48.
26 Gibson, 'Battlefield Victories and Strategic Success', 49–53.
27 Brown, 'Commander's Assessment', 31–33.
28 Burton interview with Army Captain Forrest Cook, Stanford, CA, 28 February 2007; Burton interview with Army Captain Chad Caldwell, Stanford, CA, 5 March 2007; Burton interview with Army Major Keith Walters, Stanford, CA, 26 February 2007.
29 Ricks, *Fiasco*, 416–417.
30 Burton interview with Army Lieutenant Colonel Chris Gibson, Stanford, CA, 13 March 2007; Gibson, 'Battlefield Victories and Strategic Success'.
31 Ricks, 'The Lessons of Counterinsurgency'; Ricks, *Fiasco*, 423.
32 Brown, 'Commander's Assessment', 28.
33 Ricks, *Fiasco*, 394; Cordesman, *Iraqi Force Development*, 90; 'MNSTC-I Information', Multinational Security Transition Command – Iraq. Available at: www.mnstci.iraq.centcom.mil/mission.htm (accessed 16 March 2007).
34 Cordesman, *Iraqi Force Development*, 90–91, 103–107.
35 Cordesman, *Iraqi Force Development*, 128–129, 131–132.
36 Lardner, 'Roots of Iraq Weapon Probes Date to 2004'.
37 Government Accountability Office, 'Stabilizing Iraq', 16.
38 Office of the Press Secretary, 'Fact Sheet: President Bush Received Briefing from Military Commanders'.
39 Government Accountability Office, 'Stabilizing Iraq', 17.
40 Cordesman, *Iraqi Force Development*, 423–427.
41 Metz, *Learning from Iraq*, 56–57; Michael Moss, 'How Iraq Police Reform Became Casualty of War'.
42 Metz, *Learning from Iraq*, 56–7.
43 Burton interview with Army Captain Forrest Cook, 28 February 2007.
44 Burton interview with Army Captain Chad Caldwell, 5 March 2007.
45 Hashim, *Insurgency and Counterinsurgency in Iraq*, 40–47.
46 Ricks, *Fiasco*, 413–414; O'Hanlon and Campbell; 'Iraq Index', 14.
47 O'Hanlon and Campbell, 'Iraq Index', 11; Hashim, *Insurgency and Counterinsurgency in Iraq*, 49.
48 Moss, 'How Iraq Police Reform Became Casualty of War'.
49 Metz, *Learning from Iraq*, 54–55; Tavernise, 'Iraqi Death Toll Exceeded 34,000 in 2006, UN Says'; Ricks, 'Shrine Bombing as War's Turning Point Debated'; UN Assistance Mission for Iraq (UNAMI), 'Human Rights Report'.
50 Baker and Hamilton *et al.*, *The Iraq Study Group Report*, 4.
51 O'Hanlon and Campbell, 'Iraq Index', 13; Tavernise, 'Iraq Death Toll Exceeded 34,000 in 2006, UN Says'; UNAMI, 'Human Rights Report'.
52 O'Hanlon and Campbell, 'Iraq Index', 13.
53 O'Hanlon and Campbell, 'Iraq Index', 13.
54 O'Hanlon and Campbell, 'Iraq Index', 23.
55 O'Hanlon and Campbell, 'Iraq Index', 23. In Defense Department reporting, attacks are defined as any known attack on US or Coalition forces, Iraqi security forces or civilians, and typically include improvised explosive devices (IEDs), small-arms fire and indirect weapons fire. Exact column date intervals: 29

August 2005–20 January 2006; 11 February–12 May 2006; 20 May–4 August 2006; 12 August–10 November 2006; 11 November 2006–9 February 2007.

56 Ricks, 'US Counterinsurgency Academy Giving Officers a New Mindset'; Burton interview with Army Captain Chad Caldwell, Stanford, CA, 5 March 2007.

57 Department of Defense, *Quadrennial Defense Review Report*, 41–45; Metz, *Learning from Iraq*, 58–63. 'Special Operations units include Delta Force, Army Rangers, Navy SEALs, and Army Special Forces (Green Berets)'.

58 Metz, *Learning from Iraq*, 64–65; Cone, 'The Changing National Training Center', 70–79.

59 Department of the Army, *Counterinsurgency Operations*, FMI (Field Manual Interim), 3–07.22.

60 Petraeus and Amos, 'Foreword', *FM* (Field Manual) *3–24/MCWP* (Marine Corps Warfighting Publication), 3–33.5, *Counterinsurgency*.

61 O'Hanlon and Campbell, 'Iraq Index', 20.

62 *FM 3–24*, 1–22.

63 Sanger, Gordon and Burns, 'Chaos Overran Iraq Plan in '06, Bush Team Says'.

64 Spiegel, 'Lt. Gen. Raymond Odierno Embodies "Surge" in Iraq'.

65 Sanger, Gordon and Burns, 'Chaos Overran Iraq Plan in '06, Bush Team Says'; Baker III *et al.*, *Iraq Study Group Report*.

66 Gordon, 'New Iraq Strategy Emerges'; Gordon and Sanger, 'Commander Said to be Open to More Troops'.

67 Gordon and Shanker, 'Bush to Name a New General to Oversee Iraq'; Gordon, 'A New Commander, in Step with the White House'.

68 Smith and MacFarland, 'Anbar Awakens', 44.

## Bibliography

Baker, James A. III, *et al. The Iraq Study Group Report*. New York: Vintage, 2006.

Brown, Lieutenant Colonel Ross A. 'Commander's Assessment: South Baghdad'. *Military Review* 87, 1 (January/February 2007): 27–34.

Chiarelli, Major General Peter W. and Major Patrick R. Michaelis. 'Winning the Peace: The Requirement for Full-Spectrum Operations'. *Military Review* 85, 4 (July/August 2005): 4–17.

Cone, Brigadier General Robert W. 'The Changing National Training Center'. *Military Review* 86, 3 (May/June 2006): 70–79.

Cordesman, Anthony. *Iraqi Force Development: Can Iraqi Forces Do the Job?* Center for Strategic and International Studies, Washington, DC, 23 August 2006.

Department of the Army. *Counterinsurgency Operations*, FMI (Field Manual Interim) 3–07.22. Washington, DC: Department of the Army, October 2004.

Department of Defense. *Quadrennial Defense Review Report*. Washington, DC: Department of Defense, 6 February 2006, 41–45.

Gibson, Lieutenant Colonel Chris. 'Battlefield Victories and Strategic Success: The Path Forward in Iraq'. *Military Review* 86, 5 (September/October 2006): 47–59.

Gordon, Michael R. 'New Iraq Strategy Emerges: Security, Then Politics'. *New York Times*, 20 December 2006.

Gordon, Michael R. 'A New Commander, in Step with the White House'. *New York Times*, 6 January 2007.

Gordon, Michael R. 'Occupation Plan for Iraq Faulted in Army History'. *New York Times*, 29 June 2008.

Gordon, Michael R. and D.E. Sanger. 'Commander Said to be Open to More Troops'. *New York Times*, 24 December 2006.

Gordon, Michael R. and Thom Shanker. 'Bush to Name a New General to Oversee Iraq'. *New York Times*, 5 January 2007.
Government Accountability Office. 'Stabilizing Iraq: An Assessment of the Security Situation', GAO-06–1094T, 11 September 2006. Available at: www.gao.gov/new.items/d061094t.pdf.
Hashim, Ahmed S. *Insurgency and Counterinsurgency in Iraq*. Ithaca, NY: Cornell University Press, 2006.
Keane, General (ret.). *NewsHour with Jim Lehrer*, 18 April 2006. Available at: www.pbs.org/newshour/bb/military/jan-june06/rumsfeld_4–18.html.
Lardner, Richard. 'Roots of Iraq Weapon Probes Date to 2004'. *USA Today*, 27 September 2007.
McCaffrey, General (ret.) Barry. *NewsHour with Jim Lehrer*, 30 June 2008. Available at: www.pbs.org/newshour/bb/middle_east/jan-june08/iraqreport_06–30.html.
Metz, Steven. *Learning from Iraq: Counterinsurgency in American Strategy*. Carlisle, PA: US Army War College Strategic Studies Institute, January 2007.
Moss, Michael. 'How Iraq Police Reform Became Casualty of War'. *New York Times*, 22 May 2006.
National Security Council. 'National Strategy for Victory in Iraq', The White House, 30 November 2005, 8. Available at: www.whitehouse.gov/infocus/iraq/iraq_ national_ strategy_20051130.pdf.
Office of the Press Secretary, 'Fact Sheet: President Bush Received Briefing from Military Commanders', The White House, 17 February 2006. Available at: www.whitehouse.gov/news/releases/2006/02/20060217–2.html.
O'Hanlon M.E. and J. H. Campbell, 'Iraq Index: Tracking Reconstruction and Security in Post-Saddam Iraq'. The Brookings Institution. Available at: www.brookings.edu/FP/saban/iraq/index20050318.pdf.
Packer, George. 'The Lesson of Tal Afar'. *New Yorker*, 10 April 2006.
Petraeus, Lieutenant General David, US Army and Lieutenant General James F. Amos, USMC. 'Foreword'. *FM* (Field Manual) *3–24/MCWP* (Marine Corps Warfighting Publication) 3–33.5, *Counterinsurgency*. Washington, DC: Department of the Army, December 2006.
Phillips, Michael M. 'To Woo Afghan Locals, US Troops Settle In'. *Wall Street Journal*, 9 April 2007.
Rice, Condoleezza. 'Iraq and US Policy', Opening Remarks Before the Senate Foreign Relations Committee, 19 October 2005. Available at: www.state.gov/secretary/rm/2005/55303.htm.
Ricks, Thomas E. *Fiasco: The American Military Adventure in Iraq*. New York: Penguin Press, 2006.
Ricks, Thomas E. 'The Lessons of Counterinsurgency'. *Washington Post*, 16 February 2006.
Ricks, Thomas E. 'US Counterinsurgency Academy Giving Officers a New Mindset'. *Washington Post*, 21 February 2006.
Ricks, Thomas E. 'Shrine Bombing as War's Turning Point Debated'. *Washington Post*, 13 March 2007.
Sanger, David E., Michael R. Gordon and John F. Burns. 'Chaos Overran Iraq Plan in '06, Bush Team Says'. *New York Times*, 2 January 2007.
Sepp, Kalev. 'Best Practices in Counterinsurgency'. *Military Review*, 85, 3 (2005): 8–12.
Smith, Major Niel and Colonel Sean MacFarland. 'Anbar Awakens: The Tipping Point'. *Military Review* 88, 2 (March/April 2008): 41–52.

Smith, Rupert. *The Utility of Force.* New York: Vintage Books, 2007.

Spiegel, Peter. 'Lieutenant General Raymond Odierno Embodies "Surge" in Iraq'. *Los Angeles Times,* 28 April 2008.

Tavernise, Sabrina. 'Iraqi Death Toll Exceeded 34,000 in 2006, UN Says'. *New York Times,* 17 January 2007.

UN Assistance Mission for Iraq (UNAMI). 'Human Rights Report: 1 November–31 December 2006'. Available at: www.uniraq.org/FileLib/misc/HR%20Report%20 Nov%20Dec%202006%20EN.pdf.

White House, The. 'President Bush Delivers Commencement Address at United States Air Force Academy', 28 May 2008. Available at: www.whitehouse.gov/ news/releases/2008/05/print/20080528–2.html.

Woodward, Bob. *State of Denial.* New York: Simon & Schuster, 2006.

Wright, Donald P. and Colonel Timothy R. Reese. *On Point II: Transition to the New Campaign.* Washington, DC: Government Printing Office, June 2008. Available at: http://usacac.army.mil/CAC2/CSI/OP2.asp.

# 10 Inviting the Leviathan

## External forces, war and state-building in Afghanistan

*Jan Angstrom*

War has often been thought to be intimately linked with the formation of states and the system of states. Charles Tilly and others have long argued that the formation of states in Europe during the sixteenth and seventeenth centuries happened in interaction with war.[1] Due to the costs of war, it became necessary to establish a centralised and well-functioning administration for extracting taxes. This process was made possible, and later reinforced, by the centralisation of the means for coercion. Eventually, only the state had the capabilities to wage war and war thus forced the multi-plurality of actors in late medieval Europe to copy one another and create the system of states. Today, however, the links between war and the formation of states seems less than clear-cut. Protracted wars in Central Asia, West and Central Africa have not generated strong states thus far.[2]

Instead, Western armed forces have increasingly been engaged in internationally orchestrated and supported state-building efforts in the former Third World since the end of the Cold War. There are numerous examples of this type of intervention in failed states; among others, Angola, Somalia, Afghanistan and the Democratic Republic of Congo spring to mind. However, the success rate of such enterprises has not been entirely convincing. It seems that many cases, in David Lake's words, in time come to be 'regarded not only as failed states, but as state-building failures'.[3] At the same time, in some cases where international forces have been involved, perhaps most notably Cambodia, a functioning state seems to have been reconstituted. How should this variation be explained? Why does international state-building work in some cases and not in others? To what extent can international forces contribute to creating a functioning state after state-collapse and civil war? And how should international forces operate to successfully build states in the aftermath of civil war?

The underlying assumption behind many interventions in civil wars today is that the intervening forces can do good. By intervening, it is thought, peace and stability will ensue, democracy can be established, cleptocracy will end and the overall welfare of the population will rise. Intervening forces can, the argument goes, provide security in order for the conflicting parties to demobilise and disarm without fear of renewed attacks. By supervising ceasefires, making sure that the combating parties stick to agreements and in some cases even by seizing political control over the situation, they can start to rebuild a functioning state. And intervening forces can facilitate local state-building initiatives by providing order, stability, and sustain the initially weak indigenous state, while it recuperates and starts to work. Central to the current state-building paradigm is that the intervening forces will supervise and to some extent guarantee a swift election process and democratisation. In turn, democracy, it is claimed, creates institutions that can overcome distrust, grievances and allow for political participation.[4]

However, the underlying logic of this policy agenda has been criticised. While some criticise the entire endeavour of trying to export a particular form of social order – the state – to areas that have no experience and perhaps no infrastructure to operate as states, most critics focus on how state-building is implemented in their explanations of the failure of international state-building. According to Roland Paris, the failure can be explained by an over-reliance upon liberal theory on behalf of the international community. Democracy and the market-economy are inherently competitive and in the immediate aftermath of violent conflict, Paris argues, it is by no means sure that competition is something the participants can handle without resorting to renewed armed conflict.[5] Paris suggests that one must first build a strong state – create a Hobbesian Leviathan, 'a power to keep men in awe' – before democratisation can start.[6]

Alexander Downes has pointed out some problems with the notion of external forces serving as a Leviathan. Most importantly, the local parties are aware of the fact that the international forces will be heading home at one point or another, which means that the basic security problem for the actors remains. Incentives are not to hand in any weapons and demobilise, but rather to hide the weapons and prepare for renewed conflict. The invited Leviathan, in other words, is merely an interlude before the coming war.[7] Some also suggest that the poor track record of international state-building is due to a lack of, or poorly implemented integration of, economic, political, diplomatic and military efforts.[8] Furthermore, some, most notably Edward Mansfield and Jack Snyder, have suggested that democratisation can lead to both internal and international war.[9] Finally, the stakes involved and the purpose of the war have been used to explain why some areas remain beset by protracted civil war.[10]

Other criticism can also be levelled at the liberal state-building paradigm. It does not take into account the strategic nature of state-building efforts. In effect, this means that the interactive part of state-building and legitimacy has been underdeveloped. Moreover, what seems to form the foundation of the current state-building practice is conflict theory or conflict-resolution theory. This is all very well, of course, but what is needed is a theory of the state and a theory of legitimacy.

In an important addition to the debate, which takes this criticism into account, David Lake has developed an alternative form of state-building – one drawing upon a relational understanding of legitimacy. Lake argues that the problem with the current liberal state-building paradigm is that it draws upon a flawed conception of legitimacy. It pre-supposes that legitimacy stems from the creation of institutions; after they are inaugurated, they are entrusted with creating order. Drawing upon a single case study, Somalia, Lake argues that an international state-building paradigm based upon relational legitimacy, namely a bargaining process between ruler and ruled, has a better chance of success. The relational concept of legitimacy pre-supposes that legitimacy is conditioned by the creation of order. In effect, this puts the traditional framework of state-building on its head. It is order that creates legitimacy, not the other way around.[11]

Lake's theory, however, is in need of further empirical testing. Moreover, it could be argued that Lake's model needs further development regarding the parties' strategic considerations when explaining, for example, why the parties opt to pursue strategies that diminish the value of what they strive for. And why they choose strategies that effectively prevent the disclosure of information that, if shared, would enable them to strike a more preferable bargain. Put in empirical terms, many regions of the world have best been described as conflict zones since the end of the Cold War. In Central Asia, Central as well as Western Africa, there has been near continuous violence over the past few decades. For example, there has been war in Afghanistan for 30 years. Provided that wars are costly and risky undertakings, it is surprising that the actors have not managed to strike a bargain that could provide a durable peace. In the Afghan case, estimates show that nearly 50 per cent of the indigenous population has become casualties – killed, wounded, or homeless – over the past three decades of war.[12]

Why do non-state actors go to war and continue to fight war? Why are they unable to reach a bargain in which the parties would be better off than continuing the war and turmoil? And why do warlords, rebel factions and incumbent regimes opt for strategies that undermine the asset they are trying to gain? In the state-building context, this does not only imply that wars will continue, but also critically for this chapter, so will state failure, despite the best intentions from intervening forces.

The aim of this chapter is to extend and develop Lake's new relational legitimacy state-building paradigm by examining phases of the state-building process in Afghanistan. This chapter develops an explanation of failure of state-building that draws heavily upon bargaining theory, strategic thought and the logic of Lake's relational concept of legitimacy. It is this chapter's central claim that internationally managed state-building conditions the strategic environment in which the actors find themselves. In doing so, intervening forces can foster the actors' expectations of future stability as well as co-opt them to create a homegrown Leviathan. The contribution of this study is that it further develops a theory of the role of force in international state-building efforts, based on Lake's concept of relational legitimacy, by examining state-building phases in the 30-year war in Afghanistan. In doing so, it can explain when and how intervening forces can assist in state-building efforts.

Testing a relational legitimacy model of international state-building, however, is very difficult methodologically. The reason is, as Lake recognises, that since 1990 the liberal state-building approach has been used in practically all of the cases. Following Lake, this study also uses an interrupted time-series design to solve this analytical problem.[13] By examining variation in state-building approaches in the 30-year war in Afghanistan, it is possible to discern patterns of behaviour towards state-building. Afghanistan, moreover, can also be considered as a stern test for a theory based upon a relational understanding of legitimacy, since it has experienced war for nearly 30 years without being able to generate a state. It has experienced four different attempts to build a state, and if elements of a relational state-building model have been present, one can question their utility.[14]

Afghanistan can be considered a failed state. Following Robert Rotberg, failed states are defined as being 'tense, deeply conflicted, dangerous, and contested bitterly by warring factions'.[15] Lake goes further and, following Max Weber's classical conception of the state as an actor having the legitimate monopoly of violence, claims that a failed state neither possess legitimacy nor the monopoly of violence.[16] Failed states, therefore, have poor capabilities to 'penetrate society, regulate social relationships, extract resources, and appropriate or use resources in determined ways'.[17]

The chapter continues as follows. The next section will outline the background to the problem of modern war, failed states and the problem of order. The following section will discuss Lake's theory of legitimacy and its role in state-building. The next section will analyse the phases in the state-building process in Afghanistan over the past 30 years. Finally, there will be some conclusions and suggested further areas of research.

## War, failed states and the problem of order

Scholars widely acknowledge that wars have played a central role in the emergence of the modern state-system as well the emergence of, and sometimes collapse of, individual states. The argument, in short, is that the constant presence of external threats makes the capability to conduct war a key to survival for a political system. Due to some internal advantages, mainly an increasing administrative capability, which made extraction of taxes efficient and recruitment of soldiers easier, the state became a better organisation for war than competing systems of political order in the sixteenth and seventeenth centuries. External wars, in turn, increased the state's administrative powers and, therefore, Charles Tilly concluded that states made war and war made the state.[18]

Following Tilly, there has been a new generation of scholars that examined the state-formation processes in other regions of the world. In an important qualification of the state-formation and war nexus, Victoria Tin-bor Hui argued that the creation of standing armies need not generate a state system, as standing armies composed of, for example, mercenaries weaken state cohesion. Centralisation of the means of coercion, therefore, does not necessarily lead to the creation of states.[19] Furthermore, in his study of African state-formation, Jeffrey Herbst pointed out that the political structure, geography and demography of large parts of Africa never provided incentives for large standing armies as a means for security.[20] Miguel Centeno noted a similar logic in his study of Latin American state-formation, where he suggested that the inability to wage total war produced limited state power.[21]

Anna Leander has, in an important text in this context, argued that the Tillyean processes do not have the same effect in today's Third World as it did in seventeenth-century Europe. Tilly argued that the centralisation of coercion and capital, increased administrative capacity and a grand bargain with the nobility formed the state in Europe. Indeed, Leander argues, it is precisely these factors that today prevent the state from forming. They might even propel a failing state towards complete collapse.[22]

Leander argues that the same Tillyean processes work in the opposite direction in today's small wars. There cannot be a centralisation of either coercion or capital, because the state's reach is too small, which means that the provider of (self-)preservation becomes the local leader. Similarly, since the incentives of donor nations, intervening forces and non-governmental organisations (NGOs) are to show the home audience that they are doing good, aid and support goes directly to the local leaders because it is assumed that the central government is corrupt or incapable of delivering aid to those who need it quickly. This effectively prevents individuals from having any allegiance to the state. War, in this context, does not lead to centralisation, but rather the opposite.

Furthermore, privatisation of the use of force does not create incentives for a central bureaucracy. War, in this context, does not lead to greater administrative capacity, but rather to further criminalisation of bureaucracy, as war may be thought of as a source of income.

Finally, there is no tacit bargain with civilian society, as the state cannot offer to provide the most important public good of all, security. If a bargain is struck, it is made between the local leaders and external powers so there are no incentives to strike a deal and reinforce the central state level. War, in this context, separates people from the state.[23] State death is, however, exceedingly rare, as Tanisha Fazal has demonstrated, and Robert Jackson points to the role of external powers, as well as international organisations, and norms in the concept of quasi-statehood.[24] Comparing war and state-formation in Vietnam and Afghanistan, Brian Taylor and

Roxanna Botea argue that the lack of a common ethnicity in Afghanistan has hampered state-formation, in comparison with Vietnam.[25] However, this argument underestimates the role of how war is conducted. In Vietnam, the decisive factor was the capacity to wage large-scale modern battle. In Afghanistan, as we shall see, victory in war is not dependent upon this factor. War, it seems, is no longer a part of the solution regarding state-formation, but rather a part of the problem. It perpetuates the situation of state-failure and it creates protracted problems of self-preservation for the individuals.

State-building in failed states and the aftermath of civil war can fruitfully be understood as a collective action problem. Explaining social phenomena with so-called social dilemmas is a long-standing tradition. Political philosophers ranging from Aristotle in ancient Greece to European Enlightenment thinkers such as David Hume, Thomas Hobbes and Jean-Jacques Rousseau all drew on the logic inherent in social dilemmas such as the tragedy of the commons to explain what we now would call sub-optimal outcomes.[26] Starting the analysis of state-building in failed states from the perspective of a state of nature or from anarchy in the sense of no authority above the local actors, therefore, seems fruitful. A structural explanation, moreover, can serve to identify and highlight the importance of strategic choices made by the local actors. Finally, state-building is undoubtedly primarily a question of political order.

Contemporary war in failed states is inextricably linked to political order. 'Whatever else they may be', Harvey Waterman observes, 'civil wars are conflicts over political order'.[27] The problem of order, therefore, seems to be at the heart of war and no more so than in warfare in failed states. Modern war has been described as 'neither low-intensity nor some bastard offspring of war. Rather, it is WARRE in the elemental, Hobbesian sense of the word, by far the most important form of armed conflict in our time.'[28] Thomas Hobbes famously depicted a metaphorical state of nature in which life was 'nasty, brutish, and short', and, due to the fact that all individuals had the capacity to inflict harm on one another and that no institution could guarantee anyone's security, 'a war of all against all'.[29] The connection to Hobbes is also relevant in a different context. According to Raymond Millen, it is via Hobbes' insistence on self-preservation and potential threats against the individual that we can argue that the contract between the individual and the state is first about security. Self-preservation also means, Millen argues, that in a conflict situation, the general population will not wholeheartedly back any of the sides in the conflict out of fear of repercussions from the eventual winner.[30]

War, failed states and the problem of order are therefore inextricably linked to each other. Essentially, this approach suggests that the bargain that the parties failed to reach is about political order, and specifically, about the nature of the state. Rather than greed, ideology or hatred, parties in failed states strive for order – for a Leviathan.[31] To end the violence in the conflicts – or win the wars – one has to understand the rationale of continued fighting. This is not necessarily the same set of causes that initially set off violence. The political context for decision-makers changes once war has begun. Politics does not stop, as nineteenth-century Prussian theorist Carl von Clausewitz and generations of scholars thereafter would insist on, but war changes the decisions to be made and it raises the stakes involved.[32] This means that once war is initiated, the decision on how to conduct the war and the decision to end the fighting is not necessarily dependent on the decision to start the war. In the aftermath of civil war and in the midst of state-failure, the state can solve some of the most basic social dilemmas present in the situation. Following Hobbes, a relational understanding of legitimacy implies that, in order to avoid mutual security threats, fear and possible violent death, the parties are driven to create a state that can keep each at bay.[33]

## Inviting the Leviathan: the role of external military power in relational legitimacy state-building

If we can understand war in failed states as essentially about the recreation of political order and the making of a Leviathan, then a theory of the state, of legitimacy and of strategy becomes imperative; this in light of the need to understand the role of military power in the new environment and the trajectory of modern state-building. The liberal state-building paradigm starts from the assumption that legitimacy follows from institutions and law. The relational legitimacy state-building theory, instead, as Lake argues, assumes that legitimacy follows from order. In international state-building based on the liberal state-building paradigm, it becomes imperative to hold elections quickly and to build government institutions based on election results. The decisions made by the newly elected government and implemented by its agencies are considered – and pre-supposed to be – legitimate. Legitimate institutions can then create political order. Relational legitimacy turns this logic on its head. In international state-building that is built upon a relational understanding of legitimacy, it becomes imperative to create order quickly. When order is seen to be maintained, namely that the state is able to provide the public good of security, it will earn legitimacy. Once it possesses legitimacy it can gradually centralise means of coercion. Legitimacy is, in Lake's words, the 'alchemist's dross' that makes the state's coercive powers turn into righteous political authority.[34]

Relational legitimacy is inherently strategic. An aspiring central leader must strike a bargain with local actors and the population. The state wants to centralise capital and extract taxes from the population. In return, it should provide the population with security. If the state fails to deliver security from warring factions, the population will see this as the state breaking this bargain, and it will not consent to its rule. Instead it will opt for supporting some local actor. And if the state is too harsh in enforcing order, it will be seen as a threat to the security of the population, which duly will turn its attention elsewhere. If the state does not get the taxes, moreover, it will not have the possibility to acquire the means for coercion to provide security. This can be understood as a classical social dilemma. Although it would be rational for the population to avoid paying taxes to the government, if it does so the state will not be able to provide security, thus making the population worse off. Relational legitimacy is thus critically about striking a bargain. At the end of war, this is a difficult bargain to reach, as Leander has argued.[35] This makes bargaining theory relevant for explaining the logic of why some actors consent to a new state's government, while others do not.

The bargaining theory of war has recently emerged as a nearly all-encompassing theory of war.[36] It has been used to explain the origins of war, the duration of war, and the problems inherent in peace negotiations and conflict resolution. The bargaining perspective starts with the assumption that war is essentially a form of communication in the midst of a bargaining process between at least two parties. Even during war, fighting is not the only form of communication, since the parties can also continue diplomatic efforts while fighting. Bargaining theory gains its explanatory powers from focusing on alternative routes of action. By assuming that things could have been different – and that something as costly and risky as war is sub-optimal – advocates of bargaining theory can explain why wars break out using a rationalist framework.

Bargaining theory is not compatible with constructivist theories utilising, for example, psychological processes to make sense of war. The rationalist epistemology, however, makes bargaining theory compatible with a Clausewitzean understanding of war and the conduct of war.[37] Although bargaining theory has received much attention and has proven to be a valuable tool with which to understand the causes and duration of war, hardly anyone has used bargaining theory to gain insights into the internal logic of war. Moreover, bargaining theory is usually used in quantitative studies or using formal method, which has resulted in some criticism. Reiter even claims that the ability of quantitative work 'to test the bargaining model appears fundamentally limited'.[38] By using a case-study approach, this study avoids what Reiter calls 'crude proxy variables' to test the core propositions of bargaining theory.

The bargaining theory of war starts from the assumption that wars are always restrained – or, at least, that they can be understood to be restrained. Even the Second World War – according to some, a total war fought with the objective of unconditional surrender – can be understood as a bargaining process, as Thomas Schelling has demonstrated.[39] In case of wars in failed states, this suggests that the bargaining the actors cannot agree on is order. It is the drive for order that places the actors in the original Hobbesian social dilemma: if I give up my capability to inflict harm upon my neighbour, I will be at his mercy. Therefore, it is individually rational to maintain arms. However, if all actors think the same way, the outcome will be collectively irrational. No single power will be able to become a Leviathan, and therefore war and turmoil can continue. State-building theory and bargaining theory are therefore both necessary to understand protracted state-failure.

Why then should this behaviour be more widespread and have worse consequences in internal armed conflicts? As the state – an institution with capacity to facilitate a correct flow of information between the parties – breaks down, the opponents will become more like states in the international system; namely, manipulating information and keeping information from each other. The incentives to pursue this behaviour will be widespread as the potential gains of cheating or defecting during the post-state bargaining process will be massive. Consider, for example, that this negotiation process will determine the future balance of power between the groups. Furthermore, if large-scale violence has begun, secrecy will be of utmost importance for the parties in the conduct of the war. Also, the parties have incentives to distort information about alleged atrocities since it may create international support and, at the same time, create an image of the other group as an enemy, which itself is a necessity for the fighting. Creating and maintaining enemy images also contributes to uniting one's own ethnic group. Barry Posen even argues that cohesion is a major determinant of victory in ethnic conflicts, especially since the parties are reliant upon small arms to wage the war.[40]

The parties may also have incentives to distort information to gain international support without necessarily resorting to assertions about atrocities. Consider that, during the Cold War, parties in internal conflicts could benefit logistically – for example, from the acquisition of weapons – by portraying themselves as either communist or anti-communist depending on which side in the Cold War could offer them the most advantageous deal. Arguably, the prime example of this behaviour is the UNITA rebels in the Angolan civil war, who changed allegiances between China and the United States. Hence, these incentives also obscure our understanding of civil violence.

Framing the state-failure problematique in terms of a social dilemma implies that any explanatory power is derived from the actors' rationalist decision-making. All of the main actors in wars in failed states – remnants of an incumbent state, intervening forces, local factions and the population at large – face different dilemmas, which inadvertently leads to sub-optimal outcomes.

The main influence of the invited Leviathan is that it changes and conditions a new strategic environment for the local actors – individuals as well as the warring factions. Through its greater economic and coercive powers, the invited Leviathan can create expectations of increased stability and lend its coercive powers to the home-grown Leviathan in the making.

It is through this process, Lake argues, that external powers' military forces can uphold order and thereby gain legitimacy for themselves and for the newly appointed state. In the process of gaining legitimacy for itself, however, the invited Leviathan may inadvertently weaken the legitimacy of the home-grown Leviathan, as the population may be more willing to consent to the rule of international forces. In this way, external military forces may undermine the state. In much the same vein, international aid and NGOs may undermine the legitimacy of the new central state by wanting acclamation for their human relief actions.

International forces thus face a dilemma. If they are to be successful at all and avoid being shot at by local actors, they must acquire legitimacy. However, in this process, they undermine the local state, thereby making their mission more difficult. The invited Leviathan therefore fails when it takes over the role of the incumbent state and implicitly strikes a bargain with the population. In a similar logic, if international forces stay too long, they are in danger of being interpreted as occupants, while if they stay too short, they do not significantly alter the strategic situation for the warring factions. NGOs face a similar dilemma in so far as if they 'claim' the honour of the aid to the population, the NGOs will earn the legitimacy; if they do, however, they undermine the state, thus prolonging the war and suffering. But if they do not, they will be seen to not deliver in the West, thus preventing greater support that can do miracles elsewhere.

The incumbent state, similarly, also faces a dilemma. If it acts too harshly, the general population will turn elsewhere for its security, in much the same vein as the invited Leviathan. If it fails to deliver security (and fails to keep its part of the bargain), it too will become obsolete and the general population will turn to local faction leaders for security needs. It also faces a dilemma towards the invited Leviathan in so far as it is seen not to control the outside forces, the state runs the risk of becoming obsolete, since it may appear that it is not the incumbent regime, but the outside forces that have the means for coercion.

In summary, there are multiple trajectories towards prolonged state-failure and failure in international state-building efforts. Before an external Leviathan becomes part of the conflict situation, the old state, local factions and the population have various strategic choices. Due to the stakes being high in security-related matters, the absence of credible commitments from the actors, distrust and uncertainty of others' intentions, the parties fail to strike a bargain that can prevent continued violence. And, because the violence itself further diminishes the capabilities of the state, any successful holder of the state will not be permanent; since neither the means for coercion, nor the capital is centralised, and can be allocated to potential supporters in return for a prolonged stay in power.[41]

When the international community attempts to intervene to end the bloodshed and build a state, the intervening forces – the invited Leviathan – alter the strategic make-up of the conflict. Thereby they can work as an enabler to solve the original social dilemma present in the immediate post-conflict situation. However, the intervening forces face a problem of 'relative cross-audience utility', borrowing a term from Ivan Arreguin-Toft, which leads to a renewed social dilemma.[42] Ideally, the intervening forces' strategy should make the warring factions disarm and centralise the means for coercion, make the incumbent state honour its bargain of providing public goods to the population, make the population pay taxes and defer legitimacy from the local faction leader to the state, make the international community strengthen the incumbent state, and make the population at home satisfied with a seemingly marginalised role in the state-building process. However, as described above there are plenty of trade-offs when it comes to a real situation – and these trade-offs in the worst case cause prolonged state-failure and prolonged war. The next section will test this theory against phases of state-building in Afghanistan.

## Afghanistan and the failure of invited Leviathan

The theory developed above suggests that international state-building efforts that are built upon an absolute concept of legitimacy will fail because such interventions fail to solve the basic social dilemma in which the local actors – be they warlords, incumbent governments or bands of militia – are situated. This means that the parties' strategic choice will be to continue to fight and, in turn, continue to diminish the public good they strive for – social order and security. After a brief introduction and overview of the 30-year war in Afghanistan (based in no small part on Tanner's seminal book),[43] the chapter will analyse state-building in Afghanistan with special reference to the role of an external Leviathan.

### *The Afghan 30-year war*

There have been four major attempts to create a viable state in Afghanistan over the past 30 years. The first phase consisted of the Soviet-supported regimes of Babrak Karmal and Mohammed Najibullah between 1979 and 1992. The second attempt involved the Mujahidin's rule under Burhanuddin Rabbani between 1992 and 1996 with little or no international support. The third attempt involved the Taliban-led, and Pakistani-supported, state between 1996 and 2001. The fourth attempt is the current process in which the United States leads a broad coalition in attempting to create a legitimate state in Afghanistan. If the theory can explain the various phases of state-building in the 30-year war in Afghanistan, certain empirical trends must be discerned. Either, the population and the local leaders invest significantly in the state-building process if this process follows the logic of relational legitimacy; or, the population and the local leaders do not invest significantly in the state-building process.

Before starting an analysis of Afghan state-building, one must recognise its structural constraints and historical heritage. To some extent, in the context of Afghan history it would be surprising if a strong state did emerge. The country is fragmented along ethnic lines (Pashtuns, Uzbeks, Tajiks and Hazaris), religious lines (although Islam dominates, there are multiple interpretations that to varying degrees have merged with local traditions), ideological lines, centre–periphery, and geographical and topographical constraints.[44] These conditions are hardly conducive to the formation of either a common nationality or a strong state.

It is also necessary to place the state-building efforts into a context. Even prior to Afghanistan's independence in 1919, Russia (and later the Soviet Union) intervened repeatedly in Afghan politics. During the 1940s–1970s, Soviet aid and assistance built a large part of the Afghan road network as well as several key airports. After a communist coup in 1978, in reaction to then-leader Mohammad Daoud's attempts to increase contacts with the West, the Soviet Union invaded Afghanistan in 1979 to assist the communist government, which had found itself in a precarious situation after its rise to power. Several factions openly fought against communist rule and Moscow intervened to stabilise the situation and create a Soviet-friendly state.

After having to deal with fierce resistance for a decade, the Soviet army left Afghanistan in 1989. At first sight, it may seem controversial to interpret the Soviet invasion of Afghanistan in 1979 as an attempt to build a functioning legitimate state in Afghanistan. However, the original motive of the Soviets was to assist the existing government in Afghanistan and support its 'counter-revolutionary' operations. The Soviet army struggled to come to terms with the security fears of the factions as well as conquering the terrain. As the war wore on, it eventually became clear that the Soviet army would not beat the Mujahidin and, in 1986, the Soviets initiated negotiations about a withdrawal from Afghanistan.[45]

It became clear early on that the already-splintered Mujahidin would not be able to overcome the social dilemma of post-Soviet Afghanistan. Security threats, fear and mutual distrust made the situation unsustainable. Although Afghanistan has always been fraught with in-group disagreements and fragmentation, as the war against the Soviet wore on, the Mujahidin split in different factions that sometimes collaborated and sometimes fought against each other, thus exacerbating the weakening of an already weakened Afghan national community. Although the resistance against the Soviet invaders served as a focal point and instigated a jihad, the pattern of disunity did not break after the Soviet withdrawal and, surprisingly to many observers, the Afghan leader Najibullah, who had succeeded Babrak Karmal in 1986, survived for nearly three years in power after the Soviet retreat in 1989.

After an interlude of Mujahidin rule in Afghanistan from 1992, the split between the most powerful warlords and their forces – then-President

Rabbani, Ahmed Shah Massoud, then-Prime Minister Gulbuddin Hekmatyar and Abdul Rashid Dostum – became apparent for everyone.[46] From the rubble of civil war among the former Mujahidin allies rose the Taliban movement. Between 1994 and 1996, the Taliban scored a swift military success and managed to conquer the major cities of Afghanistan. Led by the clerk Mullah Omar, the Taliban movement originated in the southern city of Kandahar and the implications of Taliban rule soon became clear in the seized areas. A strict sharia-based judiciary system was created.[47] Although the Taliban controlled some 90 per cent of the country by mid-2000, the resistance never stopped entirely. In particular, the so-called Northern Alliance led by Massoud continued the war. One could also question the ability of the Taliban state to penetrate society, since there were considerable differences between the cities and the countryside in adhering to sharia.

Following in the footsteps of the Taliban, the terrorist network al-Qaeda soon established itself in the country. Using Afghanistan as a base for recruitment, for training and for planning, al-Qaeda was able to launch several terrorist attacks around the world. Paradoxically, the most spectacular triumph of al-Qaeda – the attacks on New York and Washington, DC on 11 September 2001 – also marked the moment of its demise as an internal factor in Afghanistan. Following the terrorist attacks, the United States and its allies soon began a counter-offensive and ousted the Taliban government quickly. Since then, US-led state-building and counter-terrorist operations have been carried out. The next section will analyse the war and state-building in Afghanistan.

### *Explaining the failure of the external Leviathan*

Why has Afghanistan become not only a failed state, but also a long-lasting state-building failure? And to what extent have intervening forces been able to play the role of an invited Leviathan? The failed state in Afghanistan and the continuous state-building failures over the past 30 years can easily be incorporated into the theoretical framework developed above. This analysis will discuss the four phases of attempted state-building and analyse it through the actors' choices in relation to centralisation of coercion and capital as well as cross-audience utility.

Soviet state-building efforts failed miserably, as was evident in Afghanistan during the 1980s. The Soviet invasion effectively turned all the major sub-state Afghani actors against the regime in Kabul and against the Soviets. The original Soviet plan was to withdraw its forces in six months and during that period remodel the Afghan state structure according to the Soviet institutions with a strong party cadre at the top. In state-building terms, Barnett Rubin argues, the Soviet reluctance and inability to centralise the means for coercion (mobilising a total of 100,000 Soviet soldiers for the government's protection) meant that the state itself could not offer protection to its citizens.[48]

The Soviet state-building efforts were also hampered by the change in Soviet counter-insurgency tactics in 1983.[49] Acting too harshly meant that the Soviet forces did not gain legitimacy for themselves, nor for the Afghan state. Similarly, by not devising and communicating an exit strategy (the Soviet army therefore appeared as an occupying force rather than an invited Leviathan), the Mujahidin were allowed to play on Afghan nationalism to create legitimacy for themselves. The Mujahidin could also draw upon external support, mainly from the United States and Pakistan. This made it possible for the local level to provide security as well as create bonds of loyalty by handing out social benefits (in particular to widows of fallen Mujahidin fighters). The population, meanwhile, was as usual stuck between a rock and a hard place. According to Millen, the need for self-preservation in such a situation meant that the population tried to meddle in between, and not pledge their loyalty to any actor.[50] This can explain why none of the parties in Afghanistan later succeeded in establishing a legitimate state. Since no actor has been able, in any sustained fashion, to ensure security, they did not get the consent of the governed. By continuously shifting support (which is very rational for the individual) and supporting rivals to incumbent leaders, individuals undermine what they seek the most, security. As an 'invited' Leviathan, the Soviet Union failed to strike and uphold a bargain with the population and it failed to co-opt the local elites (the nobility), who turned into warlords. The Soviet version of a Leviathan certainly changed the strategic situation, but not into one conducive for the growth of a home-grown Leviathan. Instead, it set the conditions for continued state-failure among the local Afghan leaders.

Although the problem of warlordism was not new for the Soviet period, the war against the Soviets facilitated the warlords' acquisition of more power. In time, they effectively began to operate as states, centralising the means of coercion in their private armies and unifying sub-clans, controlling and taking a cut of the black-market trade in return for protection, and providing its populace with security and penetrating the social structures of society. Although it had some military advantages during the war against the Soviets, this system for political order, more similar to medieval princes and fiefdoms than a modern state, set the conditions for prolonged state-failure after the Soviet withdrawal.

The war against the Soviets had split the Afghan opposition according to ethnic, religious and/or regional lines. In this way, war did not contribute to forging a state, but rather set the conditions for prolonged state failure and mutual security fears. As the Mujahidin failed to unify during the war, this failure also pre-determined the post-war split. Similarly, the long war against the Soviets had put a war economy in place, heavily reliant on the poppy trade. Since drug-dealing was illegal, it made it difficult for the Afghan state in the early 1990s to extract taxes from the trade. Thereby, it refrained from a potential income that could have been used to provide public goods. Moreover, by calling for a jihad – a holy war

– against the invaders, the Afghan factions unleashed an unlimited war, instead of the more common, highly regulated war among the tribes.[51] This wrecked the, until then, dominating conventions of warfare in Afghanistan, and further deepened mutual distrust among the rival factions.

Still, strategy and warfare are important in explaining Afghan state failure, just as Tilly would have to us believe. However, what made the formative states in early modern Europe successful at war is not the same thing as what made the actors successful in war in Afghanistan. Richard Shultz and Andrea Dew suggests that one of the main military advantages of the Afghan tribal society, and an important reason why the utterly poor and relatively ill-trained soldiers have managed to successfully fend off invaders from Alexander the Great to the Soviet Union, is its dispersed and de-centralised character.[52] This is a key element in explaining why the Afghans fail to forge and maintain a state. War, rather surprisingly, therefore has still an important place in the state-building conundrum, but instead of generating states, what is successful in war in Afghanistan is its fragmentation. War in Afghanistan thus generates and regenerates the tribal system. Military power and its use, therefore, contribute to state-failure in so far as they contradict centralisation of coercion. The warlords and faction leaders have witnessed how ancient tribal tactics of limited war and guerrilla warfare have paid dividends, there are no incentives (and no external pressure) for copying the state. Instead, all the incentives are in favour of copying the other tribes.

As the Mujahidin leaders did not manage to strike a bargain or have a common political platform upon which to build in the post-Soviet period, it was hardly surprising that state failure persisted in the early-to-mid-1990s. The prolonged war, in addition, has also created mutual security fears. The failure to credibly commit to peace treaties has generated situations with mutual suspicion among the factions. The precarious balance of power between the Afghan peoples during the 1960s and 1970s, as a result, was not restored. As the civil war progressed, the warlords found it more difficult to control central government in Kabul for a sustained period of time. The absence of an invited Leviathan did not forge a home-grown one either.

In the midst of civil war, the Taliban movement transformed from a local force in southern Afghanistan to a near nation-wide force. By the time it took control over Kabul, its military power was superior to its competitors. As the movement grew and gradually came to control increasingly larger parts of Afghan territory, it also became clear that it instilled order through a very strict interpretation of sharia laws.[53] The Taliban leadership made no significant attempt to co-opt the former Mujahidin leaders and incorporate them in a grand bargain among the elites. Instead, it opted to try to control the state itself. Although there are some reports from NGOs that suggest that Taliban control of everyday life in the countryside was limited, it had control over the major cities and major airfields between 1998 and 2001. Even if Pakistan initially supported the Taliban movement, it did not work as an invited Leviathan, nor did it, in any substantial way, influence the creation of Taliban government institutions or procedures. Taliban rule was very much a home-grown wannabe Leviathan.

Although the rule seemingly produced order, at least in the south, one could argue that it did not generate a lasting effect on Taliban legitimacy. The reason is that soon after the United States and its allies initiated bombings after the 9/11 attacks, a newly formed 'southern alliance' did not have any substantial problems in recruiting supporters, even in the Pashtun heartland.[54] Meanwhile, in the northern regions, the Northern Alliance, consisting of the former allies and foes of the war against the Soviets as well as the civil war, continuously resisted the Taliban, even if it was hard pressed along the border towards Tadjikistan. Moreover, considering the short duration of Taliban rule in Afghanistan, one could possibly argue that the population, warlords and elites did not consolidate support for the Taliban, especially considering the country's recent history.

The latest attempt to build a state in Afghanistan and to find a viable political order was a direct result of the terrorist attacks on 11 September 2001. The terrorist network al-Qaeda had established itself gradually in Afghanistan under Taliban rule. When it struck against targets in the United States, it inadvertently also put an end to Taliban rule. Within two months of the attacks in New York and Washington, DC, forces from the Northern Alliance, supported by US airpower and special forces, had seized Kabul and had the Taliban retreating all over the country. Amidst the ongoing war, the United States also aimed to alter state structures in Afghanistan and create democracy. Instead of opting to control the government itself, the US chose to co-opt elites in Afghanistan – Pashtuni leaders as well as other ethnic groups' leaders – to strike a lasting bargain. The Bonn agreement from 2001 stated that all Mujahidin and all central Afghan forces should be controlled by the government in Kabul. In effect this was an attempt to centralise the means for coercion as well as strike a grand bargain among the elites. However, as has become evident subsequently, the means for coercion are still dispersed, with armed clashes exceeding one per day throughout the country. This situation becomes even more exacerbated as reports indicate that Afghanistan has no working police, judiciary system or army, which are slowly being reconstituted.[55]

The economic reconstruction and centralisation of capital has made some progress, but is – by Western standards – still painstakingly slow. As Rubin points out, one of the main problems has been to turn black market activity into the taxable domain.[56] Capital is also intimately linked with coercion. Those with a black-market livelihood tend to seek protection, not from the police, but from the local warlords, thus making themselves dependent on the local power structure instead of the central one. Centralisation of coercion and capital is thus still not implemented in Afghanistan.

In combination with the security sector reform of police forces and armed forces of Afghanistan, there are also two major intervening forces present in Afghanistan. There is, therefore, still reason to argue that there is an invited Leviathan present in the country. The intervening forces have, however, quite different tasks and different modus operandi. The intervening forces in northern Afghanistan have different approaches to this dilemma. While German aid efforts centralise in the area in which German forces operate, Swedish aid goes to the central government in Kabul. This has several short-term drawbacks for the Swedish forces – such as weakened legitimacy among the population – but it may prove effective in the long run. In the German-controlled area, the situation may be opposite. In the short-term, aid directed towards the area of the German forces may serve to build legitimacy for the German forces. However, in the long run there are problems related to the invited Leviathan stealing the central government's thunder and thus failing its mission of assisting the state-building process. In the south, mainly US and British forces are fighting side-by-side with Afghani government forces against resurrected Taliban forces. Accordingly, it is by no means sure that the population in the south will consider the Kabul government to be upholding its end of the bargain.

In terms of cross-audience utility, some of the intervening forces – the invited Leviathan – have also made a trade-off, it would seem, giving priority to their own legitimacy before that of the Afghan state. This is quite understandable given the fact that there is already a debate in several of the contributing states on whether or not it is a good idea to participate in operations in Afghanistan. Should the number of body bags sent home increase, this debate will surely increase and risk the political support for Afghan operations. Similarly, the forces must be seen to do good 'back home'.

In summary, the continuous state-failure in Afghanistan seems to be intimately linked with war, with absence of a viable order, and with absence of a grand bargain – a social contract – among the elites and among the population. As the analysis above has indicated, the data are so far anecdotal that international state-building efforts have taken different shapes during the 30 years of war. The degree of involvement in creating state institutions has varied, the degree of control of means of coercion has varied, the degree of control of capital has varied, and the degree of co-optation of the Afghani elites in a bargain has varied. Yet, state-failure has persisted. In shaping the strategic choices for the local actors, intervening forces during the 30 years of war have not always had a positive effect on the Afghan state. The fact that local factions could rely on and strike bargains with external powers, be they Pakistan, Iran the United States or the Soviet Union and its successor states, also prevented the local actors from finding a bargain that would last. It was always easier not to compromise on its goals and yet achieve support, than it was to compromise on the faction's goals and thereby build a viable political community.

## Conclusions

In this chapter, I have argued that international forces – an invited Leviathan – shape the strategic situation as they intervene in a failed state. However, these interventions need not have a positive effect on the resurrection of the state from state-collapse. Due to the research design issues (no international state-building success has been included), it is impossible to conclude whether or not an invited Leviathan can have a positive effect, or how it should act to reach success. Better data – in particular the share of aid directed to Kabul versus aid directed to the population by donating country and by NGO, as well as the degree of control of means of coercion and degree of co-optation – is also needed to test a relational legitimacy model more rigorously.

What can be concluded, though, is that war still seems to be linked to state-building, albeit possibly in a negative way. Due to the peculiarities of warfare in Afghanistan – large armies are not necessarily translated into political power – it seems that war regenerates the tribal system of local patronage. Moreover, what seems to be lacking in the relational model of legitimate state-building is a connection to the regional dynamics. The grand bargain of the state, in Europe once struck between the king and the nobility, is not simply translated to the elites and the Kabul government when it comes to Afghanistan. Instead, local warlords have been able to strike deals with external powers, effectively draining the legitimacy of the state. As it stands now, Afghanistan's 30 years of war and state failure still continue, and a grand bargain among the elites, the population and the invited Leviathan to hand over power and the responsibility to provide security in return for consent has not yet been struck.

Considering the analysis above, there are grounds for questioning the utility of force in external state-building operations. Home-grown use of force need not produce a strong Afghan state, and an invited Leviathan will be locked into hard-to-solve dilemmas. It will need to tread carefully to avoid alienating the local actors and therefore, by association, weaken the incumbent government, while at the same time, avoiding taking too much credit for its actions, and still generate order and security.

## Notes

This chapter has been previously published as Jan Angstrom, 'Inviting the Leviathan: External Forces, War, and State-Building in Afghanistan', *Small Wars and Insurgencies*, 19, 3 (2008): 374–396. We would like to thank Taylor and Francis for permission to reprint it here.

1 See, for example, Tilly, *Coercion, Capital, and European States, AD 990–1990*; Porter, *War and the Rise of the State*; van Creveld, *The Rise and Decline of the State.*
2 Holsti, *War, the State, and the State of War.*
3 Lake, 'Building Legitimate States After Civil Wars'.
4 As Lake, 'Building Legitimate States After Civil Wars', admits, there is not an explicit theory behind this model, but practice since the end of the Cold War implies such a model. Two recent RAND reports paint a similar picture of the practice of external state-building, see Dobbins *et al.*, *America's Role in Nation-Building*, as well as Dobbins, *The UN's Role in Nation-Building.*
5 Paris, *At War's End.*
6 Paris, 'Bringing the Leviathan Back In', 425–440.
7 Downes, 'The Problem with Negotiated Settlements to Ethnic Civil Wars', 230–279.
8 Fearon and Laitin, 'Neotrusteeship and the Problem of Weak States', 5–43.
9 Mansfield and Snyder, *Electing to Fight*; Snyder, *From Voting to Violence.*
10 Hartzell, 'Explaining the Stability of Negotiated Settlements to Intrastate Wars', 3–22; Walter, *Committing to Peace*; Doyle and Sambanis, *Making War and Building Peace*; Fortna, *Peace Time.*
11 Lake, 'Building Legitimate States After Civil Wars', 5–32.
12 Goodson, *Afghanistan's Endless War*, 5.
13 Lake, 'Building Legitimate States After Civil Wars'.
14 See also George and Bennett, *Case Studies and Theory Development in the Social Sciences.*
15 Rotberg, *State Failure and State Weakness in a Time of Terror*; Rotberg, *When States Fail.*
16 Lake, 'Building Legitimate States After Civil Wars', 6–10.
17 Goodson, *Afghanistan's Endless War*, 11.
18 Tilly, *Coercion, Capital, and European States, AD 990–1990.* See also Giddens, *The Nation-State and Violence*; Spruyt, *The Sovereign State and Its Competitors*; Davis and Pereira, *Irregular Armed Forces and their Role in Politics and State Formation.*
19 Tin-bor Hui, *War and State Formation in Ancient China and Early Modern Europe.*
20 Herbst, *States and Power in Africa.*
21 Centeno, 'Limited War and Limited States', 82–95.
22 Leander, 'Wars and the Un-Making of States', 69–80.
23 Leander, 'Wars and the Un-Making of States', 69–80.
24 Fazal, 'State Death in the International System', 311–344; Jackson, *Quasi-States: Sovereignty, International Relations, and the Third World.*
25 Taylor and Botea, 'Tilly Tally', 27–56.
26 See, for example, Kollock, 'Social Dilemmas', 183–214.
27 Waterman, 'Political Order and the Settlement of Civil Wars', 292.
28 Van Creveld, *The Transformation of War*, 22.
29 Hobbes, *Leviathan*; Doyle, *Ways of War and Peace*, 111–136; Tuck, 'Hobbes', 107–238.
30 Millen, 'The Hobbesian Notion of Self-Preservation Concerning Human Behavior During an Insurgency', 4–13.
31 See also Kalyvas, *The Logic of Violence in Civil War*; Weinstein, *Inside Rebellion.*
32 Duyvesteyn and Angstrom, *Rethinking the Nature of War.*
33 See also Sorensen, *Changes in Statehood*; Hobson, *The State and International Relations.*
34 Lake, 'Building Legitimate States After Civil Wars', 13.
35 Leander, 'Wars and the Un-Making of States'.
36 One should note that, although the theory has gained ground over the past decade, early work in this tradition was done by Schelling, *The Strategy of Conflict*; Pillar, *Negotiating Peace*; Blainey, *The Causes of War*; Iklé, *Every War Must*

*End*. A second generation of bargaining theorists emerged in the late 1990s. See, for example, Fearon, 'Rationalist Explanations of War', 379–414; Wagner, 'Bargaining and War', 469–484; Powell, 'Bargaining Theory and International Conflict', 1–30; Filson and Werner, 'A Bargaining Model of War and Peace', 819–838; Smith and Stam, 'Bargaining and the Nature of War', 783–813; Powell, 'War as a Commitment Problem', 169–203.

37 Reiter, 'Exploring the Bargaining Model of War', 27–43.
38 Reiter, 'Exploring the Bargaining Model of War', 37.
39 Schelling, *Arms and Influence*, 126.
40 Posen, 'The Security Dilemma and Ethnic Conflict', 79–102. See also Angstrom and Duyvesteyn, 'Evaluating Realist Explanations of Internal Conflict', 186–218.
41 Laitin, 'Somalia', 146–180.
42 Arreguin-Toft, 'How to Lose a War on Terror', 142–167.
43 Tanner, *Afghanistan*.
44 Goodson, *Afghanistan's Endless War*, 12–22.
45 See also Guistozzi, *War, Politics, and Society in Afghanistan*.
46 Margolis, *War at the Top of the World*, 52–76.
47 See also Gohari, *The Taliban Ascent to Power*.
48 Rubin, *The Fragmentation of Afghanistan*.
49 Arreguin-Toft, 'How to Lose a War on Terror', 142–167.
50 Millen, 'The Hobbesian Notion of Self-Preservation', 4–13.
51 Schultz, Jr and Dew, *Insurgents, Terrorists, and Militias*, 178.
52 Schultz, Jr and Dew, *Insurgents, Terrorists, and Militias*, 156.
53 See also Gohari, *The Taliban Ascent to Power*; Tanner, *Afghanistan*.
54 Tanner, *Afghanistan*, 308.
55 Rubin, 'Rebuilding Afghanistan', 165–170. See also Katzman, 'Afghanistan'; Jones, 'The State of the Afghan Insurgency'; Dahl Thruelsen, *NATO in Afghanistan*; and the 2007 SENLIS Afghanistan Report.
56 Rubin, 'Rebuilding Afghanistan', 165–170.

## Bibliography

Angstrom, Jan and Isabelle Duyvesteyn. 'Evaluating Realist Explanations of Internal Conflict: The Case of Liberia'. *Security Studies* 10, 3 (2001): 186–218.

Arreguin-Toft, Ivan. 'How to Lose a War on Terror: A Comparative Analysis of a Counter-Insurgency Success and Failure'. In *Understanding Victory and Defeat in Contemporary War*, eds Jan Angstrom and Isabelle Duyvesteyn. London: Routledge, 2007.

Blainey, Geoffrey. *The Causes of War*. New York: Free Press, 1988.

Centeno, Miguel Angel. 'Limited War and Limited States'. In *Irregular Armed Forces and their Role in Politics and State Formation*, eds Diane E. Davis and Anthony W. Pereira. Cambridge: Cambridge University Press, 2003, 82–95.

Dahl Thruelsen, P. *NATO in Afghanistan: What Lessons Are We Learning and Are We Willing to Adjust?* DIIS Report, 2007, 14.

Davis, Diane E. and Anthony W. Pereira, eds. *Irregular Armed Forces and their Role in Politics and State Formation*. Cambridge: Cambridge University Press, 2003.

Dobbins, James *et al. America's Role in Nation-Building: From Germany to Iraq*. Santa Monica, CA: RAND, 2003.

Dobbins, James *et al. The UN's Role in Nation-Building: From the Congo to Iraq*. Santa Monica, CA: RAND, 2005.

Downes, Alexander. 'The Problem with Negotiated Settlements to Ethnic Civil Wars'. *Security Studies* 13, 4 (2004): 230–279.

Doyle, Michael W. *Ways of War and Peace: Realism, Liberalism, and Socialism.* New York: W.W. Norton, 1997.

Doyle, Michael and Nicholas Sambanis. *Making War and Building Peace: United Nations Peace Operations.* Princeton, NJ: Princeton University Press, 2006.

Duffy Toft, Monica. *The Geography of Ethnic Violence.* Princeton, NJ: Princeton University Press, 2003.

Duyvesteyn, Isabelle and Jan Angstrom, eds. *Rethinking the Nature of War.* London: Frank Cass, 2005.

Fazal, Tanisha M. 'State Death in the International System'. *International Organization* 58, 2 (2004): 311–344.

Fearon, James D. 'Rationalist Explanations of War'. *International Organization* 49, 3 (1995): 379–414.

Fearon, James D. and David D. Laitin. 'Neotrusteeship and the Problem of Weak States'. *International Security* 28, 4 (2004): 5–43.

Filson, Darren and Suzanne Werner. 'A Bargaining Model of War and Peace: Anticipating the Onset, Duration, and Outcome of War'. *American Journal of Political Science* 46, 4 (2002): 819–838.

Fortna, Virginia Page. *Peace Time: Cease-Fires and the Durability of Peace.* Princeton, NJ: Princeton University Press, 2004.

George, Alexander L. and Andrew, Bennett. *Case Studies and Theory Development in the Social Sciences.* Cambridge, MA: MIT Press, 2004.

Giddens, Anthony. *The Nation-State and Violence.* Berkeley, CA: University of California Press, 1987.

Gohari, M.J. *The Taliban Ascent to Power.* Oxford: Oxford University, 2000.

Goodson, Larry P. *Afghanistan's Endless War: State Failure, Regional Politics, and the Rise of the Taliban.* Seattle, WA: University of Washington Press, 2001.

Guistozzi, Antonio. *War, Politics, and Society in Afghanistan.* Washington, DC: Georgetown University Press, 2000.

Hartzell, Caroline A. 'Explaining the Stability of Negotiated Settlements to Intrastate Wars'. *Journal of Conflict Resolution* 43, 1 (1999): 3–22.

Herbst, Jeffrey. *States and Power in Africa: Comparative Lessons in Authority and Control.* Princeton, NJ: Princeton University Press, 2000.

Hobbes, Thomas. *Leviathan* [1651]. Oxford: Oxford University Press, 1996.

Hobson, John M. *The State and International Relations.* Cambridge: Cambridge University Press, 2000.

Holsti, Kalevi J. *War, the State, and the State of War.* Cambridge: Cambridge University Press, 1996.

Iklé, Fred. *Every War Must End.* New York: Columbia University Press, 1991.

Jackson, Robert. *Quasi-States: Sovereignty, International Relations, and the Third World.* Cambridge: Cambridge University Press, 1990.

Jones, Seth G. 'The State of the Afghan Insurgency'. RAND Testimony, 10 December 2007.

Kalyvas, Stathis. *The Logic of Violence in Civil War.* Cambridge: Cambridge University Press, 2006.

Katzman, Kenneth. 'Afghanistan: Government Formation and Performance'. CRS Report for Congress, 4 January 2008.

Kollock, Peter. 'Social Dilemmas: The Anatomy of Cooperation'. *Annual Review of Sociology* 24 (1998): 183–214.

Laitin, David D. 'Somalia: Civil War and International Intervention'. In *Civil Wars, Insecurity, and Intervention,* eds Barbara F. Walter and Jack J. Snyder. New York: Columbia University Press, 1999.

Lake, David A. 'Building Legitimate States After Civil Wars: Order, Authority, and International Trusteeship', unpublished manuscript, dated 17 April 2007. Available at: http://chapters.ssrn.com/sol3/chapters.cfm?abstract_id¼1004425#ChapterDownload.

Leander, Anna. 'Wars and the Un-Making of States: Taking Tilly Seriously in the Contemporary World'. In *Contemporary Security Analysis and Copenhagen Peace Research,* eds Stefano Guzzini and Dietrich Jung. New York: Routledge, 2004.

Mansfield, Edward D. and Jack Snyder. *Electing to Fight: Why Emerging Democracies Go to War.* Cambridge, MA: MIT Press, 2005.

Margolis, Eric S. *War at the Top of the World: The Struggle for Afghanistan, Kashmir, and Tibet.* New York: Routledge, 2002.

Millen, Raymond. 'The Hobbesian Notion of Self-Preservation Concerning Human Behavior During an Insurgency'. *Parameters* 36, 4 (Winter 2007): 4–13.

Paris, Roland. *At War's End: Building Peace after Civil Conflict.* Cambridge: Cambridge University Press, 2004.

Paris, Roland. 'Bringing the Leviathan Back In: Classical vs Contemporary Studies of the Liberal Peace'. *International Studies Review* 8, 3 (2006): 425–430.

Pillar, Paul R. *Negotiating Peace: War Termination as a Bargaining Process.* Princeton, NJ: Princeton University Press, 1983.

Porter, Bruce D. *War and the Rise of the State: The Military Foundations of Modern Politics.* New York: Free Press, 1994.

Posen, Barry. 'The Security Dilemma and Ethnic Conflict'. In *Ethnic Conflict and International Security,* ed. M.E. Brown. Princeton, NJ: Princeton University Press, 1993.

Powell, Robert. 'Bargaining Theory and International Conflict'. *Annual Review of Political Science* 5 (2002): 1–30.

Powell, Robert. 'War as a Commitment Problem'. *International Organization* 60, 1 (2006): 169–203.

Reiter, D. 'Exploring the Bargaining Model of War'. *Perspectives on Politics* 1, 1 (2003): 27–43.

Rotberg, Robert I., ed. *State Failure and State Weakness in a Time of Terror.* Washington, DC: Brookings, 2003.

Rotberg, Robert I., ed. *When States Fail: Causes and Consequences.* Princeton, NJ: Princeton University Press, 2004.

Rubin, Barnett R. *The Fragmentation of Afghanistan.* New Haven, CT: Yale University Press, 2002.

Rubin, Barnett R. 'Rebuilding Afghanistan: The Folly of Stateless Democracy'. *Current History* (April 2004): 165–170.

Schelling, Thomas D. *The Strategy of Conflict.* Cambridge, MA: Harvard University Press, 1960.

Schelling, Thomas D. *Arms and Influence.* New Haven, CT: Yale University Press, 1966.

Schultz, Jr, Richard H. and Andrea J. Dew. *Insurgents, Terrorists, and Militias: The Warriors of Contemporary Combat.* New York: Columbia University Press, 2006.

Smith, Alistair and Allan C. Stam. 'Bargaining and the Nature of War'. *Journal of Conflict Resolution* 48, 6 (2004): 783–813.

Snyder, Jack. *From Voting to Violence: Democratization and Nationalist Conflict.* New York: W.W. Norton, 2000.

Sorensen, Gregory. *Changes in Statehood: The Transformation of International Relations.* London: Palgrave Macmillan, 2001.

Spruyt, Hendrik. *The Sovereign State and Its Competitors.* Princeton, NJ: Princeton University Press, 1994.

Tanner, Stephen. *Afghanistan: A Military History from Alexander the Great to the Fall of the Taliban.* Cambridge, MA: Da Capo Press, 2002.

Taylor, Brian D. and Rowana, Botea. 'Tilly Tally: War-Making and State-Making in the Contemporary Third World'. *International Studies Review* 10, 1 (2008): 27–56.

Tilly, Charles. *Coercion, Capital, and European States, AD 990–1990.* Oxford: Blackwell, 1992.

Tin-bor Hui, Victoria. *War and State Formation in Ancient China and Early Modern Europe.* Cambridge: Cambridge University Press, 2005.

Tuck, Richard. 'Hobbes'. In *Great Political Thinkers,* ed. Keith Thomas. Oxford: Oxford University Press, 1992, 107–238.

van Creveld, Martin. *The Transformation of War.* New York: Free Press, 1991.

van Creveld, Martin. *The Rise and Decline of the State.* Cambridge: Cambridge University Press, 2001.

Wagner, Robert Harrison. 'Bargaining and War'. *American Journal of Political Science* 44, 3 (2000): 469–484.

Walter, Barbara F. *Committing to Peace: The Successful Settlement of Civil Wars.* Princeton, NJ: Princeton University Press, 2002.

Waterman, Harvey. 'Political Order and the Settlement of Civil Wars'. In *Stopping the Killing: How Civil Wars End,* ed. Roy Licklider. New York: New York University Press, 1993.

Weinstein, Jeremy. *Inside Rebellion: The Politics of Insurgent Violence.* Cambridge: Cambridge University Press, 2007.

# 11 Between reluctance and necessity

## The utility of military force in humanitarian and development operations

*Robert Egnell*

The utility of military force in the contemporary strategic context is far from self-explanatory. Military operations with the political aims of stability, democratisation, economic development and respect for human rights have been the most prominent since the end of the Cold War, and are likely to continue to be so in the near future. The military continues to play an important role in these operations, but General Sir Rupert Smith accurately argues that military force is too often considered a solution 'for a wide range of problems for which it was not originally intended or configured'.[1] The way contemporary armed forces have adjusted to the new array of tasks is therefore of interest, and this contribution focuses on the use of military force in humanitarian and development type activities. To what extent are armed forces involved and useful in these types of activities? How does the military relate to the humanitarian and development communities? What are the humanitarian and security consequences of military involvement in these new roles?

Traditionally, the civil–military relationship in peace operations, involving military, humanitarian and development actors, has been premised on the idea of separate roles and a sequenced interaction. Military forces would provide security in the wake of a peace settlement and thereby create suitable conditions for relief and development actors to provide the aid. However, the strategic and tactical realities of operations displayed in Liberia, Afghanistan and Iraq paint a very different picture of overlapping tasks and interdependence between military and humanitarian actors. The military is increasingly engaging in humanitarian activity and the humanitarian community is increasingly employing assistance in terms of security and logistics from the military.

Lessons learned from recent operations show that uncoordinated humanitarian and military activities achieve little effect or risk being counter-productive. A clear trend is therefore to develop multifunctional approaches to operations, seeking to coordinate or integrate the activities of military, humanitarian and development actors into coherent operations with a common strategic purpose.[2] However, this contribution notes that the new trend should also be critically examined in order to avoid assumptions about effectiveness stemming from increased integration.

Humanitarian activity on the part of the military is another recent reality, partly because of tactical necessities in the field, partly because it may have positive strategic and tactical effects in hearts-and-minds operations. In counter-insurgency operations, in which the emphasis is on winning hearts and minds and increasing political legitimacy for the government, so-called civil–military operations, civil affairs units, quick impact projects and Provincial Reconstruction Teams, are thought to be imperative for success. Thus, in contemporary multifunctional operations, the activities of military and civilian actors are not only thought necessary to be coordinated or integrated for increased effectiveness, the military may also have to, or choose to, perform traditional humanitarian relief and development tasks.

This contribution has three aims: to describe the extent of civil–military interaction in peace-support operations; to analyse certain consequences of this interface; and to outline policy implications and areas for future research. In order to achieve these aims, the contribution is introduced by a review of the civil–military interface in the field of operations. This involves mapping the 'tactical realities' in the field, as well as reviewing the current trends in conceptual development regarding civil–military integration and coherence. Thereafter, the chapter analyses some potential consequences of increased military involvement in humanitarian activities, as well as increased civil–military integration. The focus of this analysis is the idea of 'militarisation of humanitarian aid' and the reverse 'humanitarianisation of the military'. What does this mean for the humanitarian principles and the idea of humanitarian space? Is integration and humanitarianisation leading to increased effectiveness and better outcomes in peace-support operations?

The main argument of the contribution is that the benefits of civil–military integration cannot be taken for granted, and that careful analysis of the involved actors and their interests are needed in order to determine the need for integrated structures; namely which actors should be integrated and which are better left outside? How should different tasks and responsibilities be divided and distributed and coordinated among the different actors involved? There is a great need for increased understanding of the causal mechanisms that lead to effectiveness in complex peace-support operations. Finally, the basic question of what can actually be achieved in terms of political change by external intervention must also be revisited.

## The extension of the civil–military interface

As touched upon in the introduction, the starting point of this contribution is the acknowledgement of a new strategic context involving a reversal of interest from traditional large-scale warfare between states to different forms of peace-support operations or small wars.[3] One of the main features of contemporary conflict is the far-reaching and complicated aims of operations, which often mean that Western armed forces will be operating in contexts involving a combination of counter-insurgency, post-conflict reconstruction, humanitarian assistance, as well as economic development and state building. This section highlights two important trends: (1) the changing tactical realities in the field, leading to increasing overlaps between military and humanitarian actors; and (2) the strategic trend towards multifunctional approaches seeking coherence and integration of all involved actors.

Tony Blair has advocated the use of armed force as 'an imperfect instrument for righting humanitarian distress'.[4] Anthony Forster argues that it is not at all uncommon for politicians and scholars to conceptualise security as a precondition of democracy, 'too readily seeing the deployment of troops as a means of achieving both security and democracy'. He further notes that state-building has become the primary raison d'être of military intervention in successive cases – Bosnia and Herzegovina (1995), Kosovo (1999) and Macedonia (2001), and, more recently, in Afghanistan and Iraq. Forster nevertheless argues that the use of the military in state-building and humanitarian operations is far more complex than tabloid headlines about using the armed forces 'as a force for good' suggest.[5] Amitai Etzioni similarly cautions us about the benefits of intervention and nation-building by arguing that many commentators greatly overestimate the transformative powers of even the most powerful nations and organisations when it comes to changing and reengineering the regimes of the other nations. Instead, he argues:

> [t]he tragic fact of the international reality is that both approaches to long-distance, large-scale social engineering have failed in most places. Liberal democracy is a delicate plant that grows only slowly under favourable conditions; it needs to be cultivated carefully by those who aim to live under it rather than by those who wish it for them.[6]

Page Fortna accurately highlights the fact that despite the vast literature on peace support operations, the area remains under-theorised and under-tested. She notes two serious gaps in the existing literature. First, the causal mechanisms through which peacekeepers keep peace have not been systematically spelled out. Second, empirical testing of whether peace is more likely after international involvement through different forms of peace operations is very limited.[7]

The use of armed forces in tasks previously owned by the international development community, such as providing reconstruction and critical infrastructure, is what Forster has come to term the 'humanitarianisation of the military'. He accurately argues that this is an overlooked but parallel process to the often-debated 'militarisation of humanitarian assistance'.[8]

The military is far from the only involved actor in contemporary peace-support operations, and will play but one part in operations that are likely to include a wide range of actors, such as other civilian government departments and agencies, international organisations, private security companies and non-governmental organisations (NGOs), host government agencies and security forces. A World Bank study firmly establishes the link between security and development by arguing that war retards development; and, conversely, development retards war.[9]

Olson and Gregorian similarly argue that in situations of incomplete or fragile peace, the interlinked nature of security and development is inescapable. Security is necessary to enable progress on development, and basic relief and longer-term development gains are necessary to achieve long-term stability by removing common obstacles to peace. 'Whether donors, militaries, development and humanitarian relief agencies choose to work together or not, they must work "side by side" in the same settings and the effects of their efforts are inevitably intertwined.'[10] At the tactical level, in the field of operations, the nature of the civil–military interface is therefore an important factor for mission success.

A recent report on civil–military interaction in Liberia and Afghanistan shows that activities and mandates often overlap as military forces engage in aid provision, and development actors participate in the security sector reform spectrum – especially in so-called demobilisation, disarmament and reintegration of ex-combatants. Overlaps are also obvious in the areas of post-conflict reconstruction and institutional development of justice systems, rule of law and governance.[11] Moreover, the failure of the sequenced approach to post-conflict management and civil–military interaction is obvious in both Liberia and Afghanistan, where great basic humanitarian needs, and in the case of Afghanistan, also security needs, still exist while already moving ahead with development and state building.[12] In Iraq, the early, and continuing, failures of economic reconstruction and political progress are even described as causes of the insurgency – or at least its scale.[13] A comprehensive and non-sequenced approach to the post-conflict situation in Iraq may therefore have limited the scale of the insurgency and improved the security situation.

The tactical realities in the field thereby show the limits of traditional separated and sequenced understandings of the civil–military divide. From a military perspective, US Marine General Charles Krulak's idea of 'three block war' shows the complex reality of soldiers by describing how troops are confronted by the entire spectrum of tactical challenges in the span of a few hours and within the space of a few city blocks.[14]

> In one moment in time, our service members will be feeding and clothing displaced refugees, providing humanitarian assistance. In the next moment, they will be holding two warring tribes apart – conducting peacekeeping operations – and, finally, they will be fighting a

> highly lethal mid-intensity battle – all on the same day … all within three city blocks.[15]

The military is thereby, out of tactical necessity, increasingly becoming involved in humanitarian and development processes, despite a traditional reluctance to do so. As an example, the US State Department has showcased the fact that, in 2006, the US military planned 556 humanitarian projects in 99 countries, arguing that

> the US military has the capability to help after natural disasters and can promote the health and economic well-being of suffering populations. That capability and that commitment are fuelling military humanitarian aid in places where US forces have had no historic role.[16]

One thing is clear: in situations of an incomplete or fragile peace, the interlinked nature of security and development is inescapable. Security is necessary to enable progress on basic relief and development, while longer-term development gains are necessary to solidify the peace process by giving people a stake in the new stability and by removing obstacles to peace. Equally, parts of the humanitarian community are increasingly looking for security support from military or private security actors in order to be able to operate in volatile areas.[17]

As noted, there are also particular tasks, like security sector reform (SSR) and demobilisation, disarmament and repatriation (DDR) that are inherently joint civil–military undertakings. As Olson and Gregorian put it: 'no matter how well the demobilisation and disarmament of ex-combatants is done by security-focused agencies, it will be a failure unless the rehabilitation and reintegration part is also done well by development agencies.' In the same way, new infrastructure and successful programmes in agriculture, health and education are directly linked to security efforts, as they are destroyed if a region reverts to war. The result is that military and civilian actors are in a relationship of what Olson and Gregorian call 'involuntary interdependence'.[18]

## Multifunctional conceptual and doctrinal developments

The increasing interface between military and humanitarian actors is, however, not merely a tactical necessity due to changing circumstances in the fields of operation. In light of the experience gained in Sierra Leone, Afghanistan and Iraq, the British Ministry of Defence argues that coherence between the different instruments of power is essential, and that it can only be achieved if strategic processes, planning and objectives were harmonised across all instruments and agencies.[19] A pressing challenge is to determine effective relationships amongst security, relief, aid and civil society-building efforts. The purpose is to allow collapsed states that were suffering from underdevelopment even prior to a devastating conflict, or in Olson and Gregorian's words: to 'develop the capacity to effectively absorb a mind-boggling array of interventions.'[20] Beyond the tactical-level interdependence between civilian and military actors, there is, in other words, a development towards increasing civil–military cooperation and integration at the strategic and operational levels. The last few years have seen the development of strategic concepts in this direction. This section briefly looks at a number of these concepts and changes in order to further highlight the strategic context in which military, humanitarian and development actors operate.

At the operational level, the armed forces in many Western states are currently developing doctrine in the direction of effects-based approaches to operations (EBAO). The US Joint Forces Command defines EBAO as 'a process for obtaining desired strategic outcome or effect on the enemy through the synergistic and cumulative application of the full range of military and nonmilitary capabilities at all levels of conflict'. 'Effect' is described as the physical, functional or psychological outcome, event or consequence that results from specific military or non-military actions.[21] Additional defining elements of EBAO include seeing the operations as a process beginning with developing knowledge of the adversary, viewed as a complex adaptive system, the environment, and own assets and capabilities. Knowledge of the enemy will enable the commander to determine the effects needed to convince or compel the enemy to change behaviour.[22] Although there is some disparity in the different national developments of effects-based approaches to operations, below is a list of the most common features of the concept:

- focus on effects and outcomes rather than tasks;
- use of several instruments to achieve the intended effects;
- cooperation and coordination between the different instruments;
- emphasis on mutual aims of all involved actors, based on common analysis of the situation;
- comprehensive view of the different phases of conflict – from prevention to post-conflict reconstruction;
- emphasis on comprehensive understanding of the conflict situation;
- follow-up and evaluation are central when planning for certain effects;
- belief in system theory and understanding.[23]

Although a defining feature of EBAO is the need to coordinate the different instruments of power towards mutual effects, the concept has not been successful outside the military sector. EBAO proved to be a hard sell to civilian actors and early ambitions to raise it to the strategic level have therefore been replaced by a more pragmatic view of EBAO as a military operational-level concept with links to strategic level ideas of concept development of multifunctional and comprehensive approaches discussed below (pp. 244–248).[24]

The European Union (EU) has come to acknowledge that the organisation's growing ambitions and responsibilities in multifunctional peace operations need additional procedural support to be realised effectively.[25] In response, the EU adopted the British conceptual development of Comprehensive Approaches, which serves to create 'holistic' views of conflicts/post-conflict situations, a treatment of the instruments at the organisation's disposal as 'tools in a toolbox', emphasis on integrated civil–military planning, and focus on the desired outcomes rather than processes of operations.[26] In general terms, the main objective of the Comprehensive Approach is to identify what needs to be done, when and by whom, in order to create a lasting solution in complex theatres.

The more specific development of Comprehensive Planning is intended to define a practical framework for effective planning coordination of the different EU actors involved in crisis management. It is defined as:

> a systematic approach designed to address the need for effective intra-pillar and inter-pillar coordination activity by all relevant EU actors in crisis management planning. [...] It includes identification and consideration of interdependencies, priorities and sequence of activities and harness resources in an effective manner, through a coherent framework that permits review of progress to be made.[27]

Some of the requirements of comprehensive planning are identified as joint assessment and analysis of the situation, including root causes, key actors, etc., as well as the formulation of clear measurements of success for each desired outcome and planned task.

A similar development is taking place within the United Nations (UN) structure, focusing on creating operational coherence between the different UN agencies and organs. The idea of 'Integrated Missions' started to evolve by the end of the 1990s, strongly influenced by the complexity encountered in the peace operations of that decade and the shortcomings in Somalia, Rwanda and the former Yugoslavia. These experiences indicated that the separate efforts by different UN actors did not have enough impact, and that more coherent multidimensional operations were needed. The so-called Brahimi Report provided an early initiative to integrate different UN organs and agencies in UN missions. Among other recommendations, the report proposed that 'Integrated Missions Task-Forces' for planning and implementation of multifunctional operations should be set up.[28]

The Integrated Mission concept provides a framework for multi-actor contributions by setting a common strategic aim, and by allowing the UN agencies to strive for the common aim in line with their respective procedures and processes. The shared strategic objectives should be the result of a deliberate effort by all elements of the UN system to achieve a shared understanding of the mandates and functions of the various elements of the UN presence in the field, and to 'maximise UN effectiveness, efficiency, and impact in all aspects of its work'.[29] Integrated Missions should also involve a structure derived from an in-depth understanding of the specific operational context, the political, humanitarian, human rights and development imperatives in that particular context, and of the particular mix of instruments needed and available to achieve the desired aims.[30] As in the case of the EU comprehensive approach, a process for planning, implementing and monitoring integrated missions has been established:

> The Integrated Mission Planning Process is designed to facilitate achievement of the common understanding among the different agencies by establishing a planning process that engages the capacities of all parts of the UN System relevant to achieving impact in a given country setting. The IMPP does not aim to take over all other planning processes. It aims to ensure that the right people are at the table, that the right issues are being considered, and that the appropriate authorities and accountabilities are in place to motivate flexible, creative, and integrated strategic and operational thinking and planning.[31]

The ambitious conceptual development of multifunctional approaches to war has nevertheless not been accompanied by an equally impressive implementation of these concepts. At the national strategic level, the interagency structure often fails to achieve coherence because of bureaucratic cultures that emphasise competition and distrust across ministerial and agency boundaries. Equally, in the civil–military interface at this level, even within the ministries of defence, there is often an administrative stovepipe structure that limits the extent of joint civil–military analysis and planning of operations.[32]

There is limited empirical data on the effectiveness of multifunctional approaches to operations. In essence, integrated or comprehensive operations have yet to be implemented in full, although some aspects have been tried in the contexts of Afghanistan, Iraq, Liberia and Sudan. This raises the question of whether humanitarian and military activities necessarily benefit from integration and produce some added value compared to separated activities. In the field of operations, the humanitarian community has questioned the very premise of integrating for increased operational effectiveness. Clearly, many NGOs see little value in being coordinated or subordinated to larger political aims of operations. Isabelle Duyvesteyn highlights the fact that quantifiable data from peacekeeping operations produces evidence that interventions in ongoing conflicts to promote peace are 'highly problematic'. Military and humanitarian-type activities in particular often produce counter-productive results.[33] The failures of past isolated military, diplomatic and humanitarian efforts mean that much hope is placed in the multifunctional and integrated types of intervention. However, the perceived need for increased civil–military coherence and integration is often based only on an assumption that the different actors and strands of activity are both inherently compatible and more effective when coordinated than when employed individually.[34]

As already noted, Fortna has highlighted the lack of research on the causal mechanisms between certain behaviour in peace-support operations and effectiveness. In fact, the jury on the effectiveness of peace-support operations is still out. While Fortna has isolated enough variables in her quantitative work to argue with some confidence that interstate peacekeeping is beneficial, General Sir Rupert Smith presents a bleaker view of intervention effectiveness – especially when considering the ambitious political aims of contemporary operations.[35] Moreover, the leap between grasping the effects of intervention in general and understanding the causal effects of different strategies and methods in complex and multifunctional peace-support operations is still large.

One step has been taken by Doyle and Sambanis who, through extensive quantitative analysis of UN peace operations, have highlighted the importance of creating strategies for peace operations that fit 'the ecologies' of contemporary conflicts in general, and the specific parameters of each conflict zone more specifically. This generally means comprehensive approaches that include the entire toolbox of national power – military, judicial, economic, constabulary, etc.[36] Nevertheless, without further empirical evidence, and with weak theoretical underpinnings, the concepts of integration and multifunctionality largely rely on assumptions of increased effectiveness from integration. More research in this area is necessary.

Before looking more specifically at consequences of the extended interface between humanitarian and military actors, the following section highlights the humanitarian principles. The purpose is to understand the reluctance of the humanitarian community to work with military organisations, and to inform the following analyses of potential consequences of integration.

## The humanitarian principles

Erin Weir argues that the humanitarian imperative is 'fundamentally incompatible with political objectives and military force'. To come to this conclusion, she compares the philosophies and principles of humanitarian action and peace-support operations.[37] Peace-support operations are inherently political and value laden, often based on Just War theory and international legal frameworks. The foundation of such operations is the use of force in order to establish or maintain security, and thereby to create a condition from where to establish long-term economic and political development.[38]

Humanitarian action, on the other hand, is based on the principle that receiving humanitarian assistance, and offering such assistance, are fundamental rights of all people and an obligation of the international community respectively. Humanitarian activities are therefore ideally motivated by the desire to alleviate human suffering. It should not be 'a partisan or political act and should not be viewed as such'.[39] In the implementation of this humanitarian imperative, aid organisations are guided by four 'core principles' of humanity, impartiality, neutrality and independence, designed to differentiate humanitarian activity from different forms of politically motivated activities.[40] Humanity indicates the duty to alleviate human suffering – a principle that is the source of all other principles. Impartiality means that aid priorities should be calculated on the basis of need alone and without regard to race, religion, creed or nationality of the recipients.[41] The principle of neutrality originally asserted that humanitarian organisations should not take sides or engage in controversies of a political, racial, religious or ideological nature. The principle, traditionally interpreted, means that organisations are not allowed to speak out against even the worst regimes or crimes against humanity. Many organisations, most notably Médecins Sans Frontières (MSF), have therefore completely reinterpreted this principle to include an obligation to document and speak out against human right abuses. Neutrality is instead maintained by speaking out against all actors involved in abuse in an impartial manner.[42] Finally, independence means that in order to be impartial and neutral in the provision of aid, to avoid becoming a tool of politics, humanitarian organisations must have complete independence from all actors that can be seen as exerting political influence over the distribution of aid. The main factor is the source of funding for humanitarian activities.[43] Where does the funding come from and how does that affect the provision of aid?

It is important to emphasise the fact that what this chapter refers to as the humanitarian community, or civilian aspects of peace-support operations, is actually a highly disparate group of actors with different aims, mandates and ways of operating. Describing and analysing this disparity is beyond the scope of this contribution, but for the sake of clarity Weir has provided three categories of humanitarian actors: the International Committee of the Red Cross (ICRC), UN humanitarian agencies and NGOs. While they all subscribe to the same humanitarian principles, 'there are significant differences in the interpretation and application of these principles in the field'.[44]

A further complicating factor in the civil–military interface is the cultural differences between humanitarian and military organisations. Donna Winslow is but one scholar who describes tensions between the military and humanitarian actors as a result of differing organisational cultures.[45] These differences influence the ability to interact and cooperate in the field. The cultural differences include differences in organisational goals, including values and basic assumptions, but also the functional imperatives of the respective organisations (alleviate suffering versus preparation for war). It also includes differences in organisational composition – the fact that there are often gender, age and ethnicity gaps between military and humanitarian organisations. Other differences are the approaches to violence, nationalism and decision-making styles. While the military emphasises controlled use of violence, has strongly nationalistic values and hierarchical structures, the humanitarian side is strongly opposed to the use of violence, is internationalist and often works in decentralised forms.[46]

The conclusion of Weir's comparative analysis is that the humanitarian imperative is fundamentally incompatible with the political objectives and use of force in peace-support operations.[47] Coupled with the cultural disparities between military and humanitarian actors, cooperation and coordination of these actors in the field of operations is inherently difficult. At the same time, the reality of the contemporary strategic context of complex peace-support operations is that the military is increasingly involved in humanitarian activities, and humanitarian and military organisations are working closer together, partly due to tactical realities in the field, and partly because of strategies that emphasise increased civil–military coherence and integration. Weir also notes that the one thing that binds peacekeepers and humanitarian workers together is the common desire to do what is 'morally right' in the face of brutality and hardship. The military and humanitarian activities thereby have the possibility to be complementary.[48] However, in order to determine suitable levels of integration, if at all, the consequences of the current trends must be discussed in greater detail. This is the purpose of the following section.

## Consequences of military humanitarian activity and integration

Although the use of military units for humanitarian projects is a hotly debated topic, it is too often presented in an unproblematic way. As an example, a US government information site highlights the words of Army Major Philip Spangler who argues that '[w]e want to add to, not detract from, what NGOs are doing'. The Major recently worked on water-sanitation problems in Chad, Niger and Mali, where he said NGOs have put in wells but not always maintained them. His command could fix water pumps, ambulances and other donated equipment.[49]

From a military perspective, there is potential for strategic value in conducting humanitarian work. As an example, Reuben Brigety argues that in the context of instant global media and the war on terrorism, 'demonstrating the capacity to care for civilians in the midst of warfare actually helps to achieve the strategic objectives for which the US uses force in the first place – namely, to deter hostile intent toward America and its citizens'. It is, in other words, a way to achieve legitimacy and force protection, as well as to win the hearts and minds of the local population in pursuit of the larger strategic objectives.[50] Thomas Henriksen at the Hoover Institution also describes humanitarian activity as a necessary measure: 'If we don't do this, we leave an opening to al-Qaida.'[51]

To place this line of argumentation in a theoretical context, General Sir Rupert Smith argues that in the struggle for the hearts and minds of the local population, the number of battlefield victories or reconstruction projects completed matter little if the local population and the wider audiences of international opinion think you are not winning, or visibly improving people's situation. Instead, the success is achieved by communicating with the people through the media and other outlets, getting the right narrative out there and changing perceptions.[52]

Johnson and Tierney argue that people often evaluate the outcomes of conflict on the basis of factors that are largely independent from the events in the field of operations. Instead, people's pre-existing beliefs, the symbolism of events, and manipulation by elites and the media affect the perception of the outcome.[53] Among several historical examples they mention the 'success' of the emergency evacuation of British forces from Dunkirk in the early days of the Second World War. The withdrawal operation was certainly a success, but it focused attention away from the fact that the British and French suffered a tremendous defeat when facing the onslaught of the German Blitzkrieg.[54] The important thing is that the defeat was transformed into some form of success in the minds of soldiers and policy-makers. Johnson and Tierney therefore accurately argue that the perceptions of victory and defeat, resulting from competition between the reality on the ground and the biases mentioned above, is of great importance as they 'not only affect the account book of history but also shape the fate of leaders, democratic processes, support for foreign policies, and the lessons used to guide decisions in the future'.[55]

Lawrence Freedman has developed the concept of 'strategic narrative'. He describes narratives as 'compelling story lines which can explain events', and says that they are intentionally designed to structure 'responses of others to developing events'.[56] The strategic feature of narratives lies in the fact that they are not spontaneous, but 'deliberately constructed or reinforced out of the ideas and thoughts that are already current'.[57] Freedman argues that the idea of the narrative opens up the possibility that military operations might need to be focused on undermining the narratives on which the insurgent bases its appeal and which guides its activists towards actions. He notes that 'success will depend on how a particular irregular war's purpose, course and conduct is viewed by the public opinion at home as well as within the theatre of operations'.[58] The biases and interpretations of events should, in other words, be actively influenced by intervening powers to not simply rely on the 'facts' from the ground, but also to create the all-important perception of winning or producing successes.

Also underlying this line of argumentation is the strategic context of what Martin Shaw calls 'the global surveillance mode of warfare'. Shaw's important contribution is made by arguing that all armed actors must reckon with comprehensive surveillance by global state institutions, law, markets, media and civil society. In this context of global surveillance, conflicts are fought under the critical gaze not only of the local population, and the people in intervening nations, but of the world as a whole. Such surveillance means that some tactics of the past are not politically acceptable in today's context where conveying the right message and winning the hearts and minds of the local and the global populations are key features. Casualties must be minimised, laws adhered to and media managed. In particular, Western governments must pay constant attention to, and obviously also play by the rules of, global surveillance, as these states generally make and enforce the rules.[59]

Thus, the idea is that within this context of global surveillance, military humanitarian activities can help achieve stability by improving the daily lives of the local population, increasing the legitimacy of the intervention, and winning the hearts and minds of the local population. Force protection can thus be increased, and intelligence gathering simplified and improved. However, in the global surveillance age of instant media coverage, the Internet and mobile phones, controlling a story or a narrative is not easy.[60] In the fight for hearts and minds of local and global populations, assumptions regarding the benefits of military involvement in humanitarian and development activities are therefore potentially dangerous. Calculations of local interests have proved notoriously difficult to achieve with any level of accuracy, not least because of cultural misunderstandings and language barriers in the field of operations.

The difficulty of communicating the intended narrative applies to humanitarian activities as well. There is, in other words, a risk that humanitarian activities performed by military units may not be interpreted by the local actors as intended, and thereby have unintended consequences. Moreover, even the best intentions and activities in the humanitarian, development and security spheres may be distorted into a different narrative in the hands of insurgents with far better local understanding of, and access to, the local population.[61] David Kilcullen convincingly argues that 'in modern counterinsurgency, the side may win which best mobilises and energises its global, regional and local support base and prevents its adversaries doing likewise'.[62]

Lieutenant General Sir Robert Fry, one-time Deputy Commander of Multinational Force Iraq, has argued that one of the greatest problems in Iraq was the failure to translate accurate tactical behaviour into operational effect in the pursuit of strategic goals.[63] Conducting appropriate and effective tactical operations and translating them into strategic aims, and then getting that message across to the local and global audience in a coherent strategic narrative, is arguably amongst the most important challenges in current and future peace-support operations. Understanding the potential impact on the humanitarian community and its ability to operate effectively must certainly be part of such calculations. Placed in a larger perspective of comprehensive and effects-based approaches to operations, the potential positive tactical effects of military humanitarian activity may prove to be Pyrrhic victories that undermine the effectiveness of the humanitarian and development communities, and thereby the larger political aims of the operation.

## The militarisation of humanitarian activity

A common argument is that civil–military integration is dangerous to the humanitarian community as it blurs the lines between politics and humanitarianism, and threatens the perception of humanitarian neutrality, impartiality and independence.[64] For example, Raj Rana argues that:

> The distinction between humanitarian, political and military action becomes blurred when armed forces are perceived as being humanitarian actors, when civilians are embedded into military structures, and when the impression is created that humanitarian organisations and their personnel are merely tools within integrated approaches to conflict management.[65]

During interventions, integrated multifunctional approaches may force the humanitarian community not only to associate with intervening powers, but also to be subordinated to the political and military agendas of the larger intervention. Clearly, the humanitarian space – access to suffering communities on both sides of the confrontation line, based on the humanitarian principles – thereby risks being eroded. Moreover, increasing humanitarian tasks performed by the military may cause both recipients of aid, as well as the conflicting parties, to find it difficult to distinguish between providers of assistance and combatants. Brigety argues that this confusion may not only apply to mere physical identification, but that it can also apply to the motives of the different actors:

> While, as stated above, military forces may wish to be identified with humane purposes for both tactical and strategic reasons, humanitarian groups almost invariably do not wish to be identified either physically or politically with any party to an armed conflict.[66]

The confusion of different actors' identity may compromise the humanitarian principles that the humanitarian agencies depend on for operational security in the field, as well as for political independence in funding and policy circles.[67]

Not only the military organisations, or the multifunctional strategies, are guilty of blurring the lines. Many NGOs have adopted what Stephen

Cornish calls a 'maximalist approach' to humanitarian aid, which means that they do not only seek to reduce short-term suffering, but to address the root causes of suffering. Maximalist NGOs therefore increasingly engage in long-term development, governance assistance and security-sector reform.[68] In any case, complex peace-support operations with the far-reaching aims of stability, democracy and economic development require military, development and humanitarian types of activity from the onset of operations. The number of tasks to be performed within the different areas is vast and they will require enormous resources in the forms of funding and personnel. As noted, the humanitarian space is a provider of security and access for humanitarian organisations that may be eroded if such organisations are associated with the larger military or political operations, or if the line between military and humanitarian actors is blurred because of military personnel conducting humanitarian activity. Eroding the humanitarian space based on the traditional principles of impartiality, neutrality and independence may therefore also create the need for an alternative form of security, provided for by military or private security actors. Military provision of security for humanitarian activities from all humanitarian risks drains already overstretched military organisations and limits the troops available for their core tasks of security. For the same reason, solving the problem of security for humanitarian activity by simply letting the military, with the capacity to provide for its own security, to conduct humanitarian operations is not an option.

As noted above, the blurring of the lines between the military and humanitarian communities is to some extent a result of new contextual realities in the field of operations and not simply a conceptual trend that can be reversed by different level decisions. Moreover, as Erin Weir notes, 'integration is not simply the invention of an overzealous UN bureaucrat, but the reaction to a number of startling failures on the part of the international community'. She argues that these failures have created a 'necessary push' to achieve better coordination and cooperation between all involved actors – a necessary development that nevertheless puts the humanitarian imperative at risk.[69]

In Afghanistan, the humanitarian space and the security of humanitarian organisations based on consent from the local population have eroded, forcing many organisations to abandon areas they have operated in for a very long time. The integrated approaches meant that all external actors were associated with the intervention. A Taliban spokesperson even stated that aid organisations were working for the Americans and thereby were legitimate targets of the insurgency.[70] The result is an ever-expanding area of the country suffering from an acute humanitarian emergency, as the private contractors and military provincial reconstruction teams have not been able to maintain the functions of the NGOs.[71]

It should be noted, however, that the very notion that the erosion of the traditional humanitarian space is caused by association with the military, or by military humanitarian activities that blur the lines between the two groups of actors, is questionable. The problem of the humanitarian space has been most prominent in contexts of forced Western intervention in Afghanistan and Iraq, where insurgents have deliberately targeted both NGOs and the UN. Arguing that this is simply the effect of military involvement in humanitarian activities, or the perception of humanitarian non-independence because of association, would be stretching matters too far. Instead, insurgents seem to be fighting not only against the intervention, but against all Western influence and presence in the region. As Karen Guttieri expresses it:

> The simple status of being an 'outsider' generates a political signature.... When militaries step into an environment previously dominated by relief organisations, as in Somalia in the 1990s and Afghanistan after September 11th – the humanitarian relationships with communities, and also with political and military actors shifts.[72]

There is simply no room for humanitarian space in some conflicts – regardless of the approaches adopted by the intervening powers and the humanitarian community.

Another argument against military involvement in humanitarian activities is that military forces, although well organised, efficient and with plenty of manpower, are generally not well trained for humanitarian activity. They have limited understanding of the humanitarian principles at best, and have an organisational culture and ethos not adjusted for such tasks. Paolo Tripodi even argues that professional soldiers with training that emphasises patriotism and a warfighting ethos have caused problems in military peacekeeping operations, which instead require the characteristics of patience, restraint and flexibility.[73] While peacekeeping is also different from traditional war-fighting, Tripodi's argument can surely also be made using humanitarian activities, which are obviously even more different from the cultural essence of military organisations than peacekeeping. Humanitarian organisations, on the other hand, often have long experience in complex emergencies, and have generally been in the field of operations for years, if not decades, before the military units show up. Humanitarian organisations therefore have the necessary skills, with needs assessment, water purification, humanitarian de-mining and human-rights monitoring as a few examples. They also often have in-depth local knowledge of the area of operations, as well as of local cultures and customs.

Let us also assume that both military and humanitarian organisations adjust their professional cultures and competences according to their respective functional imperatives or main tasks.[74] The disparity between military and humanitarian culture and competence should therefore be an indication that it is just as unlikely that a humanitarian worker would do well as an amateur soldier, as a soldier showing extreme talent in humanitarian work. Put simply, the military is not as competent as civilian organisations are in the provision of humanitarian assistance. Thus, military involvement in humanitarian-type activities, for which they are comparatively incompetent, may undermine the quality of aid delivery and 'the ultimate success of the humanitarian effort'.[75] As an example, Stephen Cornish notes that development projects in Afghanistan, funded through external support and directed through private contractors or so-called Provincial Reconstruction Teams (PRTs), have been 'costly, wasteful, lacking in quality and often not taking into account community needs'.[76]

Related to the view that the military does not have a suitable professional culture and ethos for humanitarian types of activities is the idea that officers and soldiers lack the ultimate commitment to humanitarian tasks, and that the recipients of aid may suffer as a consequence. This may not necessarily be caused by a lack of commitment and interest in humanitarian activity. Instead, the fact is that the focus of military organisations, and rightly so, is being directed elsewhere – security, in the traditional military sense. Humanitarian activity will thereby often be a secondary activity for the military. Moreover, the principal motivation of military humanitarian tasks is to achieve military interest such as stability and force protection. It is not governed by the principle of humanity – the right of suffering people to receive aid based on need rather than military objectives. Brigety refers to recipients of aid being subjected to the 'whims' of military interests rather than the humanitarian principles:

> When the interests change, the amount or nature of the aid may change as well. Humanitarian agencies, however, see themselves as dedicated to responding to humanitarian suffering for as long as such suffering exists, regardless of the affiliation of those in need, for as long as they have the capacity to respond, and as long as security conditions permit. All other political and strategic considerations are, generally speaking, irrelevant.[77]

Louise Heywood has described the fickle nature of military humanitarian and development interest in Iraq. She argues that with every unit rotation, with every new commander in a region, interests and methods to achieve different military and civilian aims change.[78] Humanitarian activities should primarily be based on need rather than the interests of military commanders, and development and reconstruction projects need stability and continuity for effective results.

In sum, there are arguments against increased military humanitarian activity, and civil–military integration. While they are important, they are not particularly constructive as they do not acknowledge that the development towards military humanitarian activity and civil–military integration are partly the result of a changing strategic context rather than new ideas and strategies in peace-support operations. Weir accurately argues that the criticism of integration from the humanitarian using the humanitarian imperative is a bit like evoking a right. It does not leave room for compromise and requires a clarity that is not possible in the contemporary strategic context of complex peace-support operations. Instead, she argues:

> The reality is that humanitarian action and peacebuilding are no longer clearly separable concepts, and the most humanitarian actors have to compromise every day – choosing to use armed escorts, military transport, or targeted funding – in order to get the job done for the people who need their help.[79]

Weir's conclusion is that the overlap between military and humanitarian efforts requires a structure that can unify the activities of these disparate actors. Integration is, in other words, the only way forward.[80] However, the complaints raise important questions that must be considered in the development of multifunctional approaches to operations. It also hints towards the problematic notion that integration may not be the conclusive solution to the problem of effectiveness in peace-support operations. Before discussing these matters further in the concluding section, let us take a look at the accompanying trend called the humanitarianisation of the military.

## The humanitarianisation of the military

There may also be danger in military involvement in humanitarian activities. Anthony Forster notes that the use of British armed forces in Afghanistan's Helmand province provides an example of blurring lines between military and humanitarian work, and a case of what he calls the humanitarianisation of the military.[81] As these troops were deployed to Afghanistan, a Ministry of Defence briefing document noted that they would always have the primacy of the reconstruction effort in mind, and that they were not there to wage war but to help to rebuild. Moreover, then Defence Secretary John Reid suggested that the British troops might not even have to fire their weapons by stating that '[w]e would be perfectly happy to leave in three years' time without firing a shot because our job is to protect reconstruction'.[82] This quickly proved to be a serious underestimation of the security situation in the province, and within days of their deployment the British forces were engaged in sustained and intense fighting against Taliban forces. Planned British reconstruction and development projects have also been on hold because of the security situation, which does not allow civilian actors access in the Helmand province. Leo Docherty has described the failure of the British to apply comprehensive approaches in the Helmand province. Although the British Department for International Development and the Foreign and Commonwealth Office had no practical measures in place in the northern parts of the province, military teams were sent into the region without any means to win hearts and minds. The teams quickly became the targets of local fighters and were drawn into high-intensity war-fighting.[83]

As a consequence of the distance between political statements and realities on the ground, combined with a considerable overstretch of the armed forces – being required to do too many tasks with too few troops – retired senior officers have criticised the political leadership in Britain for failing to give proper support to the armed forces in these deployments, and for confused military objectives that are putting lives in danger.[84]

There may also be a risk in that troops deploy for operations without accurate understanding of the dangers, and without the proper mindset and training to face those dangers. Christopher Dandeker has written on the need for military culture and ethos to be different from that of civil society, albeit within certain limits, in order to perform effectively in combat situations.[85] During the 1990s, a debate existed regarding the potential negative consequences of deploying troops for peacekeeping tasks – or tasks beyond traditional soldiering. As an example, General Colin Powell, then Chairman of the Joint Chiefs of Staff, argued against the deployment of US troops in Bosnia by citing the need to retain a warrior culture for the primary military mission of fighting and winning the nation's wars.[86] The idea that peacekeeping would damage combat readiness and warrior ideals has somewhat died down with the increasing combat intensity of the 'peace operations' in Afghanistan and Iraq, and with the fact that increasing numbers of officers and analysts are considering these types of operations the most likely and important in the near future. But perhaps the issue should be raised again in relation to humanitarian-type activities. How does such activity affect military units' culture and combat readiness? How many different tasks is it possible to train soldiers to do at an acceptable level?

While answering those questions is beyond the scope of this chapter, it should be noted that a pre-deployment training not focused enough on combat may leave military forces unprepared for serious situations and thereby risk the effectiveness of the operation, their own lives, and the lives of the local population and humanitarian workers they may need to protect. However, simply focusing on combat training will leave soldiers unprepared for other important and inescapable tasks in contemporary operations. A healthy balance needs to be struck. Yet, clearly, the learning capacity of soldiers is not unlimited, and in order to perform effectively, an emphasis is necessary on the core military tasks.

## Conclusion: drawing new lines in the sand

This contribution has sought to achieve three aims: to describe the extensive civil–military interface in peace-support operations, to analyse some consequences of integrated approaches to operations and military activities, and the remaining task, to outline various policy implications. The descriptive section highlighted two important trends. First, the increasing strategic emphasis on civil–military integration and coherence in peace-support operations. Second, an increasing military involvement in humanitarian activities based on the assumption that within the operational context of global surveillance, conducting so-called 'hearts-and-minds operations' may have a positive strategic impact.

The findings of the analysis of potential consequences of these trends argued that, despite the fact that both trends are often seen as necessary developments for increased effectiveness in complex peace-support operations, there are serious issues that should also be considered when employing such strategies. With increased integration, the humanitarian principles are watered down and the humanitarian space risks being eroded. The cost of providing an alternative form of security for humanitarian activities may well overshadow the potential benefits of coherent multifunctional operations. The benefits of military involvement in humanitarian activities may be questioned on several counts. In sum, armed forces, while being expensive to train, equip, maintain and deploy, have little humanitarian expertise and commitment. Military organisations are therefore likely to achieve less at a higher cost – while at the same time endangering the work of humanitarian organisations.

Military humanitarian activities are often performed to achieve tactical or strategic effects such as improving force protection and stability, or creating legitimacy for the larger political transformations. However, calculating the psychological effects on the local population is inherently difficult, not least because of poor understanding of local cultures and customs. Thus, while the humanitarian impact of military involvement in such activities is often limited at best, the unintended effects may be serious if the activities are 'misunderstood', or if an alternative narrative is presented by insurgents. Moreover, the humanitarianisation of military affairs means that yet another traditionally non-military group of tasks are expected of the armed forces. There are limits to the amount of tasks that soldiers and officers can be trained to perform effectively, and adding humanitarian activities to the list may undermine necessary military competencies such as war-fighting and peacekeeping.

In sum, we are left with two 'necessary' trends and some serious side-effects. It is therefore time to leave the existential debates about civil–military integration and instead look at ways to implement such approaches with maximum effect and some level of damage control.

First, the theoretically weak assumption of increased effectiveness from civil–military integration should be studied further. One step would be to further highlight the outcome-based thinking of multifunctional approaches. What are the precise aims of operations, how do they translate into operational and tactical objectives, and how can different instruments of power – civilian and military – be used to achieve those aims? It may not simply be a problem of finding the right method. In some cases, increased integration into coherent comprehensive approaches may be the more effective solution. In other cases, a clear line of demarcation between certain actors and their activities may be the answer. The important thing is that such calculations are made, and that they are holistic in that they involve both short- and long-term effects, intended and unintended, at all levels of operations. As an example, the unintended consequence of civil–military integration in Afghanistan is that it has eroded the humanitarian space in the southern parts of the country and forced many humanitarian organisations to leave. While the humanitarian situation has worsened, the support for the Taliban insurgency is growing. The intended strategic effect of stabilisation, economic development and strengthening of the central government may therefore have been damaged.

Second, further structural, cultural and doctrinal adjustment of the armed forces is necessary in the direction of traditionally non-military tasks in order to become more effective. However, a delimitation regarding the tasks that should be asked of the military is equally important. As argued above, while armed forces are always practical to use because of their availability, as well as their hierarchical and subordinate structure and culture, they are not capable of performing any number of tasks with good effect. In essence, it comes down to a political choice regarding the use of the military instrument. It can be used as a blunt instrument for any type of task and without professional expertise in any field, or it can be sharpened to perform a more limited number of tasks with greater professionalism, but at the same time requiring investments in civilian operational capacity to perform tasks beyond the military delimitation. A military tool as sharp as traditionally configured armed forces for conventional warfare is probably not a useful option as the strategic context is unlikely to provide opportunities for the effective use of such a tool.

Third, the use of armed forces for humanitarian activities needs to be reconsidered. Again, the mere assumption that such activities will lead to increased support from the local population is an oversimplification, and serious thinking is necessary in order to better understand the consequences of military activity in the humanitarian field. The provision of humanitarian aid must be provided for during conflict in order to establish or maintain the legitimacy of the operation – in the eyes of the local population as well as globally. This contribution suggests that maintaining the humanitarian space based on impartiality, neutrality and independence of humanitarian organisations is more effective and efficient than either providing military security in support of humanitarian activities, or by military and private-contractor involvement in such activities. To achieve that, some form of divide between humanitarian and political actors such as development organisations, and especially the armed forces, should ideally be maintained. Importantly, the demarcation line is not between military and civilian actors, but between those who operate according to humanitarian principles and those who are part of a political operation – involving both development and security policy.

Despite the fact that this contribution advocates a divide between the humanitarian and political activities in the field, it should be acknowledged that this demarcation line can never be too distinct. As noted in the analysis above, the nature of contemporary operations means that military, humanitarian and other civilian actors will always need to interact and cooperate to some extent. The humanitarian organisations will from time to time need security assistance, regardless of whether a better-delineated humanitarian space is maintained. Similarly, the stabilisation and state-building processes may be seriously hampered by humanitarian disasters that can limit the legitimacy of the operation and host government. Communication and some level of cooperation, albeit preferably driven by humanitarian principles, will therefore be necessary, but lines in the sand should be drawn in order not to further blur the contrasts between the different actors and their activities.

How such a divide can be practically established and communicated to the people in the field of operations is an important future research question. Nonetheless, in a substantial number of conflicts – especially in interventions with questionable international mandates and limited local legitimacy – the humanitarian space is an ideal that is not likely to become a reality as all external actors are placed in the same category. In these cases, alternatives for the provision of humanitarian aid must be established.

Finally, in more general terms, the current development of integrated and comprehensive approaches to operations contain a 'can-do' mentality that disregards the current evidence that interventions in armed conflicts are always highly problematic and that we cannot even be certain that they have a positive effect. Thus, there is a need to consider what can actually be achieved by intervening in complex war zones and post-conflict societies. A conclusion of a recent conference on transatlantic approaches to post-conflict management was that, given the many failures in achieving far-reaching political aims in post-conflict management operations, there must either be a substantial increase in the resources used in each intervention, or the aims must be lowered.[87] While peace-support operations are not simply driven by cost–benefit analyses, such recalculations regarding efforts and possible outcomes may well tip the scale towards less optimism and involvement in peace operations.

Yet, again, the causal mechanisms describing how different strategies and tactics – not least the alternative uses and levels of integration of the available political instruments of power – affect outcome in peace-support operations need further study. The substantive number of harmful consequences of civil–military integration means that such approaches should not be seen as a silver bullet in the struggle for global peace and security. Nevertheless, with better understanding of the causal mechanisms between certain behaviour and operational outcomes, and with better understanding of the necessary delimitations for such integration, it may well prove to be a step in the right direction.

## Notes

This chapter has been previously published as Robert Egnell, 'Between Reluctance and Necessity; The Utility of Military Force in Humanitarian and Development Operations', *Small Wars and Insurgencies* 19, 3 (2008): 397–422. We would like to thank Taylor and Francis for permission to reprint it here.

1 Smith, *The Utility of Force*, xii.
2 See, for example, Smith, *The Utility of Force*; Weir, *Conflict and Compromise*; UK Ministry of Defence, *The Comprehensive Approach.*
3 Smith, *The Utility of Force*; van Creveld, *The Transformation of War*; Freedman, *The Transformation of Strategic Affairs.*
4 Blair, 'Doctrine of the International Community'.
5 Forster, 'Breaking the Covenant', 1045.
6 Etzioni, *Security First*, 3.
7 Fortna, 'Interstate Peacekeeping', 482.
8 Forster, 'Breaking the Covenant', 1045.
9 Collier *et al.*, *Breaking the Conflict Trap*, 1.
10 Olson and Gregorian, 'Side by Side or Together?', 81.
11 Olson and Gregorian, 'Side by Side or Together?', 80.
12 Olson and Gregorian, 'Side by Side or Together?', 80.
13 Metz, 'Insurgency and Counter-Insurgency in Iraq', 26–28; Ricks, *Fiasco*, 191.
14 Krulak, 'The Strategic Corporal'.
15 Krulak, 'The Three Block War', 139–142.
16 Kelleher, 'US Military Humanitarian Efforts Planned for 99 Nations Forces Respond to Disasters and Deliver Longer-Term Aid Overseas'.
17 Olson, 'Fighting for Humanitarian Space', 15.
18 Olson and Gregorian, *Side by Side or Together?*, 81.
19 UK MoD, *The Comprehensive Approach*, 1–1.
20 Olson and Gregorian, *Side by Side or Together?*, 81.
21 US Joint Forces Command, *Effects-Based Operations*, 5.
22 US Joint Forces Command, *Effects-Based Operations*, 5.
23 Carlsen, Carling and Eriksson, *EBO och effektbaserat tänkande i internationell konflikthantering*, 21.
24 Derblom, Egnell and Nilsson, *The Impact of Strategic Concepts and Approaches on the Effects-Based Approach to Operations*, 9.
25 United Kingdom, Austria and Finland non-paper, *Enhancing EU Civil–Military Coordination*, 1.
26 United Kingdom, Austria and Finland non-paper, *Enhancing EU Civil–Military Coordination*, 5.
27 Council of the European Union, *Draft EU Concept for Comprehensive Planning.*
28 United Nations, *Report of the Panel on United Nations Peace Operations*, xiii.
29 United Nations, *Integrated Missions Planning Process*, 3.
30 United Nations, *Integrated Missions Planning Process*, 3.
31 United Nations, *Integrated Missions Planning Process*, 3.
32 Egnell, 'Explaining US and British Performance in Complex Expeditionary Operations', 1050.

33 Duyvesteyn, 'The Effectiveness of Intervention Instruments in Armed Conflict'.
34 Gordon, 'Military–Humanitarian Relationships and the Invasion of Iraq'.
35 Fortna, 'Does Peacekeeping Keep Peace?', 288; Smith, *The Utility of Force*.
36 Doyle and Sambanis, *Making War & Building Peace*, 334–337.
37 Weir, *Conflict and Compromise*, 26.
38 Weir, *Conflict and Compromise*, 20–26.
39 IRIC, 'The Code of Conduct for the International Committee of the Red Cross and the Red Crescent Movement, and NGOs in Disaster Relief', 1.
40 Terry, *Condemned to Repeat*, 26.
41 IRIC, 'Code of Conduct'.
42 Médicins Sans Frontières Switzerland, 'Le Témoignage'.
43 Weir, *Conflict and Compromise*, 22.
44 Weir, *Conflict and Compromise*, 27–30.
45 Winslow, 'Strange Bedfellows in Humanitarian Crisis', 113–128.
46 Winslow, 'Strange Bedfellows in Humanitarian Crisis', 116.
47 Weir, *Conflict and Compromise*, 26.
48 Weir, *Conflict and Compromise*, 26.
49 Cited in Kelleher, 'US Military Humanitarian Efforts'.
50 Brigety, 'From Three to One'.
51 Kelleher, 'US Military Humanitarian Efforts'.
52 Smith, *The Utility of Force*, 391.
53 Johnson and Tierney, *Failing to Win*, 5.
54 Johnson and Tierney, *Failing to Win*, 9.
55 Johnson and Tierney, *Failing to Win*, 18.
56 Freedman, *The Transformation of Strategic Affairs*, 22.
57 Freedman, *The Transformation of Strategic Affairs*, 22.
58 Freedman, *The Transformation of Strategic Affairs*, 93.
59 Shaw, *The New Western Way of War*, 75. The difficult choices faced by Western states in an era of increasing humanism and humanity in warfare are also described by Coker in his *Humane Warfare*.
60 Shaw, *The New Western Way of War*, 47, 56.
61 Egnell, 'The Missing Link', 36.
62 Kilcullen, 'Counter-Insurgency Redux', 121.
63 Fry, 'Expeditionary Operations in the Modern Era', 62.
64 Weir, *Conflict and Compromise*, 45.
65 Raj Rana, 'Contemporary Challenges in the Civil–Military Relationship', 586.
66 Brigety, 'From Three to One'.
67 Brigety, 'From Three to One'.
68 Cornish, 'No Room for Humanitarianism in 3D Policies', 12.
69 Weir, *Conflict and Compromise*, 45.
70 Owen and Travers, '3D Visions', 46.
71 Cornish, 'No Room for Humanitarianism', 38.
72 Guttieri, 'Humanitarian Space in Insecure Environments'.
73 Tripodi, 'Peacekeepers of the Twenty-First Century', 71–72.
74 This assumption is based on traditional civil–military relations theory that states that professional armed forces are influenced by a functional and a societal imperative. See, for example, Huntington, *The Soldier and the State*, 2.
75 Brigety, 'From Three to One'.
76 Cornish, 'No Room for Humanitarianism', 30.
77 Brigety, 'From Three to One'.
78 Interview with Captain Heywood, 26 January 2007. Many of her comments are published in Heywood, 'CIMIC in Iraq', 36–40.
79 Weir, *Conflict and Compromise*, 45.
80 Weir, *Conflict and Compromise*, 45.

81 Forster, 'Breaking the Covenant', 1045.
82 Albone *et al.*, 'Soldiers Who Went to Build Bridges Fight For Their Lives'.
83 See Docherty, *Desert of Death*, 185–192.
84 Forster, 'Breaking the Covenant', 1045–1046.
85 Dandeker, 'On the Need to be Different', 4–9.
86 Powell, 'U.S. Forces', 32–45.
87 'Transatlantic Approaches to Post-Conflict Management', SWP–NIPP Workshop, Stiftung Wissenschaft und Politik (SWP), Berlin, 28–29 February 2008.

## Bibliography

Albone, Tim, Richard Beeston, David Charter and Michael Evans. 'Soldiers Who Went to Build Bridges Fght For Their Lives'. *Times Online*, 1 July 2006, www.timesonline.co.uk/tol/news/world/middle_east/article681590.ece html (accessed 6 January 2010).

Blair, Tony. 'Doctrine of the International Community', speech at Economic Club of Chicago, Hilton Hotel, Chicago, 22 April 1999, www.number-10. gov.uk/output/Page1297.asp (accessed 8 January 2005).

Brigety II, Reuben E. 'From Three to One: Rethinking the "Three Block War" and Humanitarian Operations in Combat'. The Joint Services Conference on Professional Ethics, Springfield, VA, 29 and 30 January 2004.

Carlsen, Henrik, Christian Carling and Anders E. Eriksson. *EBO och effektbaserat tänkande i internationell konflikthantering. FOI-R–1990–SE, 200621* (Stockholm: FOI, 2006).

Coker, Christopher. *Humane Warfare*. London: Routledge, 2001.

Collier, P., V.L. Elliott, H. Hegre, A. Hoeffler, M. Reynal-Querol and N. Sambanis. *Breaking the Conflict Trap: Civil Wars and Development Policy*. Washington, DC: World Bank and Oxford University Press, 2003.

Cornish, Stephen. 'No Room for Humanitarianism in 3D Policies: Have Forcible Humanitarian Interventions and Integrated Approaches Lost Their Way?'. *Journal of Military and Strategic Studies* 10, 1 (2007): 1–48.

Council of the European Union. *Draft EU Concept for Comprehensive Planning*. Document 13983/0, 3, November 2005.

Dandeker, Christopher. 'On the Need to be Different: Military Uniqueness and Civil–Military Relations in Modern Society'. *RUSI Journal* 146, 3 (2001): 4–9.

Derblom, Markus, Robert Egnell and Claes Nilsson. *The Impact of Strategic Concepts and Approaches on the Effects-Based Approach to Operations: A Baseline Collective Assessment Report*. FOI-R–2394–SE. Stockholm: FOI, 2007.

Docherty, Leo. *Desert of Death: A Soldier's Journey from Iraq to Afghanistan*. London: Faber, 2007.

Doyle, Michael W. and Nicholas Sambanis. *Making War & Building Peace: United Nations Peace Operations*. Princeton, NJ: Princeton University Press, 2006.

Duyvesteyn, Isabelle. 'The Effectiveness of Intervention Instruments in Armed Conflict: Conflict Resolution is the Only Solution?'. In *Peace, Security and Development in an Era of Globalization: The Integrated Security Approach Viewed from a Multidisciplinary Perspective*, eds. G. Molier and E. Nieuwenhuys. Leiden: Martinus Nyhoff, 2009.

Egnell, Robert. 'Explaining US and British Performance in Complex Expeditionary Operations: The Civil–Military Dimension'. *Journal of Strategic Studies* 29, 6 (December 2006): 1041–1075.

Egnell, Robert. 'The Missing Link: Civil–Military Aspects of Effectiveness in Complex Irregular Warfare'. PhD dissertation, King's College, London, 2007.

Etzioni, Amitai. *Security First: For a Muscular Moral Foreign Policy*. New Haven, CT: Yale University Press, 2007.

Forster, Anthony. 'Breaking the Covenant: Governance of the British Army in the Twenty-First Century'. *International Affairs* 82, 6 (2006): 1043–1057.

Fortna, Virginia Page. 'Does Peacekeeping Keep Peace? International Intervention and the Duration of Peace After Civil War'. *International Studies Quarterly* 48, 2 (2004): 269–292.

Fortna, Virginia Page. 'Interstate Peacekeeping: Causal Mechanisms and Empirical Effects'. *World Politics* 56, 4 (2004): 481–519.

Freedman, Lawrence. *The Transformation of Strategic Affairs*. Adelphi Paper, No. 379. London: Routledge for IISS, 2006.

Fry, Lieutenant General Robert. 'Expeditionary Operations in the Modern Era'. *RUSI Journal* 150, 6 (December 2005): 60–63.

Gordon, Stuart. 'Military–Humanitarian Relationships and the Invasion of Iraq (2003): Reforging Certainties?', *Journal of Humanitarian Assistance* (July 2004).

Guttieri, Karen. 'Humanitarian Space in Insecure Environments: A Shifting Paradigm'. *Strategic Insights* 4, 11 (2005), www.ccc.nps.navy.mil/si/2005/Nov/guttieriNov05.asp (accessed 12 March 2008).

Heywood, Louise. 'CIMIC in Iraq'. *RUSI Journal* 151, 6 (December 2006): 36–40.

Huntington, Samuel P. *The Soldier and the State: The Theory and Politics of Civil–Military Relations*. Cambridge, MA: Harvard University Press, 1957.

IRIC (International Committee of the Red Cross). 'The Code of Conduct for the International Committee of the Red Cross and the Red Crescent Movement, and NGOs in Disaster Relief', www.icrc.org/Web/eng/siteeng0.nsf/html/p1067 (accessed 15 March 2008).

Johnson, Dominic and Dominic Tierney. *Failing to Win: Perceptions of Victory and Defeat in International Politics*. Cambridge, MA: Harvard University Press, 2006.

Kelleher, Elizabeth. 'US Military Humanitarian Efforts Planned for 99 Nations Forces Respond to Disasters and Deliver Longer-Term Aid Overseas'. USINFO.state.gov, 13 July 2006, http://usinfo.state.gov/xarchives/display.html?p¼washfile-english&y¼2006&m¼July&x¼20060712172520berehellek0.4737207.

Kilcullen, D. 'Counter-Insurgency Redux'. *Survival* 48, 4 (2007): 111–130.

Krulak, Charles C. 'The Three Block War: Fighting in Urban Areas'. *Vital Speeches of the Day* 64, 5 (1997): 139–142.

Krulak, Charles C. 'The Strategic Corporal: Leadership in the Three Block War'. *Marines Magazine* (January 1999).

Médicins Sans Frontières Switzerland. 'Le Témoignage', www.msf.ch/temoignage.45.0.html (accessed 12 February 2008).

Metz, Steven. 'Insurgency and Counter-Insurgency in Iraq'. *Washington Quarterly* 27, 1 (2003): 25–36.

Olson, Lara. 'Fighting for Humanitarian Space: NGOs in Afghanistan'. *Journal of Military and Strategic Studies* 9, 1 (2006): 15.

Olson, Lara and Hrach Gregorian. 'Side by Side or Together? Working for Security, Development & Peace in Afghanistan and Liberia'. A report on the 30 & 31 March 2007 'Workshop Coordinated Approaches to Security, Development and Peacemaking: Lessons Learned from Afghanistan and Liberia', held by the Centre for Military and Strategic Studies (CMSS), University of Calgary, 2007.

Owen, Taylor and Patrick Travers. '3D Visions – Can Canada Reconcile its

Defence, Diplomacy, and Development Objectives in Afghanistan?'. *The Walrus* (July/August 2007).

Powell, General Colin. 'US Forces: Challenges Ahead'. *Foreign Affairs* 71, 5 (1993): 32–45.

Rana, Raj. 'Contemporary Challenges in the Civil–Military Relationship: Complementarity or Incompatibility?'. *International Review of the Red Cross* 86, 855 (2004): 565–591.

Ricks, Thomas E. *Fiasco: The American Military Adventure in Iraq*. New York: Allen Lane, 2006.

Shaw, Martin. *The New Western Way of War: Risk Transfer War and its Crisis in Iraq*. Cambridge: Polity, 2005.

Smith, Rupert. *The Utility of Force: The Art of War in the Modern World*. London: Allen Lane, 2005.

Terry, Fiona. *Condemned to Repeat: The Paradox of Humanitarian Action*. Ithaca, NY: Cornell University Press, 2002.

'Transatlantic Approaches to Post-Conflict Management'. SWP–NIPP Workshop, Stiftung Wissenschaft und Politik (SWP), Berlin, 28–29 February 2008.

Tripodi, Paolo. 'Peacekeepers of the Twenty-First Century: A Comparison Between Professional Soldiers and Draftees in Peace Support Operations'. *Small Wars & Insurgencies* 14, 2 (Summer 2003): 71–86.

UK Ministry of Defence. *The Comprehensive Approach*, Joint Discussion Note 4/05 (January 2006).

United Kingdom, Austria and Finland non-paper. *Enhancing EU Civil–Military Coordination* (May 2005).

United Nations. *Report of the Panel on United Nations Peace Operations*. UN doc. A/55/305, S/2000/809, 2000.

United Nations. *Integrated Missions Planning Process (IMPP)* – Guidelines Endorsed by the Secretary-General on 13 June 2006.

US Joint Forces Command. *Effects-Based Operations*. White Paper Version 1.0. Norfolk, VA: Concepts Department J9, 2001.

van Creveld, Martin. *The Transformation of War*. New York: Free Press, 1991.

Weir, Erin A. *Conflict and Compromise: UN Integrated Missions and the Humanitarian Imperative*. KAIPTC Monograph, No. 4, June 2006.

Winslow, Donna. 'Strange Bedfellows in Humanitarian Crisis: NGOs and the Military'. In *Twisting Arms and Flexing Muscles: Humanitarian Intervention and Peacebuilding in Perspective*, eds. N. Mychajlyszyn, and T.D. Shaw, 113–128. Aldershot: Ashgate, 2005.

# 12 War, what it is good for …

*Isabelle Duyvesteyn*[1]

Imagine, you are a military commander tasked by your political masters to devise a military strategy. The political end goal of this strategy is to be the exit from the battle theatre. In practice this means that your opponent – who you generally intend to impress by force – will also be able to read your intentions: you are on your way out. From a perspective of military strategy this does not make any sense.[2] It gives your opponent the advantage of knowing that your staying power is limited. He automatically has the advantage of time, and simply has to wait until you leave in order to gain the upper hand. If the political objective is to exit, what is the military end state that leads to this? Odd as it might seem, this is exactly the situation General George Casey found himself in Iraq in 2006. His political masters, the United States government, formulated as its main political goal the withdrawal of US forces from Iraq, with local forces and authorities taking over responsibility. One of the main puzzles this situation poses is: how did it arise? Did the military commanders, Casey included, fail to make clear to the political leadership that this is not how military strategy works or how force should be employed? Or did the politicians fail to grasp the essential workings of strategic logic and the utility of military force?[3]

Imagine, you are a military commander in charge of a large multinational force. Every contingent has been sent by their national government to show the world what they are capable of as a force for good. At the same time every contingent is concerned about safeguarding domestic support for the operation and limit casualties and collateral damage. One of the instruments used to facilitate this is to negotiate a series of exceptions or caveats when and where force may be employed. With every participating state stipulating exceptions, the list has becomes so large that in every situation where force has to be employed a careful balance has to be struck between these interests and limitations and military necessity. Odd as it might seem, this is exactly the situation General Ricardo Sanchez found himself in Iraq in 2004. In his memoirs he writes:

> The restrictions were so complex that I had to carry a five-page spreadsheet listing all the countries, their rules of engagement, and who was

> authorized to do what [. . .] almost always, they had to go back to their national political leadership to ask permission.[4]

How did this situation arise? Did the political leadership not understand that this is not how military strategy works? The timely use of force and a sufficient dose of surprise are critical ingredients when it comes to the application and utility of force.

Imagine, you are a military commander tasked by your government to devise a winning strategy for an irregular war in a far-away country. Your analysis shows that the center of gravity is the population and the central government in the capital city. Recent elections have bestowed dubious democratic credentials on that government due to election irregularities. Polls show that, despite the democratic process and huge expense and effort, few people actually see the government as legitimate or serving their interests. Military force of whatever shape has very little to offer when it comes to perfecting this less-than-perfect local political situation. With the weight attached to democratic elections and respect for the functioning of allegedly democratic government, what is it that military force can do? Odd as it might seem, this is exactly the situation General Stanley McChrystal found himself in, in Afghanistan in the autumn of 2009. While his analysis is undoubtedly correct that the government in Kabul is the center of gravity, it is also the factor that he is least able to influence by the military force under his command. It is not a military but a political problem, which demonstrates again a limited understanding of the utility of military force.

These three little vignettes illustrate some of the very pressing and practical problems and limitations of the utility of force. In the preceding pages of this volume we have tried to empirically verify how and under what circumstances force has indeed utility. This collection of essays is a first attempt to take a closer look at when force has actually made a difference. It is time, we argue, to explore in full the utility of force, and to empirically (in)validate the claim that military force might not be so useful to deal with all kinds of politically perceived problems other than traditional conventional war. The aim of this conclusion is to bring the most important findings together, tie them to the more general discussion in the relevant literature, and identify further avenues for research.

## Force as part of the solution or as part of the problem?

To what extent is force, in fact, part of the solution of political problems? What empirical evidence can we point to when demonstrating that force has utility? Several uses of force can be identified in order to shape this discussion. We will use the conceptualization as proposed by Rupert Smith. In his view, force has four functions: destruction, coercion and deterrence, containment, and amelioration.[5]

### *Destruction*

The destructive function of force has undoubtedly proven itself in the past. We can point to the old paradigm of industrialized conventional war and there are numerous examples to illustrate this case. By using death and destruction, the perpetrators of violent acts raised the cost of the confrontation to such an extent that the opponent's choice became reduced to surrender or death. As Smith notes in the foreword to this volume, the battle is not necessarily the conflict, and the battle can be very violent and destructive in wars amongst the people. The essence of this destructive function is a concentration of force in space and time directed at the weak point of the enemy. A corollary of this approach is the warrior ethos. Recently, Robert Kaplan has argued that only the United States army has the right mentality and training to deal with present threats in the world.[6] Even American Secretary of Defense Robert Gates has criticized allies for being weak-kneed when it comes to confrontations such as in Afghanistan and Iraq.[7] This could also be read as an American explanation of the lack of results in these wars; that not enough force has been employed to achieve the desired ends.[8] The warrior posture, uniquely present in the American armed forces, contributes to the destructive function of the most powerful arsenal in the world.

The American conventional might has reached what Edward Luttwak has called its culminating point of success.[9] The all-powerfulness of the United States has caused it opponents to reach for ways of circumventing its overwhelming conventional overweight. Terrorism and insurgency are logical choices. The essence of the strategies of terrorism and insurgency is not primarily aimed at destruction. The link between destruction and political objectives on both sides is indirect. Intervening factors are necessary to deliver the political desirables.

There are several authors who award the destructive function of armed force great importance in many contemporary conflicts. Edward Luttwak, for example, has argued that we need more force to beat the terrorists at their game.[10] Similarly, Ralph Peters has argued that attrition is the recipe for success: 'the only way you can decisively cripple the command networks of terrorist organizations is by killing terrorists.... It is not a matter of whether attrition is good or bad. It's necessary.'[11] Martin van Creveld, in his study on the changing face of war, states that, in fact, it is more violence that is warranted in insurgency-type struggles. In case intelligence is lacking and you find yourself in a situation of disadvantage, what is necessary is 'to strike as hard as possible within the shortest possible time; better to strike too hard than not hard enough'.[12] While these points of view come from eminent strategic thinkers, such as Luttwak and van Creveld, and they receive a lot of attention, they form a distinct minority voice in the debate about the use of force.

Counterinsurgency has recently become the central concept in strategic studies and the dominant frame within which to study conflict. The historical development of American COIN thinking has been outlined in the contribution by Thomas Mockaitis. In the new American COIN field manual, peace operations, previously the dominant theme, have been implicitly incorporated in the concept of counterinsurgency, which it is now supposed to encompass it.[13]

> Modern COIN ... incorporates stability operations, also known as peace support operations, reconstruction, and nation building. Just recently, these were considered a separate category of military activity closely associated with multinational or United Nations peacekeeping operations in which force is rarely used.[14]

COIN operations are of a highly political nature. In their chapter, Burton and Nagl pay homage to the long tradition of scholarship recognizing this state of affairs. While David Galula has claimed in the past that COIN is 80 percent political and 20 percent military, David Kilcullen has recently raised the percentage to 100 percent political.[15] Although war in general can be interpreted as political, following Clausewitz, these comments signal how Western armies in general perceive their role in counterinsurgency: separate from the political dimension, rather than as an integral part of politics.

In the context of COIN operations, the destructive capacity of armed force often works counterproductively, detailed in the chapter by James Corum. Despite this insight, there is a utility of force in COIN. Mockaitis has pointed out that, in the highly unique context of decolonization, conventional forces were the instrument of choice of the Western states: 'Unusual circumstances in each country allowed the Americans and their allies to defeat the insurgents using conventional means, especially the application of firepower.' Furthermore, as Burton and Nagl have shown: 'As the most successful operations and tactics by US forces in Iraq demonstrated, the use of force to secure and control the population and separate it from the insurgents is crucial to success in "war amongst the people".'[16] The problem, however, was that '[i]t was a lesson that was learned fairly rapidly among operational – and tactical – leaders, but painfully slowly by strategic commanders'. Force can be necessary, as they write: 'In counterinsurgency, military force to "capture" the will of the people through security has an essential utility. The failure of Iraqi forces to perform this function required US troops to perform this role.' The main problem they identify as 'the continuing failure to translate strategy into operations and tactics or integrate effective operations and tactics into strategy meant that battlefield gains did not lead to any sustainable progress toward either transition or security in Iraq'.

Apart from this tactical–strategic disconnect, several suggestions have been offered in the past as to why these kinds of operations are so difficult: faulty strategic concepts, the limitations of democratic states in the use of armed force, too few troops, the political agenda, culture and outside support.[17] However, in a recent book, Jeffrey Record argues that, contrary to popular belief, terrorism and insurgency strategies are not assured recipes for success. Only in a minority of cases does 'David' actually win from 'Goliath'.[18] In fact, 'Goliaths usually win, but their exceptional defeats are almost always a combination of inferior political will, ineffective strategic interaction with the enemy, and failure to isolate the weaker side from foreign help'.[19] In particular, Record awards great importance to this last factor. In both Iraq and Afghanistan, the role of neighboring states, Iran and Pakistan, deserves more attention in light of this argument. In COIN, these studies suggest, it is not destruction but other variables that hold the key to effective operations.

A variation on the theme of destruction, and part of a series of recent tactical successes, is the use of local forces. A combination of precision airstrikes and Western Special Forces, combined with local forces, has been effective, for example, in Afghanistan.[20] This practice, however, has a very long pedigree, as noted in both James Corum's and Christopher Spearin's contributions in this volume. Another scholar, Robert Cassidy, has argued that the most important lesson of all these past cases is that 'the deliberate and early employment of indigenous forces in a counterinsurgent role can be an effective method in helping to achieve success'.[21] An advantage of using local forces is their knowledge of the local culture and environment. This facilitates the procurement of intelligence, which is deemed crucial in counterinsurgency campaigns, as discussed by James Corum. As John Nagl has written elsewhere:

> It is perhaps only a slight exaggeration to suggest that, on their own, foreign forces cannot defeat an insurgency; the best they can hope for is to create the conditions that will enable local forces to win it for them.[22]

This is what happened in the case of Afghanistan in 2001–2002. The way in which the Taliban regime was toppled produced, according to Stephen Biddle, an '"Afghan Model" of warfare in which indigenous allies replace American conventional ground troops by exploiting US airpower and a handful of American special operations forces'.[23] While this Afghan Model has received a lot of criticism, not the least because it relied on very unique circumstances, it has led to an interesting debate about the utility of Special Forces combined with local forces to create overweight. Hy Rothstein argues that, even though Special Forces do an excellent job, they cannot do it all, and neither is it feasible to transform the whole of the armed services into Special Forces.[24]

In the case of dealing with terrorism, I have argued in my contribution that the use of force, ranging from assistance to the civilian power, to full-scale warfare, can be linked to escalation.[25] When we follow the logic of

Smith's new paradigm, the strategy of terrorism is a foremost example of the clash of wills aimed at creating a condition. When confrontations involve the use of force to change the behavior of the opponent, violence is not very effective.[26] The only cases of terrorist activity that have been linked to some measure of success are nationalist or decolonization struggles. This follows the logic that force can create a condition, which can be followed up by clearly linked political action. In terrorist campaigns with Marxist-inspired agendas, for example, force has been aimed primarily at changing the opponent's behavior. This turned out to be extremely difficult.

Destruction, so far, seems to be a problematic function of force in the contemporary context, except in highly circumscribed situations. COIN and counter-terrorism are highly political undertakings and rely on overwhelmingly tactical uses of force, with very strongly infused political operational and strategic plans. The opposite, insurgency and terrorism, similarly rely on a highly politicized approach. Destruction can make a difference in 'industrial war' and as part of operations of Special Forces in conjunction with local forces.

### *Coercion and deterrence*

The role of force in cases of deterrence has proven its worth. In his authoritative study on the subject, Lawrence Freedman describes its utility as follows:

> deterrence works best when the targets are able to act rationally, and when the deterrer and the deterred are working within a sufficiently shared normative framework so that it is possible to inculcate a sense of appropriate behaviour in defined situations that can be reinforced by a combination of social pressures and a sense of fair and effective punishment.[27]

This punishment is clearly to be meted out by the use of force.

A similar type of function is identified by Burton and Nagl in COIN and by Kreps and Larsdotter for peace operations. Deployment of force possesses utility in terms of signaling, credibility and legitimacy. For the COIN activities, even in light of their highly political nature, command of force is necessary to give the operation credibility. In a similar vein, Sarah Kreps and Kersti Larsdotter stress credibility and decisive posture in their analyses of peace operations. Kreps, in her comparison of the operations in the DRC and Sierra Leone, awards great importance to the attitude of the British Special Forces in the case of Sierra Leone, which made them highly effective. Larsdotter notes that force in peace operations is generally seen as 'a means of communication'.[28] Her investigations are based on interviews with participants in Provincial Reconstructions Teams (PRT) in

Afghanistan. She formulates a testable hypothesis, based on her preliminary investigations, that a robust posture might be linked to more effective hearts-and-minds operations, since she observes a reduction in attacks in areas where a 'show of force' approach was adopted.

Coercive measures are used to make an opponent change its course of action. This concept has been notably incorporated in the practices of peace operations. There are many who consider these types of operations as a highly successful instrument in international affairs. There are several studies that have tested these suppositions on the basis of empirical material. Virginia Page Fortna, for example, comes to the conclusion that 'peacekeeping works, particularly after the Cold War when most of the attempts to keep peace after civil wars have been made'.[29] There are others who share her conclusions.[30] Robert Egnell, in this volume, argues that peace operations clearly have a function and fulfill a need.

Despite these positive appraisals, peacekeeping operations are not without problems. Peacekeeping might be, par excellence, the tactical tool Smith describes in his new paradigm. These uses of force are extremely dependent on clear links with political aims. What is it that the military is supposed to achieve and contribute toward the political goal? Military commanders, writing about their experiences, have pointed at what they see as the root of the problems in contemporary operations. Explaining the difficulties are, among others: fuzzy political concepts that are difficult to translate militarily and a lack of political will to carry through the operation, a persistent warrior culture and a lack of resources. In a 1992 analysis, Steven Metz pinpointed one of the main problems: 'By using a strategy and doctrine devoid of clear thinking on the end states of insurgencies, we aim mightily for nothing.'[31] Unfortunately, based on our current knowledge, we do not really know what our priority list should be to remedy these problems and increase the effectiveness of peace operations. Many prescriptions have been formulated.

First, there are those who stress security. The military instrument is first and foremost suited to create a secure situation.[32] Linked to this provision of security is the demilitarization of society.[33] However, demilitarization relies on a logic that the former belligerents are ready to lay down weapons before resolution has taken place.[34] This is often a fallacy. Integration of militias into regular state armed forces is, in practice, more often seen as an economic provision of employment rather than a conflict-resolution mechanism.[35] Security sector reform (SSR) is currently the most popular term to describe the set of activities that need to be performed to transform a conflict society. Few empirical studies exist to date to validate its claims.

Second, others stress the need for stability in order to conduct successful intervention operations.[36] When stability is achieved, it is more likely that the warring parties will discard their weapons, peaceful exchanges will take place and political dialogue can be started up. It is true that increases in state capacity have been statistically linked to a decrease in the chances of a resumption of war.[37] However, it is unlikely that stability on its own shapes a post-conflict environment. Furthermore, Jan Angstrom has argued in his contribution to this volume that outside forces as providers of stability and order, in fact, hamper efforts of nation-building.

Third, democracy has been hailed as a cure for all ills.[38] However, repeated research has shown that in transitions from authoritarianism to a democratic state, more violence is likely.[39] Furthermore, when it comes to intervention, those states that are already making advances independently toward democratization are more likely to be the subject of an intervention operation.[40] These efforts to hasten democratic development with outside assistance only seem to pay off in developed states, which have their own previous democratic history, e.g. Japan and Germany.[41] What seems to be produced most often, in the words of Fareed Zakaria, is 'illiberal democracy'; a state with a façade of democratic credentials with disguised patrimonial or pre-modern political practices.[42]

Fourth, economic development has also been suggested as a bridge toward peace. We know from statistical analyses that economic incentives in post-conflict states work within the context of an established democratic polity. Patrimonial or authoritarian states, however, manage to make the economic development work toward strengthening the state and the parochial interests of the elite. Economic power in this case reinforces (non-democratic) political power. The opposite, however – a lack of economic development – has statistically been linked to an increase in democratization.[43] Furthermore, when economic growth is promoted, the risk of a resumption of war decreases;[44] again, this applies only to democratic states.[45]

These suggestions and prescriptions are not mutually exclusive; rather, they meet in the overall ambition of state- and nation-building concepts.[46] The comprehensive approach, the whole of government approach, effects-based operations, a 3D approach (Defense, Diplomacy and Development) and full spectrum operations are all terms used to describe these ambitions. We should not forget that these elements and state instruments can also be linked in a negative fashion. The failure in one area can be a cause of more conflict in another. As Egnell writes: 'in Iraq, the early, and continuing, failures of economic reconstruction and political progress are even described as causes of the insurgency.'[47] It is clear that the military instrument functions in practice as one among many. Smith argues that the links between the state instrument of force and other available instruments have become more important in the new paradigm: 'the political leadership ... [needs to ensure that] the military actions are coherent with the actions of its other levers of power.'[48] When the effect is limited to the tactical level, other instruments are expected to deliver the desired result on the political plane. Here we encounter another set of problems. When looking at the effectiveness of other instruments of state for international political results, some skepticism is warranted. Due to an increasing call for evidence-based policy, a whole series of investigations have been conducted into the effectiveness of foreign-policy instruments in dealing with conflict.[49]

In the 1990s, there was a lively debate about the effectiveness of economic sanctions. This attention coincided with a steep increase in the use of this foreign-policy instrument.[50] The main issue of disagreement seems to be, in hindsight, whether it was around 40 percent of economic sanctions that led to a change in behavior of the targeted state or only 5 percent.[51] The fact that sanctions primarily affect the population in a negative way instead of the (political) elite places further question marks around the utility of this instrument.[52]

Similarly, the effectiveness of weapons embargoes has been the subject of a study by the Stockholm International Peace Research Institute. It concluded, based on a comparison between the 27 embargoes instituted by the United Nations since the 1990s, that it is not very effective in changing the behavior of the targeted state.[53] In particular, the complicity of neighboring states involved in breaking the embargo goes a long way in explaining this lack of effectiveness.

Another important category of foreign-policy levers are diplomatic interventions and political instruments. In particular the effects and effectiveness of negotiated settlements has started to become the subject of debate. Several studies have recently pointed out that a negotiated settlement provides a very shaky foundation for bringing about desired political ends, such as peace and democratization. Conflicts terminated through a negotiated settlement tend to last much longer compared to alternative resolution strategies. Importantly, negotiated settlements since the 1990s possess a very high propensity to break down and lead to renewed fighting.[54]

Of course, there are many more instruments a state can employ to affect the behavior of another state, such as multilateral pressure through the United Nations, humanitarian operations, pressure from human-rights organizations and indictments from the International Criminal Court. However, these instruments deserve a thorough evaluation when it comes to establishing their effectiveness in changing attitudes and behavior, which so far has not been conducted.

Coercion and deterrence possess utility as a function of force to signal intentions, and to create legitimacy and credibility of an actor in a battle theatre. Apart from these tactical uses, the available empirical evidence substantiates Smith's claim that force is largely of tactical utility. There is a large dependence on other levers to carry through its intended message and create the intended results. Problematic in this discussion is the limited utility of other state instruments, working in conjunction with force. Evidence suggests that there are clear limits to state power employed to change the behavior of opponents.

***Containment***

Containment is another function of military force. Smith and many other authors count traditional peacekeeping among its successful application: 'Kashmir and ... Cyprus are classic, and successful, examples of peace-keeping.'[55] The separation of belligerents and the observation of ceasefires and demilitarized zones can be seen as a task for which military force has high utility. However, a similar question mark as in the previous paragraph is warranted. Force has utility to create a condition; it is not the solution to the problem. The Cyprus ceasefire observation mission has led to a period of 45 years of relative peace without a resolution of the underlying problems. The politicians in this respect have not managed to build on the success of the containment function of force. Therefore, we can question whether force has, in this case, indeed served its utility.

Since private security companies in general refrain from military offensive tasks, it seems logical to discuss their contribution at this point. As Christopher Spearin has pointed out in his contribution, their employment is no panacea for increasing the utility of force. Rather, things have become more complicated with their ascendancy during the past decades. Their added value is sometimes difficult to gauge because they perform all the tasks of a regular military force. The nature of their work requires them to be increasingly closer to the regular chains of command. Furthermore, the argument that personnel problems can be relieved through the use of PSCs is problematic. As Spearin argues, they draw on the same specialist pool as the regular military forces and the recruitment of local theatre personnel does not solve the problem. The use of PSCs as a route toward greater utility of force is, therefore, contentious.

***Amelioration***

Amelioration stands out through its lack of actual employment of force. Force can be useful in its deployment rather than its employment. The difference with the previous function is that its ambition is larger than simply freezing a situation. A first example of the non-use of force to ameliorate is the training and advisory function of armed forces. In his chapter, James Corum comes to the conclusion that

> the most reasonable solution in helping an allied nation to combat an insurgency is to refrain from the temptation to take over the war effort of the host nation, and to emphasize the training and support of the host nation's security forces.

A second example of the amelioration function is the peacekeeping function. To ameliorate by employing force has also been the fundament of humanitarian activities of armed forces. Sarah Kreps points out in her contribution that for this use of the instrument of force, the conditions under which it is put into practice are extremely important. Capacity, also in a geographic perspective, is the distinguishing factor that made the peacekeeping activities in Sierra Leone a success and contributed to failures in the DRC.

The Provincial Reconstruction Teams (PRT) in Afghanistan, as another example of this kind of employment of force, are intended to kick start the rebuilding of the country and facilitate the operation of the military component by winning over the population for the intervention forces. Kersti Larsdotter suggests that a robust posture might serve the foreign forces better than a more low-key presence.[56] Her suggestions go against the large body of counterinsurgency thinking, which stress the importance of winning the hearts and minds of the population through non-violent behavior. Her thought-provoking ideas warrant more investigation.

However, regular soldiers becoming humanitarians can be problematic, and perhaps they should refrain from these activities. Robert Egnell warns us against all-too-optimistic expectations: 'the mere assumption that such activities will lead to increased support from the local population is an oversimplification and serious thinking is necessary in order to better understand the consequences of military activity'.[57] An alternative suggestion comes from Thomas Barnett, who has proposed a bifurcated military: one part specializes in conventional war and the other in peace operations.[58] Yet another suggestion, and the opposite of the ideas sketched so far, comes from Sarah Sewall. In the introduction to the new US Army *Field Manual on Counterinsurgency*, she proposes that perhaps the armed forces should widen their scope of responsibility and include all humanitarian aspects, to assure effectiveness.[59] The same has been proposed by Major General Peter Chiarelli, as discussed in the contribution by Burton and Nagl.

A third example of amelioration is the nation-building function of force, as detailed by Jan Angstrom. He argues, in fact, that this function of the armed forces is part of the problem, rather than the solution: 'internationally managed state-building conditions the strategic environment in which the actors find themselves.'[60] When foreign armed forces take over part of the state or the state-making mechanisms, they hamper indigenous efforts of building up a Leviathan.

> International forces thus face a dilemma. If they are to be successful at all and avoid being shot at by local actors, they must acquire legitimacy. However, in this process, they undermine the local state, thereby making their mission more difficult. The invited Leviathan therefore fails when it takes over the role of the incumbent state and implicitly strikes a bargain with the population.[61]

This argument is illustrated by the case of the 30 years of war in Afghanistan: the Soviets, as an invited Leviathan, failed to entice the elite to support it; similarly the Taliban regime failed. Angstrom argues that, in the Afghan case, 'Military power and its use ... contribute to state-failure in so far as it contradicts centralization of coercion', it feeds the tribal system and patronage politics leading to the failure of the state gaining legitimacy. While war in Late Medieval Europe forged a state-system, the peculiarities of warfare and military power in the Afghan case propels the state toward protracted state failure. Christopher Spearin even notes in his chapter that the competition between intervention forces, in particular private security companies and the new Afghan National Army, is very acutely felt. Reliable personnel leave the army and police forces for very lucrative jobs in the PSCs. In this respect, we should be reconsidering the utility of force and looking perhaps elsewhere for the establishment of viable and locally legitimate forms of rule that provide stability, which post-conflict societies so desperately need.

Amelioration as a function of force finds expression in reconstruction, humanitarian activity and nation-building. There is a utility of force here but the exact conditions and the link with effectiveness in operations remains less than clear. One suggestion proposes a positive effect of a forceful posture on the tactical level in these activities.

## Observations and suggestions for further research

Rupert Smith concludes his study with the claim that force does possess utility: 'military force has a role to play; but that role is not a detached one, nor one which will achieve the strategic objective by itself.'[62] Force is, in fact, first and foremost a tactical instrument. Research efforts, therefore, should focus more on the tactical level of strategy and its link with the other levels and levers of power. Kersti Larsdotter, in this volume, has pointed out that our knowledge on the tactical output of armed forces has been a forgotten subject, perhaps because it is so difficult to measure. She has suggested that we should focus more on the posture of the armed forces in contemporary conflicts.

The conclusion that the utility of force is mainly tactical combined with the precarious effectiveness of the other instruments of statecraft demands serious consideration and research effort. The use of force has become more politicized than we might think, or like to think. We could argue, at the same time, that the use of force has become more Clausewitzean, in the sense that the stress on its nature as a political instrument has increased. Politicians need to clearly understand the boundaries beyond which force does not have practical utility. Military commanders need to explain properly what these limits are. Furthermore, we need a clear strategic framework, in which the political leadership has thought out how military force is to play its role among the other instruments of state.[63] Force, as Christopher Dandeker has discussed in his contribution, is nowadays mainly used to create a condition which has been linked to success instead of geared toward achieving victory. A convincing narrative then becomes of the utmost importance. We run the risk, however, in the words of Jeffrey Record, that politicians implicitly come to see 'war as a substitute for politics'.[64] Robert Cooper argues in his review of Smith's book for the *Sunday Times* that

> Clausewitz recommended that the head of the armed forces should sit in the cabinet, not so that the politicians should receive military advice, but so that the army should understand exactly what political goals it was fighting for and could conduct the campaign accordingly. We need to find the modern equivalent.[65]

Overall, we should not forget that intervention in armed conflict has been statistically linked to a proliferation of warring factions, a longer duration of the violent interaction, geographic spill-over and an increase in the number of issues of contention.[66] This in itself warrants a critical re-evaluation of intervention practices.

What are the consequences of the observation that the utility of force is circumscribed? Drawing the extreme conclusion, Jeffrey Record has suggested that the United States in particular should refrain from getting involved with counterinsurgency or wars amongst the people. If the state does not possess the stamina to carry through what is required in a successful confrontation, the ultimate consequence should be abstinence from this category of conflicts: 'a strong case can be made for adopting a policy of refusing US involvement in foreign internal wars altogether.'[67] The conclusions in this volume do not support this extreme position of abstinence and non-involvement. However, more careful investigations into the utility of force are of paramount importance.

Several avenues present themselves for further research.

First, we should not take for granted causal links that have not been subject of proper investigation. As Egnell warns in this volume, 'we need to determine the effective relationships between security, relief, aid and civil society-building efforts'.[68] Presuming that increased civil–military cooperation produces better results is not grounded in any evidence-based research but seems to form the fundament for large-scale efforts. One could even argue that this idea has been elevated to the dominant and all-consuming ideology in the field.[69]

Second, the conditions under which force does possess utility should be more carefully investigated. War cannot become a substitute for political action because we do not fully understand its workings. As the example of the Iraq War between 2004 and 2006, discussed by Burton and Nagl, shows, the tactical utility of force was clearly understood by the lower ranks. However, these conditions were not put to an advantage and translated upwards to where it mattered in a strategic perspective. Force undoubtedly does play a role in contemporary war, but we need more refined and empirically substantiated evidence of exactly where, when and how.

Third, closely related to the conditions, is the combination of instruments that increases their usefulness. The all-inclusive approaches, such as the nation-building projects dissected in the contribution by Angstrom, rely on a presumption that combinations of instruments are better than a single well-targeted instrument. We do not know which combinations work where, when and how. We do know that when force is used sub-strategically, the dependence on other instruments grows. Politicians need to be made more aware of this state of affairs. Additionally, we need to focus our research efforts on the successful employment of combinations of foreign-policy instruments and perhaps learn more from past experiences.

Fourth, the order of the instruments deserves careful consideration. Smith argues in his book that a linear approach to armed conflict can be highly problematic.[70] In particular, the idea of using force as a last resort carries criticism. It might be possible that force needs to be considered a first resort in order to increase its utility. As argued in the contribution about the utility of force in cases of counterterrorism, when responding with force is the first option, softer measures might produce a different result compared to situations that see the reverse order. How believable is an offer of negotiation, when previously the opponent has received threats to his or her life?

Fifth, since the will of the people is the *Schwerpunkt* in wars amongst the people, we need to better understand how we can capture it. An important suggestion in this regard would be to focus research efforts toward identifying narratives and culturally sensitive interpretations of the use of force. In the words of Lawrence Freedman:

> The idea of the narrative ... opens up another possibility of military operations. Instead of being geared to eliminating the assets of the enemy, they might need to be focused on undermining those narratives on which that enemy bases its appeal and which animates and guides its activists.[71]

Dandeker, Corum and Egnell all concur. Egnell has argued in this volume:

> Conducting appropriate and effective tactical operations and translating them into strategic aims, and then getting that message across to the local and global audience in a coherent strategic narrative, is arguable amongst the most important challenge in current and future peace-support operations.[72]

Corum writes: 'The US and Western nations must do better in presenting our side of the conflict to the world media.' Kersti Larsdotter has proposed that in order to capture the will of the people, more military power and a robust approach might be needed. Elsewhere, she has argued that the use of force in peace operations can be culturally sensitive.[73] In a culture that places great importance on military prowess, carrying weapons openly might be interpreted as a sign of commitment and legitimacy. These ideas should be included in further studies into this subject.

In a contribution from 1980, Robert Art identified the main uses of force as defense, deterrence, compellence and 'swaggering'.[74] The last category included – in the context of the Cold War – shows of force and signaling. He argued that

> [t]he swagger use of force is the most egoistic: it aims to enhance the national pride of a people or to satisfy the personal ambitions of its ruler.... Swaggering is more something to be enjoyed for itself than to be employed for a specific, consciously thought-out-end.[75]

It seems that this description of a category of the use of force applies particularly well to recent times. Since the end of the Cold War, the Western world has used its military apparatus as a world police force, to show leadership in the new world order, to demonstrate continued relevance of armed force in a new context, and to show continued commitment toward pre-existing alliances. It might be argued that many of these recent expressions fit neatly into the category of swaggering. As this volume has indicated, we should be well aware that this use of force is not without problems and, importantly, unintended consequences.

## Notes

1 This conclusion is an edited version of Isabelle Duyvesteyn, 'Exploring the Utility of Force: Some Conclusions', *Small Wars and Insurgencies*, 19, 3 (2008): 423–443. We thank Taylor and Francis for permission to reprint it here.
2 For a discussion of the military fallacy of the exit strategy notion, see: Rose, 'The Exit Strategy Delusion'; Duyvesteyn, 'Understanding Victory and Defeat', 232.
3 This is, amongst others, the assessment of Alice Hills in the case of the Falluja crisis in 2004: Hills, 'Fear and Loathing in Falluja', 624. Others have argued that during the reign of Donald Rumsfeld at the Pentagon, the control of civilian leaders over military affairs developed into political micro-management: Hersh, *Chain of Command.*
4 Sanchez, *Wiser in Battle*, 245.
5 Smith, *The Utility of Force*, 320–321.
6 Kaplan, *Hog Pilots, Blue Water Grunts.*
7 Interview with Robert Gates, 'Gates Faults NATO Force in Southern Afghanistan', *Los Angeles Times*, 16 January 2008.
8 Compare this to the explanations for the failure in Vietnam: more force would have been the solution (Summers, *On Strategy*).
9 Luttwak, *Strategy.*
10 Luttwak, 'Counterinsurgency Warfare as Military Malpractice'.
11 Peters, 'In Praise of Attrition'.

12 Van Creveld, *The Changing Face of War*, 269.
13 *The U.S. Army and Marine Corps Counterinsurgency Field Manual: U.S. Army Field Manual No. 3–24.*
14 Sewall, 'Introduction to the University of Chicago Press Edition', xxiii.
15 Galula, *Counterinsurgency Warfare*, 89; Kilcullen, 'Counterinsurgency Redux', 129.
16 Burton and Nagl, 'Adapting to Counter-Insurgency', this volume.
17 Mack, 'Why Big Nations Lose Small Wars'; Merom, *How Democracies Lose Small Wars*; Arreguin-Toft, *How the Weak Win Wars*; Nagl, *Learning to Eat Soup with a Knife*; Record, *Beating Goliath.*
18 Record, *Beating Goliath*; Angstrom and Duyvesteyn, *Understanding Victory and Defeat.*
19 Record, *Beating Goliath*, xii.
20 Clark, *Waging Modern War*, xxiii.
21 Cassidy, 'The Long Small War'.
22 Nagl, *Learning to Eat Soup with a Knife*, xiv.
23 Biddle, 'Toppling the Taliban in Afghanistan', 187. This idea has a longer pedigree. Steven Metz talks about an El Salvador model of counterinsurgency in which local forces play a crucial role: Metz, 'New Challenges and Old Concepts', 20.
24 Rothstein, *Afghanistan and the Troubled Future of Unconventional War.*
25 Duyvesteyn, 'Great Expectations', this volume.
26 Sullivan, 'War Aims and War Outcomes'.
27 Freedman, *Deterrence*, 5.
28 Larsdotter, 'Exploring the Utility of Armed Force in Peace Operations', this volume.
29 Page Fortna, 'Does Peacekeeping Keep Peace?', 288.
30 Doyle and Sambanis, 'International Peace Building'; van der Lijn, *Walking the Tightrope*; Seybolt, *Humanitarian Military Intervention.*
31 Metz, 'Victory and Compromise in Counterinsurgency', 53.
32 Etzioni, *Security First.*
33 Stedman, 'Policy Implications'.
34 Duyvesteyn, 'The Effectiveness of Intervention Instruments in Armed Conflict'.
35 Glassmyer and Sambanis, 'Rebel Military Integration and Civil War Termination'.
36 Paris, *At War's End.*
37 DeRouen and Sobek, 'The Dynamics of Civil War Duration and Outcome'.
38 Fukuyama and McFaul, 'Should Democracy be Promoted or Demoted?
39 Mansfield and Snyder, 'Democratization and the Danger of War'; Mansfield and Snyder, *Electing to Fight*; Hegre *et al.*, 'Toward a Democratic Civil Peace?'
40 Gleditsch, Siljeholm Christiansen and Hegre, 'Democratic Jihad?
41 Hegre, 'Disentangling Democracy and Development as Determinants of Armed Conflict'.
42 Zakaria, *The Future of Freedom.*
43 Hegre, 'Disentangling Democracy'.
44 Collier, Hoeffler and Soderbom, *Post Conflict Risks*; Collier *et al.*, *Breaking the Conflict Trap*, 158.
45 Hegre, 'Disentangling Democracy'.
46 Dobbins *et al.*, *America's Role in Nation-Building.*
47 Egnell, 'Between Reluctance and Necessity', this volume.
48 Smith, *The Utility of Force*, 307.
49 This section is inspired by the discussion in Duyvesteyn, 'The Effectiveness of Intervention Instruments in Armed Conflict'.
50 *The Human Security Report*, 2005, 154.

51 Hufbauer and Elliot, 'Same Song, Same Refrain?'; Elliot, 'The Sanctions Glass'; Pape, 'Why Economic Sanctions Still Do Not Work'; Hufbauer, Schott and Elliott, *Economic Sanctions Reconsidered.* Even the effectiveness of smart or targeted sanctions is disappointing: Elliott, 'Analyzing the Effects of Targeted Sanctions'. There is some evidence that the present sanctions practices are more effective against democratic regimes compared to non-democratic ones: Lektzian and Souva, 'An Institutional Theory of Sanctions Onset and Success'.
52 Mueller and Mueller, 'Sanctions of Mass Destruction'; Weiss *et al.*, *Political Gain and Civilian Pain.*
53 Fruchart *et al.*, *United Nations Arms Embargoes.*
54 *The Human Security Brief* 2006, 4; Toft, *Peace Through Security*; Regan and Aydin, 'Diplomacy and Other Forms of Intervention in Civil Wars'; Collier, Hoeffler and Soderbom, *Post Conflict Risks*; Page Fortna, *Peace Time.*
55 Smith, *The Utility of Force*, 274.
56 Larsdotter, 'Exploring the Utility of Armed Force in Peace Operations', this volume.
57 Egnell, 'Between Reluctance and Necessity', this volume.
58 Barnett, *The Pentagon's New Map.*
59 Sewall, 'Introduction to the University of Chicago Press Edition', xxxi.
60 Angstrom, 'Inviting the Leviathan', this volume.
61 Angstrom, 'Inviting the Leviathan', this volume.
62 Smith, *The Utility of Force*, 394.
63 Sewall, 'Introduction to the University of Chicago Press Edition', xl–xli; Sarkesian and Connor, *The U.S. Military Profession into the Twenty-First Century*, 22. See also Burton and Nagl, this volume.
64 Record, *Beating Goliath*, xii. Ralph Peters too comments: 'The US military, otherwise magnificently capable, is an extremely inefficient tool for combat in urban environments', Peters, 'Our Soldiers, Their Cities', 43.
65 Cooper, 'The Utility of Force by General Sir Rupert Smith', *Sunday Times*, 18 September 2005.
66 Harbom *et al.*, 'Dyadic Dimensions of Armed Conflict 1946–2007', 697–710; Regan, 'Third-Party Intervention and the Duration of Intrastate Conflicts'; Regan, 'Conditions of Successful Third-Party Intervention in Intrastate Conflicts'.
67 Record, *Beating Goliath*, xii.
68 Egnell, 'Between Reluctance and Necessity', this volume.
69 Belloni, 'The Trouble with Humanitarianism'.
70 Smith, *The Utility of Force*, 376–377.
71 Freedman, *The Transformation of Strategic Affairs*, 26.
72 Egnell, 'Between Reluctance and Necessity', this volume.
73 Larsdotter, 'Culture and the Outcome of Military Intervention'.
74 Art, 'To What Ends Military Power?', 3–35.
75 Art, 'To What Ends Military Power?', 10–11.

## Bibliography

Angstrom, Jan and Duyvesteyn, Isabelle (eds), *Understanding Victory and Defeat*, London: Routledge, 2007.

Arreguin-Toft, Ivan, *How the Weak Win Wars: A Theory of Asymmetric Conflict*, Cambridge: Cambridge University Press, 2005.

Art, Robert, 'To What Ends Military Power?', *International Security*, 4, 4 (1980): 3–35.

Barnett, Thomas P., *The Pentagon's New Map*, New York: Berkley Books, 2004.

Belloni, Roberto, 'The Trouble with Humanitarianism', *Review of International Studies*, 33, 3 (2007): 451–474.

Betz, David, 'Redesigning Land Forces for Wars Amongst the People', *Contemporary Security Policy*, 28, 2 (2007): 221–243.

Biddle, Stephen, 'Toppling the Taliban in Afghanistan', in J. Angstrom and I. Duyvesteyn (eds) *Understanding Victory and Defeat in Contemporary War*, London: Routledge, 2007.

Cassidy, R.M., 'The Long Small War: Indigenous Forces for Counterinsurgency', *Parameters*, 36, 2 (2006): 47–62.

Clark, Wesley K., *Waging Modern War: Bosnia, Kosovo and the Future of Combat*, New York: Public Affairs, 2001.

Collier, P., Elliot, L., Hegre, H., Hoeffler, A., Reynal-Querol, M. and Sambanis, N., *Breaking the Conflict Trap: Civil War and Development Policy*, Oxford: Oxford University Press, 2003.

Collier, Paul, Hoeffler, Anke and Soderbom, Mans, *Post Conflict Risks*, paper for the Centre for the Study of African Economies, Department of Economics, University of Oxford, 2006. Available at: www.exlegi.ox.ac.uk/anke%20hoeffler.pdf

Cooper, Robert, 'The Utility of Force by General Sir Rupert Smith', *Sunday Times*, 18 September 2005.

Department of the Army *The U.S. Army and Marine Corps Counterinsurgency Field Manual: U.S. Army Field Manual No. 3–24*, Chicago, IL: University of Chicago Press, 2007.

DeRouen, Karl and Sobek, D., 'The Dynamics of Civil War Duration and Outcome', *Journal of Peace Research*, 41, 3 (2004): 303–320.

Dobbins, James, *America's Role in Nation-Building: From Germany to Iraq*, Santa Monica, CA: RAND 2003.

Doyle, M.W. and Sambanis, N., 'International Peace Building: A Theoretical and Quantitative Analysis', *American Political Science Review*, 94, 4 (2000): 779–802.

Duffy Toft, Monica, *Peace Through Security: Making Negotiated Settlements Stick*, Working Paper 23, McGill University: Research Group in International Security, 2006.

Duyvesteyn, Isabelle, 'Understanding Victory and Defeat; Some Conclusions', in Jan Angstrom and Isabelle Duyvesteyn (eds), *Understanding Victory and Defeat in Contemporary War*, London: Routledge, 2007.

Duyvesteyn, Isabelle, 'The Effectiveness of Intervention Instruments in Armed Conflict: Conflict Resolution is the Only Solution?', in G. Molier and E. Nieuwenhuys (eds), *Peace, Security and Development in an Era of Globalization: The Integrated Security Approach Viewed From A Multidisciplinary Perspective*, Leiden: Martinus Nÿhoff, 2009.

Elliot, K.A., 'The Sanctions Glass: Half Full or Completely Empty?', *International Security*, 23, 1 (1998): 50–65.

Elliott, K.A., 'Analyzing the Effects of Targeted Sanctions', in D. Cortright and G.A. Lopez (eds), *Smart Sanctions: Targeting Economic Statecraft*, New York: Rowman and Littlefield 2002.

Etzioni, A., *Security First*, New Haven, CN: Yale University Press, 2007.

Fishel, J., *Civil Military Operations in the New World*, Westport, CN: Praeger, 1997.

Freedman, L., *Deterrence*, Cambridge: Polity, 2004.

Freedman, L., *The Transformation of Strategic Affairs*, Adelphi Paper 379, London: Routledge, 2006.

Fruchart, D., Holtom, P., Wezeman, S.T., Strandow, D. and Wallensteen, P., *United Nations Arms Embargoes: Their Impact on Arms Flows and Target Behaviour*, Stockholm: SIPRI, 2007.

Fukuyama, Francis and McFaul, Michael, 'Should Democracy be Promoted or Demoted?, *Washington Quarterly*, 31, 1 (2008): 23–45.

Galula, David, *Counterinsurgency Warfare: Theory and Practice*, London: Pall Mall, 1964.

Glassmyer, K. and Sambanis, N., 'Rebel Military Integration and Civil War Termination', *Journal of Peace Research*, 45, 3 (2008): 365–384.

Gleditsch, Nils Petter, Siljeholm Christiansen, L. and Hegre, Havard, 'Democratic Jihad? Military Intervention and Democracy', paper prepared for the International Studies Association Conference, 17–20 March 2004.

Gow, James, 'The New Clausewitz? War, Force, Art and Utility – Rupert Smith on 21st Century Strategy, Operations, Tactics in a Comprehensive Context', *Journal of Strategic Studies*, 29, 6 (2006): 1151–1170.

Harbom, L., Melander, E. and Wallensteen, P., 'Dyadic Dimensions of Armed Conflict, 1946–2007', *Journal of Peace Research*, 45, 5 (2008): 697–710.

Hegre, Havard, 'Disentangling Democracy and Development as Determinants of Armed Conflict', paper prepared for the World Bank, 9 January 2003.

Hegre, Havard, Gates, Scott, Gleditsch, Nils Petter, *et al.*, 'Toward a Democratic Civil Peace? Democracy, Political Change, and Civil War, 1816–1992', *American Political Science Review*, 95, 1 (2001): 16–33.

Hersh, Seymour, *Chain of Command: The Road from 9/11 to Abu Ghraib*, New York: HarperCollins, 2004.

Hills, Alice, 'Fear and Loathing in Falluja', *Armed Forces and Society*, 32, 4 (2006): 623–639.

Hufbauer, Gary and Elliot, Kimberly, 'Same Song, Same Refrain? Economic Sanctions in the 1990s', *American Economic Review*, 89, 2 (1999): 403–409.

Hufbauer, Gary, Schott, J., Elliott, Kimberly Ann, *Economic Sanctions Reconsidered: History and Current Policy*, Washington, DC: Institute for International Economics, 1993.

Kaplan, Robert, *Hog Pilots, Blue Water Grunts: The American Military in the Air, at Sea and on the Ground*, New York: Random House, 2007.

Kilcullen, David, 'Counterinsurgency Redux', *Survival*, 48, 4 (2007): 111–130.

Larsdotter, Kersti, 'Culture and the Outcome of Military Intervention: Developing Some Hypotheses', in Jan Angstrom and Isabelle Duyvesteyn (eds), *Understanding Victory and Defeat in Contemporary War*, London: Routledge, 2007.

Lektzian, D. and Souva, M., 'An Institutional Theory of Sanctions Onset and Success', *Journal of Conflict Resolution*, 51, 6 (2007): 848–871.

van der Lijn, Jair, *Walking the Tightrope: Do UN Peacekeeping Operations Actually Contribute to Durable Peace?*, Amsterdam: Rozenberg, 2006.

Luttwak, Edward, *Strategy: The Logic of War and Peace*, Cambridge, MA: Harvard University Press, 1987.

Luttwak, Edward, 'Counterinsurgency Warfare as Military Malpractice', *Harper's Magazine*, 2 (2007): 33–42.

Mack, Andrew, 'Why Big Nations Lose Small Wars: The Politics of Asymmetric Conflict', *World Politics*, 27, 2 (1975): 175–200.

Mansfield, Edward and Snyder, Jack, 'Democratization and the Danger of War', *International Security*, 20, 1 (1995): 5–38.

Mansfield, Edward and Snyder, Jack, *Electing to Fight: Why Emerging Democracies Go to War,* Cambridge, MA: MIT Press, 2005.

Merom, Gil, *How Democracies Lose Small Wars,* Cambridge: Cambridge University Press, 2003.

Metz, Steven, 'Victory and Compromise in Counterinsurgency', *Military Review,* 72, 4 (1992): 47–53.

Metz, Steven, 'New Challenges and Old Concepts: Understanding 21st Century Insurgency', *Parameters,* 37, 4 (2007): 20–32.

Mueller, John and Mueller, Karl, 'Sanctions of Mass Destruction', *Foreign Affairs,* 78, 3 (1999): 43–53.

Nagl, John, *Learning to Eat Soup with a Knife: Counterinsurgency Lessons from Malaya and Vietnam,* Chicago: University of Chicago Press, 2005.

Page Fortna, Virginia, 'Does Peacekeeping Keep Peace? International Intervention and the Duration of Peace After Civil War', *International Studies Quarterly,* 48, 2 (2004): 269–292.

Page Fortna, Virginia, *Peace Time: Cease-Fires and the Durability of Peace,* Princeton, NJ: Princeton University Press, 2004.

Pape, Robert, 'Why Economic Sanctions Still Do Not Work', *International Security,* 23, 1 (1998): 66–77.

Paris, Roland, *At War's End; Building Peace after Civil Conflict,* Cambridge: Cambridge University Press, 2004.

Peters, Ralph, 'Our Soldiers, Their Cities', *Parameters,* 26, 1 (1996): 43–50.

Peters, R., 'In Praise of Attrition', *Parameters,* 34, 2 (2004): 24–32.

Powell, Colin, *My American Journey,* New York: Random House, 1995.

Record, Jeffrey, *Beating Goliath: Why Insurgencies Win,* Washington, DC: Potomac Books, 2007.

Regan, Patrick M. and Aydin, A., 'Diplomacy and Other Forms of Intervention in Civil Wars', *Journal of Conflict Resolution,* 50, 5 (2006): 736–756.

Rose, G., 'The Exit Strategy Delusion', *Foreign Affairs,* 77, 1 (1998): 56–67.

Rothstein, Hy, *Afghanistan and the Troubled Future of Unconventional War,* Annapolis, MD: Naval Institute Press, 2006.

Sanchez, Ricardo, *Wiser in Battle: A Soldier's Story,* New York: HarperCollins, 2008.

Sarkesian, Sam C. and Connor Jr., R.E., *The U.S. Military Profession into the Twenty-First Century: War, Peace and Politics,* London: Routledge, 2006.

Sewall, Sarah, 'Introduction to the University of Chicago Press Edition: A Radical Field Manual', in *The U.S. Army and Marine Corps Counterinsurgency Field Manual: U.S. Army Field Manual No. 3–24,* 2007.

Seybolt, Taylor, *Humanitarian Military Intervention: The Conditions for Success and Failure,* Oxford: Oxford University Press, 2007.

Smith, Rupert, *The Utility of Force: The Art of War in the Modern World,* London: Allen Lane, 2005.

Stedman, Stephen J., 'Policy Implications', in S.J. Stedman, D. Rothchild and E. Cousens (eds), *Ending Civil Wars: The Implementation of Peace Agreements,* Boulder, CO: Lynne Rienner, 2002.

Sullivan, Patricia, 'War Aims and War Outcomes: Why Powerful States Lose Limited Wars', *Journal of Conflict Resolution,* 51, 3 (2007): 496–524.

Summers, Harry, *On Strategy: A Critical Analysis of the Vietnam War,* Novato, CA: Presidio Press, 1982.

Toft, Monica, *Peace Through Security: Making Negotiated Settlements Stick*, Working Paper 23, McGill University: Research Group in International Security, 2006.

van Creveld, Martin, *The Changing Face of War: Lessons of Combat from the Marne to Iraq*, New York: Ballantine, 2006.

Weiss, Thomas, Cortwright, David, Lopez, Gary and Minear, Larry, *Political Gain and Civilian Pain*, New York: Rowman and Littlefield, 1997.

Zakaria, Fareed, *The Future of Freedom: Illiberal Democracy at Home and Abroad*, New York: Norton, 2004.

# Index

*Italic* page numbers indicate tables not included in the text page range. **Bold** page numbers indicate figures not included in the text page range.


1st Armored Division 188
3rd Armored Cavalry Regiment (ACR) 187–8

Abizaid, John 185–6, 201
Abrams, Creighton 151
AFDL (Alliance of Democratic Forces for the Liberation of Zaire) *see* Alliance of Democratic Forces for the Liberation of Zaire
Afghan Model 262
Afghan National Army (ANA) *see* ANA
Afghan National Police (ANP) *see* ANP
Afghanistan: 30-year war 218–20; aim of intervention 233; British-led PRT 125–8, 130, 132, 134; casualties 23, 210; civil-military interaction 234, 245; Coalition Forces 119, 126; counterinsurgency campaign 163; defining moments 27; erosion of humanitarian space 246, 251; as failed state 211; German-led PRT 128–31, 132, 134; government 27–8; humanitarianisation of the military 248–9; local forces 262; and peace 119; poppy crop 30; progress 17; PSC personnel 55–7; and PSCs 45; state-building failures 220–5, 268–9; strategic goals 26; and success 26; terrorist attacks 131–4; and US forces 154; and utility of military force 259
African National Congress 78
*An Agenda for Peace* 106, 110
Aguinaldo, Emilio 146
Aided, Mohammed Farah 153
airpower 170–1
al-Qaeda 154, 195, 220, 223
Albright, Madeleine 6
Algeria: National Liberation Front 78
Alliance of Democratic Forces for the Liberation of Zaire (AFDL) 95
amelioration: function of force 267–9
American Way of War 47–9, 55
Amnesty International 98
Amos, James F. 199
ANA (Afghan National Army) 56–7, 127, 130
Anbar awakening 155, 202
Angola 51
Annan, Kofi 104
ANP (Afghan National Police) 56–7, 127
Armed Forces Revolutionary Council (West Side Boys) 103
ArmorGroup 42, 46, 55
army doctrine 23–4
Arreguin-Toft, Ivan 218
art of war 19
Art, Robert 77, 272
assassination: counterterrorism 73, 75–6
assistance: to civilian authorities 71–2
attrition 260
audiences 28
Australia: changing military role 43
Avant, Deborah 47

Baghdad 161, 196
Bangladesh 162–3
BAPSC (British Association of Private Security Companies) 42, 47
bargaining theory of war 215–16
Barnett, Thomas 268
Bearpark, Andy 42
Bennett, Huw 126
Berlin discotheque bombing 74
Biddle, Stephen 121, 262
Bin Laden, Osama 154
Black Panther leaders 75
Blackwater USA (now Blackwater Worldwide) 44, 46, 59*n*35
Blair, Tony 233
bombings: Iraq War (2003) *199*
Bonn agreement 223
Bosnia 107, 153, 233, 249
Botea, Roxanna 213
Brahimi Report 237
Brigety, Reuben 241, 244, 247
Britain *see* UK (United Kingdom)
British approach 119
British Association of Private Security Companies (BAPSC) *see* BAPSC

British-led PRT: Mazar-e Sharif 125–8, 130, 132, 134
Brooks, Doug 42
Brophy-Baermann, Bryan 75
Bullock, Richard K. 124
Bush administration 72, 161, 186
Bush, George W. 48, 181
Byman, Daniel 66, 119, 124

Caldwell, Chad 195
Callwell, Charles 145
Canada: and Kosovo crisis 51; PSCs 45–6; Quebec Liberation Front 78
Canadian Forces 43; changing military role 43
capacity 107–9, 268
capacity argument: peacekeeping 91
CAPs (Combined Action Platoons) 151, 172–3
Casey, George 182–3, 184–5, 186, 198, 200, 201, 258
Cassidy, Robert 262
Castro, Fidel 160
casualties: acceptable 29; DCR and Sierra Leone *94*; Iraq and Afghanistan 23; and peacekeeping success 92; PSCs 51; and retaliation 74
CENTCOM (Central Command) 185, 194
Centeno, Miguel 212
Central Intelligence Agency (CIA) *see* CIA
CGS (Chief of General Staff) 23–4
Chalk, Peter 72
Chapter VI missions 106
Chapter VII missions 103, 106
Chiarelli, Peter 189, 268
Chief of General Staff (CGS) *see* CGS
CIA (Central Intelligence Agency) 151, 154
civil–military integration: consequences of 241–4; extent of 232–5; humanitarianisation of the military 248–9; implementation 250–3; implications 249–50; militarisation of humanitarian activity 244–8
civil–military relationship: in peace operations 231
Civil Operations and Rural Development (CORDS) *see* CORDS
civil wars: and military intervention 119, 208–9; and terrorism 76; USA 145
civilian casualties: Iraq War (2003) *197*
Clark, Wesley 1
Clausewitz xvii, 1, 5, 9, 16, 25, 144, 189, 214, 215, 261, 269, 270
Clinton administration 51
Coalition Forces: Afghanistan 119, 126
coercion: function of force 263–6
COIN doctrine: 1945–1960 149–50; early 147; evolution of 142, 144–5; *FM 3–24 Counterinsurgency* 162, 175–6; peace operations 261; post-9/11 154–5; post-Cold War era 153–4; post-Vietnam 152–3; soft approach 119, 121; and Vietnam 150–1
COIN operations 261, 263
Cold War 4, 20
colonial experience 144–5
'colonial model' 55
Combined Action Platoons (CAPs) *see* CAPs
Combined Joint Task Force - Seven (CTJF-7) *see* CTJF-7
complexity: of 'wars amongst the people' 24–5
Comprehensive Approaches 237
Comprehensive Planning 237
concessionary model: of counterterrorism 67–8
condition: as defining war 7; notion of 17; and success 25
conduct: of military forces *see* military conduct
conflict: confrontation and 7
conflict research 5
Congressional Research Service 67
Connor, Robert 6
consent 121
'constabulary force' 17–18
containment: function of force 267
Control Risk Group 51
conventional war 159–60, 260; *see also* industrial war
Conway, James T. 44–5
Conybear, John 75
Cook, Forrest 195
Cooper, Robert 270
'cordon-and-knock' missions 191
CORDS (Civil Operations and Rural Development) 151
Cornish, Stephen 244–5, 247
Corsair 51
counterinsurgency: and Iraq *see* Iraq War (2003); and leadership 165–9; and military force 160–1, 162, 176–7; political aspects 161; role of intelligence 163–5; and strategic studies 261; strategy 161; as a term 143; US military and 145–7, 159; *see also* COIN
counterinsurgency campaigns: Afghanistan 163; and conduct of forces 119; Iraq 163
counterinsurgency doctrine *see* COIN doctrine
counterterrorism: judiciary measures 67, 69, 70, 80; literature on 66; past approaches 67–70; use of military instrument 65–6, 70–8, 81–2
counterterrorist spiral 81
Cradock, John 33*n*46
credibility: PSCs 53
Crelinsten, Ronald 67
Crenshaw, Martha 67, 79
criminal underworld: terrorism and 80
Cronin, Audrey Kurth 79–80
CTJF-7 (Combined Joint Task Force - Seven) 182
Cyprus 160, 167–9, 267

Daoud, Mohammad 219
Dandeker, Christopher 7, 43, 249
Dannatt, Richard 32*n*31
Davis, Paul K. 67
DDR (disarmament, demobilization and reintegration) 56, 235, 264

decapitation: counterterrorism 75–6
Decker, George H. 150
defeat: meaning of 5; of terrorism campaigns 79–80
defence spending 47, 49
defensive violence 42
defining moments 27–8, 30
definitions of success: peace operations 123–4
demilitarization 264
Democratic Forces for the Liberation of Rwanda (FDLR) *see* FDLR
Democratic Republic of Congo (DRC) *see* DRC
democratization 209, 265
Denard, Robert 41
deniability: PSCs 41–2, 44
destruction: function of force 260–3
deterrence 3, 72, 73–4
deterrent: function of force 263–6
development: security and 234, 235
Dew, Andrea 222
disarmament, demobilization and reintegration (DDR) *see* DDR
division of responsibilities 49
Dobbie, Charles 121
Docherty, Leo 248
doctrine: definition 143–4
Donahue, Laura 67, 82
Downes, Alexander 209
Downs, George 92, 119, 120, 123
Doyle, Michael 92, 102, 107, 239
DRC (Democratic Republic of Congo) 95–103, 108–9, 268; UN peacekeeping mission *see* MONUC
Dunkirk 242
Duyvesteyn, Isabelle 238

EBAO (effects-based approaches to operations) 236
ECOMOG (Economic Community of West African States Monitoring Group) 103
economic assistance 102, 109
Economic Community of West African States Monitoring Group (ECOMOG) *see* ECOMOG
economic development 92, *94*, 265
economic sanctions 266
Edson, Merrit 'Red Mike' 147
effective militaries 8–9
effects-based approaches to operations (EBAO) *see* EBAO
Egyptian Islamic Jihad 77, 80
El Salvador 152–3, 173–5
end of war 21
Engene, Jan Oskar 69
Eppright, Charles 74, 77
equipment 8
escalation: risk of 76, 77
ethical dilemmas: terrorism 73, 76, 82
ethnic conflicts 216
Etzioni, Amitai 233
EU (European Union) 237
exceptions: to use of force 258–9
exit strategies 258

factions: number of 92, *94*
failed states 208, 211, 213
Falklands Conflict 21
Fallujah 44–5
Farabundo Martí National Liberation Front (FMLN) *see* FMLN
FARC (Revolutionary Armed Forces) *see* Revolutionary Armed Forces
FAST Event database 131–2
Fazal, Tanisha 212
FDLR (Democratic Forces for the Liberation of Rwanda) 101
'fire flush' operations 151
firepower 149, 159–60, 164, 172–3
First Congo War 95
First World War *see* World War I
1st Armored Division 188
'flexible response' 150
*FM 3–16 Counterguerrilla Operations* 151
*FM 3–24 Counterinsurgency* 155, 162, 175–6, 199–200
*FM 31–15 Operations Against Irregular Forces* 149
*FM 31–20 Operations Against Guerrilla Forces* 149
*FM 31–23 Stability and Support Operations* 152
*FM 100–5 Field Operations* 147, 152, 153
*FM 100–20 Low-Intensity Conflict* 152
FMLN (Farabundo Martí National Liberation Front) 174
force: discussion of utility 4–9; four functions of 259–69
force protection: Germany 128
foreign economic assistance 102, 109
foreign-policy instruments: effectiveness 266
Forster, Anthony 233, 248
Fortna, Page 92, 233, 239
Freedman, Lawrence 242, 263, 271
Fry, Robert 243
'full-spectrum operations' 189, **190**

Galula, David 189, 198, 200, 261
*The Gamble* 25
Gates, Robert 52, 54, 159, 161, 260
Gaza Strip 171–2
Gbadolite Agreement 96
geographical factors: and peacekeeping 108–9
German-led PRT: Kunduz 125, 128–31, 132
German Red Army Faction 80
Germany: and Afghanistan 224; force protection 128
Gibson, Chris 190–1, 192
Global Risk Security 51
global surveillance 243, 250
Goldwater-Nichols Defense Reorganization Act 152
Goma Agreement 101
Gow, James 2, 9
Gray, Colin 76
Gray, David R. 17–18
Greek Civil War 148
Greek Cypriot rebels 160
Gregorian, Hrach 234, 235
Guelke, Adrian 68

guerillas 144
Gurr, Robert Ted 72
Guthrie, Charles 106
Guttieri, Karen 246

Hamas 16, 171–2
Hanle, Donald 75, 77
Harding, John 167–9
Harkavy, Robert E. 121
Hartzell, Caroline 92
Hastings, Max 25
hearts and minds 22, 30, 45, 121, 132, 143, 232, 242, 250
Heiberg, Marianne 121
Henriksen, Thomas 242
Herbst, Jeffrey 109, 212
Heywood, Louise 247
high-tech intelligence 163, 164
Hillen, John 49
Hix, William 184
Hobbes, Thomas 213
Holmes, Richard 21–2
Holsti, Kalevi 4–5
hostages 68, 71
Howard, Michael 66
Hukbalahap Revolt 148
human intelligence (HUMINT) *see* HUMINT
human intelligence specialists 164–5
human rights abuses 240
human shields 170–1, 172
humanitarian activities: and military activities *see* civil-military integration
humanitarian actors: classes of 240
humanitarian operations: use of term 2
humanitarian principles 239–41
humanitarian space: erosion of 244–6, 251
'humanitarianisation of the military' 232, 233, 248–9, 250
humanity 240
HUMINT (human intelligence) 164–5
Huntington, Samuel 17, 18–19, 20
'hybrid wars' 33*n*31

identity-based wars 92, *94*
IDF (Israel Defence Forces) 16, 66
Ignatieff, Michael 22
illiberal democracy 265
image: of conflicts 169–72
impartiality 240, 244–5, 251
independence 240, 244–5, 251
indigenous personnel 55–7, 262
industrial war 21, 23–4, 260; *see also* conventional war
information distortion 216
insurgencies: as strategy 260, 262; as a term 142–3; and terrorism 76, 78–9
insurgents: provocation by 44–5
Integrated Missions 237–8
integration: concept of 8; of militias 56–7, 264
intelligence: and counterinsurgency 163–5
international aid 102, 109
international capacity 107–9
International Committee of the Red Cross: code of conduct for combatants 45
international forces: and amelioration 268–9; and legitimacy 217
International Peace Operations Association (IPOA) *see* IPOA
International Terrorism Attributes of Terrorist Events (ITERATE) *see* ITERATE
interstate war: research on 4; *see also* conventional war; industrial war
intervention 208–9, 217–18, 225
intervention effectiveness 239
intervention/stabilisation operations: preparation for 23–4
IPOA (International Peace Operations Association) 42, 46, 47
IRA (Irish Republican Army) *see* Irish Republican Army
Iraq: aim of intervention 233; erosion of humanitarian space 246; insurgency 234; media strategy 171; military humanitarian aid 247; tactical–strategic disconnect 261; and USA 154–5, 175
'Iraq Index' 196
Iraq Study Group 196, 201
Iraq War (1991) 175
Iraq War (2003): breakdown of order 160–1, 195–6; Casey's command 182–3; and casualties 23; civilian casualties *197*; counterinsurgency campaign 163; counterinsurgency preparation 181–2; counterinsurgency strategy 198–201; daily attacks *198*; developments in operations 187–92; First Battle of Fallujah 44–5; multiple-fatality bombings *199*; as postmodern conflict 21–2; Sanchez's command 181–2; security force development 193–5; and success 16–17, 25; Tall Afar experience 187–8; transition 183–7, 202–3; USA approach 201–2
Iraqi Army 194
Iraqi Police 193, 194, 195
Irgun 78
Irish Republican Army (IRA) 79, 80
irregular conflicts 159
irregular forces 144
irregular wars: literature on 4; and new paradigm 2
ISAF mission 27–8, 34*n*57, 126
Islamic Jihad 77
Israel: assassination 75; Hamas attack 171–2; invasion of Lebanon 170; and Lebanon 76; retaliation 75
Israel Defence Forces (IDF) *see* IDF
Italian Red Brigades 80
ITERATE (International Terrorism Attributes of Terrorist Events) 68–9

Jackson, Robert 212
Jakobsen, Peter V. 130
Janowitz, Morris 17–18, 19–20
Jeffrey, Jim 185
Jenkins, Brian Michael 68
Johnson, Dominic 242
Jones, Seth 131
Just War theory 239
Justice and Equality Movement 109

Kabila, Joseph 95
Kabila, Laurent 95
Kagan, Frederick 48
Kaldor, Mary 5
Kaplan, Robert 260
Karmal, Babrak 218, 219
Kasdorf, Bruno 50
Keane, John 181
Kennedy, John F. 150
Kennedy, Leslie 72
kidnapping: of PSCs personnel 45; Sierra Leone 104
Kilcullen, David 243, 261
Kimmitt, Mark 44–5
Komer, Robert 151
Kosovo 22, 51, 107, 153, 233
Krulak, Charles 29, 234–5
Krulak, H. 151
Kupchan, Charles 49
Kurdish Workers Party (PKK) 75, 79–80

Lake, David 208, 210, 217
Law of Armed Conflict 46
leadership: in counterinsurgency 165–9
Leander, Anna 212
Lebanon 76, 170
legitimacy: and counterinsurgency 175; and *FM 3–24 Counterinsurgency* 155, 175–6; and international forces 217; ISAF mission 34*n*57; and order 214; and state-building 210–11; terrorist 76; and timeless missions 27
levels of war 26–7, 29–30
liberal democracy 233
liberal interventionism 29
Liberia 234
Libya: and USA (United States of America) 74
Livingston, Neil 75
local forces 55–7, 262
Lockerbie bombing 74
Lomé Peace Accords 103
lootable resources 93, *94*, 102, 109
Low-Intensity Conflict 152
Luft, Gal 73, 75
Lum, Cynthia 72
Lusaka Peace Accord 96, 101
Luttwak, Edward 260

McCaffrey, Barry 50, 181
McChrystal, Stanley 259
Macedonia 233
McFarland, Sean 202
McKenna, Richard 169
McMaster, H.R. 187–8
MACV (Military Assistance Command Vietnam) 150–1
Makarios, Archbishop 168
Malayan insurgency 166–7
Malvesti, Michele 67, 74
manpower: US reduction 48
Mansfield, Edward 209
Manwaring, Max 175
Marine Corps 145, 146–7, 148, 151, 152, 154, 172
martyr creation 76
mass media: use of 169–72
'massive retaliation' 150
maximalist approach: to humanitarian aid 245
Médecins Sans Frontières 240
the media 28, 169–72
media surveillance 243
Memorial Institute for the Prevention of Terrorism (MIPT) database *see* MIPT
mercenaries 41, 42
Metz, Steven 264
'militarisation of humanitarian activity' 232, 233, 244–8
military activities: and humanitarian activities *see* civil–military integration
Military Assistance Command Vietnam (MACV) *see* MACV
military conduct: British-led PRT 125–8, 130, 132, 134; German-led PRT 125, 128–31, 132, 134; and peace operations 119–23
military doctrine 143–4
military effectiveness 8–9
military force: counterinsurgency and 176–7; discussion of utility 4–9; as tactical utility 2
military goals: political aims and 6–7
military instrument: terrorism and 65–6, 70–8
Military Observer Teams (MOTs) *see* MOTs
military power: research 5
Military Professional Resources Incorporated 51
military role: changing 43–4
militias 56
Millen, Raymond 213, 221
Miller, Gregory 67, 77–8, 80
minimum force approach 121, 122–3, 126, 131, 132, 134–5
MIPT (Memorial Institute for the Prevention of Terrorism) database 69, 70
'Mission Accomplished' speech 48
Mobutu, President 95
monitoring: of professional space 19
monitoring missions 106, 115*n*97
MONUC (UN Mission in the Democratic Republic of Congo) 90–1, *94*, 95–103, 107, 108–9
MOTs (Military Observer Teams) 127
Mueller, John 5
Mujahidin 219, 221–2
multifunctional approaches 231–2, 238–9, 250–1
multinational force: state exceptions 258–9
multiple-fatality bombings: Iraq War (2003) *199*
Mumbai attacks 28–9
Münkler, Herfried 5

Najibulla, Mohammed 218, 219
Narodnaya Volya 79–80
narratives of war 24–30, 30–1, 242–3
nation-building 265, 268; *see also* state-building
National Liberation Front: Algeria 78
'National Strategy for Supporting Iraq' 185

National Strategy for Victory in Iraq 186–7, 200
NATO: firepower 149; and Kosovo 22; Peace Support Operation 122
negotiated settlements 266
negotiations: in counterterrorism 68
Negroponte, John 184–5
Netanyahu, Benyamin 66
Neuman, Stephanie G. 121
neutrality 240, 244–5, 251
'new' wars 1
NGOs: maximalist approach 244–5
Nicaraguan campaign 146–7
Nicaraguan Contras 152–3
Nisoor Square incident 46
Nixon, Richard 152
non-interstate warfare 39, 159
non-state adversaries 39
Northern Alliance 220, 223
nuclear weapons 4, 18, 19, 21

'objective civilian control' 19
Olson, Lara 234, 235
OOTWA (Operations Other Than War) 153
Operation Artémis 97, 99
Operation Cast Lead 16, 34*n*52
Operation Palliser 104–5, 107–8
Operation 'Together Forward' 194–5, 196
operational level: of war 27, 29
Operations Other Than War (OOTWA) *see* OOTWA
'overlay' 40
overreaction 11, 66, 71, 76, 81

Pakistan: and the Taliban 223
Palestine Liberation Organization (PLO) *see* PLO
Palestinian Liberation Organization 79
Paris, Roland 123, 209
'Patterns of Global Terrorism' reports 67, 69
Patterson, Malcolm 41
peace agreements 91, 120–1
peace enforcement: use of term 2
peace-enforcement operations 106–7
peace operations: comprehensive approaches 239; definitions of success 123–4; force as deterrent 263–4; and military forces 119–20
Peace Support Operation: NATO 122
peace support operations 239–41
peacekeeping: literature 92–3; situational factors 92–3, *94*, 100–3; UN 106–11; and use of force 6, 267–8; use of term 2
peacekeeping missions: research 5; UN 90–1 *see also* MONUC; UNAMSIL
Peters, Ralph 48, 260
Petraeus, David 26, 46, 175, 193, 199, 201
Petraeus surge 25
Philippine insurrection 146
Philippines 148
Phoenix Program 151
PKK (Kurdish Workers Party) *see* Kurdish Workers Party
PLO (Palestine Liberation Organization) 170
political aims: and military goals 6–7
political factors: and peace 109–10
political order: war and 213, 214
poppy crop 30
Posen, Barry 5, 216
post-Cold War: strategic thought 4–5
post-Cold War conflicts 21–3
postmodern conflict 21–2
Pouligny, Beatrice 121
Powell, Colin 249
preemption: counterterrorism 72–3
Pretoria Agreement 96, 101
Prince, Eric 54
professional autonomy 18–19
professional space 19
propaganda: by insurgents 169–70
protection: counterterrorism 71–2
Provincial Reconstruction Teams (PRTs) *see* PRTs
provocation: by insurgents 44–5; by terrorists 71
proxies 55–6
PRT Kunduz 128–31, 132, 134
PRT Mazar-e Sharif 126–8, 130, 132, 134
PRTs (Provincial Reconstruction Teams) 126, 247, 263–4, 268; *see also* British-led PRT; German-led PRT
PSCs (Private Security Companies): activities 42; Afghanistan 43; British kidnapping 45; Canadian 45–6; clients 42; demand side 40; deniability 40–1, 44; and human rights standards 45; indigenous personnel 55–7; rise of 39–40; salaries 52; SOF (Special Operations Forces) 52–4; and state militaries 43, 44–7; strikes 51; studies 40; supply factors 40; trade associations 42; and USA 47–57; value of 267; and violence 41, 42
public scepticism 29
public support: for timeless missions 29–30

QRF (Quick Reaction Force) 127, 130
*Quadrennial Defense Review* 43, 50, 52, 54
Quebec Liberation Front 78
Quick Reaction Force (QRF) *see* QRF
Quinlivan, James 107

Rana, Raj 244
RAND Corporation 66
RAND-MIPT Terrorism Incident Database 131
Reagan administration 67
Record, Jeffrey 48, 55–6, 262, 270
Red Army Faction 80
Red Brigades 80
refugee crises: and use of force 5
Reid, John 248
reintegration 56–7
Reiter, D. 215
relational legitimacy 210–11, 214–15
relational legitimacy state-building theory 214–18
*The Remnants of War* 5
research: suggestions for further 269–72
responsiveness 8

retaliation: counterterrorism 74–5
Revolution in Military Affairs (RMA) 47
Revolutionary Armed Forces (FARC) 80
Revolutionary United Front (RUF) *see* RUF
Rice, Condoleeza 5–6, 43, 185, 186
Richardson, Louise 77, 79
Ricks, Thomas E. 25, 184
Rivers, Galye 77
RMA (Revolution in Military Affairs) *see* Revolution in Military Affairs
Rose, Michael 1
Ross, Jeffrey 72
Rotberg, Robert 211
Rothstein, Hy 262
Rubin, Barnett 220, 223
RUF (Revolutionary United Front) 103, 104, 107–8
Rumsfeld, Donald 50
Russia: and Afghanistan 219
Rwanda 95, 96, 101

Saladin 55
Salvadoran government 152–3, 173
Sambanis, Nicholas 92, 102, 107, 239
Sanchez, Ricardo 182, 258–9
*The Sand Pebbles* 169–70
Sandino, Augusto 147
Sankoh, Foday 104
Sarkesian, Sam 6
scepticism 29
Schelling, Thomas 4, 215
Schmid, Alex 67
Schoomaker, Peter 49
science of war 19
Second Congo War 95
Second World War *see* World War II
sectarian violence 196
security: and development 234, 235; and use of force 264
security sector reform (SSR) 235, 264
Sederberg, Peter 76, 77, 78
Sedra, Mark 125
Sendero Luminoso (Shining Path) 75, 77, 79–80
Sepp, Kalev 184
September 11 attacks 4, 223
Sewall, Sarah 268
Seybolt, Taylor 119, 124
Shaw, Martin 23, 243
Sherley, Alison S. 72
Shining Path (Sendero Luminoso) *see* Sendero Luminoso
'shock and awe' air campaign 154
'show of force' approach 122, 128, 130, 131, 132, 134–5
Shultz, Richard 222
Sierra Leone 103–6, 107–8, 268; UN peacekeeping mission *see* UNAMSIL
Silke, Andrew 77
Singer, Peter 41
skill 8
skill sets 53
small wars 144–5
*Small Wars Manual* 147
Smith, Niel 202
Smith, Rupert 1–2, 6–8, 17, 20–1, 22, 26, 48, 57, 121, 183, 231, 239, 242, 260, 265, 269
Snyder, Jack 209
SOCOM (Special Operations Command) 152
SOF (Special Operations Forces) 52–4, 152, 154
Sokolsky, Joel 49
'soldier-scholars' 1
soldiers of fortune 41
Somalia 153, 160
South Africa: African National Congress 78
South Korea 148
Soviet Union: and Afghanistan 219, 220–1
Spangler, Philip 241
Spanish–American War 146
'spasm terrorism' 69
Special Operations Command (SOCOM) *see* SOCOM
Special Operations Forces (SOF) *see* SOF
speed: expectation of 48
spoilers 93, 100–1, 109, 122
SSR (security sector reform) *see* security sector reform
stabilization operations 264–5; use of term 2
stalemate 92, *94*
START (Study of Terrorism and Responses to Terrorism) 69
state-building: and amelioration 268; approaches to 265; in failed states 213; and international forces 208, 209; and intervention 233; and legitimacy 210–11; and war 225
state-building failures: Afghanistan 220–5, 269
state formation: and war 208, 211–13
Stedman, Stephen 92, 93, 119, 120, 123
Steele Foundation 51
Strategic Hamlet program 151
strategic level: of war 26–7, 29, 161
'strategic narrative' 242
strategic patience 26–7
strategic thought: post-Cold War 4–5
strikes: PSCs 51
Study of Terrorism and Responses to Terrorism (START) *see* START
sub-strategic use of force 6–7
substitution: terrorism 81–2
success: definitions of 123–4; pursuit of 24–30; and victory 17–20
successful endings: terrorist campaigns 78–9
Sudan 109
suicide bombings 195
the 'surge': Iraq 17, 21, 25, 26, 155, 183, 201
surveillance, global 243, 250
swaggering 272
Sweden: and Afghanistan 224
Symbionese Liberation Army 75

tactical level: of war 27, 29
Taliban 26, 27, 33*n*32, 220, 222–3, 245, 251
Tall Afar experience 187–8, 192
targetted killings *see* assassination
Taylor, Brian 212–13
TCNs (Third Country Nationals) *see* Third Country Nationals

technology: emphasis on 47–8
Templer, Gerald 166–7
terror: as a term 143
terrorism: and criminal underworld 80; endings 78–80; instruments used against 65–6; and military instrument 65–6; as strategy 260, 262; UN and 65; and use of force 262–3; *see also* counterterrorism
Terrorism in Western Europe database (TWEED) *see* TWEED
terrorist attacks 4, 74–5, 131–4, 223
terrorist campaigns: successful endings 78–9; and violent spectacle 28–9
terrorists: provocation by 71
Third Country Nationals (TCNs): and PSCs 55–7
3rd Armored Cavalry Regiment (ACR) 187–8
'three block war' 44, 234–5
Tierney, Dominic 242
Tilly, Charles 208, 211, 212, 222
timeless missions 26–7, 29–30, 48
Tin-bor Hui, Victoria 212
transformation: from terrorism 79
transition: Iraq War (2003) 183–7, 202–3
treaty/peace agreement *94*
Tripodi, Paolo 246
Trubowitz, Peter 49
Tupamaros terrorism 80
TWEED (Terrorism in Western Europe database) 69, 70

UCMJ (Uniform Code of Military Justice) *see* Uniform Code of Military Justice
Uganda 95, 96
UK (United Kingdom): Afghanistan 224, 248–9; army doctrine 23–4; changing military role 43; colonial experience 145; conduct of military forces 119; and counterinsurgency 145; Cyprus 160, 167–9, 267; Malaya 166–7; Ministry of Defence 235; Sierra Leone 91, 104–5, 106, 107–8; *see also* British-led PRT
UK–US alliance 23–4
UN (United Nations): Chapter VII missions 106; Integrated Missions 237–8; and terrorism 65
UN Children's Fund 102
UN Mission in Sierra Leone (UNAMSIL) *see* UNAMSIL
UN Mission in the Democratic Republic of Congo (MONUC) *see* MONUC
UN peacekeeping 106–11
UN peacekeeping missions 90–1
UN Resolutions 96, 99, 126
UNAMSIL (UN Mission in Sierra Leone) 90–1, 103–6, 107–8
Uniform Code of Military Justice (UCMJ) 46
United States Special Operations Command (USSOCOM) *see* USSOCOM
Uruguay 80
US Army 145, 147, 260
US military: and COIN doctrine 147, 149–55; and Cold War 148; and counterinsurgency 145–7; and World War II 148
US Protection and Investigation 55
USA (United States of America): Afghanistan 154, 223; and allies 49–50; Angola 51; army doctrine 23–4; Bosnia 249; changing military role 43; and COIN doctrine *see* COIN doctrine; conduct of military forces 119; and counterinsurgency 159; counterterrorism 67; El Salvador 172–5; Fallujah 44–5; human intelligence system 165; Iraq 154–5, 175 (*see also* Iraq War (2003)); Libya 74; MIPT database 70; PSCs 46, 47–57; SOF (Special Operations Forces) 52–4; Vietnam War 172–3
USSOCOM (United States Special Operations Command) 52
*The Utility of Force* 1–3, 183

van Creveld, Martin 4, 39, 82, 260
victory: concept of 20; elusiveness of 16, 39, 48; meaning of 5; success and 17–20
Vietnam War: and COIN doctrine 149–51; USA 172–3
violence: and peace operations 122; and PSCs 41, 42
virtual war 22
Volckmann, Russell 149
von Clausewitz, Carl 9, 214

Walter, Barbara 92
war: conceptualization of 5; as counterterrorism 76–8; levels of 26–7, 29–30; and political order 213, 214; state formation and 208, 211–13
war duration 92, *94*
'war of ideas' 22
war outcomes 92, *94*
Wardlaw, Grant 71–2
warlordism 221, 222, 225
warrior ethos 249, 260, 264
'wars amongst the people' 1, 2, 7, 8, 17, 21–3, 24, 121, 183, 203, 260, 261, 270
Waterman, Harvey 213
weapons 8
weapons embargoes 266
Weber, Max 19, 211
Weir, Erin 239, 240, 241, 245, 247–8
Wellesley, Arthur 144
West Side Boys 103
Western strategic thought: post-Cold War 4–5
Westmoreland, William 151, 172–3
Wilkinson, Paul 69, 80
Winslow, Donna 240
World War I 146
World War II 148, 215, 242

Young, Arthur 167

Zakaria, Fareed 265
al-Zarqawi, Abu Musab 195, 200